The
original edition
of this book con-
tained oversized maps.
PDF files of these maps
can be downloaded from
www.elibron.com/maps

JOHN ROSE TROUP

WITH

STANLEY'S REAR COLUMN

Elibron Classics
www.elibron.com

Elibron Classics series.

ISBN 1-4212-3952-3 (paperback)
ISBN 1-4212-3951-5 (hardcover)

This Elibron Classics Replica Edition is an unabridged facsimile of the edition published in 1890 by Chapman and Hall, Limited, London.

WITH

STANLEY'S REAR COLUMN.

MAJOR E. M. BARTTELOT

COMMANDER OF STANLEY'S REAR COLUMN.

WITH
STANLEY'S REAR COLUMN

BY

J. ROSE TROUP

LATE TRANSPORT OFFICER OF THE EMIN PASHA RELIEF EXPEDITION

With Illustrations

LONDON: CHAPMAN AND HALL,
LIMITED.

1890.

Richard Clay & Sons, Limited,
London & Bungay.

Dedicated

TO

MY WIFE,

WHO HAS SHARED THE ANXIETIES AND LABOURS CONNECTED WITH THE PREPARATION OF THIS VOLUME.

PREFACE.

On my return from Africa in September 1888, my health being very much broken by what I had so recently gone through in the service of Mr. Stanley, I was much pained by reading most inaccurate accounts of what had taken place at Yambuya, and I was even more troubled to find the Emin Pasha Relief Committee—with my report in their possession—failed to correct the false statements made in the press, though I wrote requesting them to do me and my comrades this act of justice, because we were forbidden by our contracts to publish anything. Failing to obtain from them any satisfaction, I felt that to defend the honour of myself and comrades became my duty, especially when our action was condemned in a letter of Mr. Stanley's published April 1889. I knew of the paragraph in the contract by which I had agreed not to publish anything for six months after the appearance of the official publication of our leader, but I believed this was binding only on condition that he had fulfilled his part of the contract, and this he had failed to do, as he never supplied me with the tent, bed, Winchester rifle, canteen, and a due share of European provisions. Mr.

Stanley's failure to fulfil his contract appeared to me to nullify it; so that in a letter to the newspapers, pointing out Mr. Stanley's errors in his letter from Bonalya, I called upon the Committee once again to do me justice, and added, in case they did not do so, "I further propose, as he (Mr. Stanley) will never be able, unless steps are taken, to form a right opinion of the state of affairs at that place, to publish, at the earliest possible date, a detailed account of our end of the Expedition." The Committee paid no attention to this, and I felt that the duty devolved upon me of making public the whole matter. My book was therefore prepared, but on the eve of its publication Mr. Stanley's representative in London took steps to obtain an injunction.

I will not burden this account with a description of the varying courses that ended eventually in bringing the matter into the Chancery Court. The judge granted the *interim* injunction under the impression that I could have obtained the specified articles by simply asking for them, but this I had done unsuccessfully again and again; therefore, in order to bring out clearly the impossibility of getting them, and that I never did receive them, I found it necessary to bring a counter-claim for damages against Mr. Stanley for his breach of contract. Before the action could be brought to uphold the *interim* injunction came on, Mr. Stanley had returned to England, and after a time—possibly through his desire not to appear in court—he suggested compromising the matter. After a good deal of talking and arranging, the following agreement was signed by both of us, and the action in the Chancery Court was withdrawn.

"Stanley *v.* Troup.

"Memorandum of agreement made the twentieth day of May, 1890, between the plaintiff and defendant, whereby it is agreed as follows, viz.—

"(1) The plaintiff shall pay to defendant, his solicitors, proper costs, charges, and expenses, as between solicitor and client in reference to the action, and bear his own costs, charges, and expenses.

"(2) The defendant to be at liberty to publish his book on or after the fifteenth of October, 1890, but not before.

"(3) All proceedings in the action to be stayed.

"(4) None of the other officers of the Expedition employed by the plaintiff to be allowed to publish before the fifteenth of October, 1890."

It will be seen from the above that Mr. Stanley was to pay all my costs, and it is usually inferred from such an arrangement that he admitted that the action did not have his sanction, or that some other powerful reason influenced him to make a considerable sacrifice to end the controversy. His book was not then published, but I had argued that the report sent to her Majesty's Government, and issued by them as a Parliamentary paper, was the official publication; but if, as he maintained, his two volumes, for which he was to receive high remuneration, were the official publication, then I was granted leave to publish three and a half months, instead of six months, afterwards.

It is in accordance with this agreement that I now publish

this volume. I must apologize to the reader for having adopted a rather unpopular form in presenting the account of events in quotations from my diary. It appeared to me that the record written at the time, while everything was still fresh in my memory, would give the public a clearer idea of what happened, how one event led on to another, and of the impressions I then received. When five officers, almost total strangers to each other, are shut up for a considerable length of time in a camp in Central Africa, isolated, cut off from communication with the outer world, and deprived of the ordinary occupations of life, it is not remarkable that a certain amount of disagreement should arise, but I see no reason why such matters should be made public property, so I have excised from my diary reference to affairs of a purely personal nature, as well as others that I thought it best to omit. Certain actions I at the time deemed inexpedient, and therefore remonstrated against them; some of these in the light of subsequent knowledge bear a different interpretation, and many were due to the faults of others.

With this explanation I lay this volume before my readers.

CONTENTS.

LIST OF ILLUSTRATIONS.

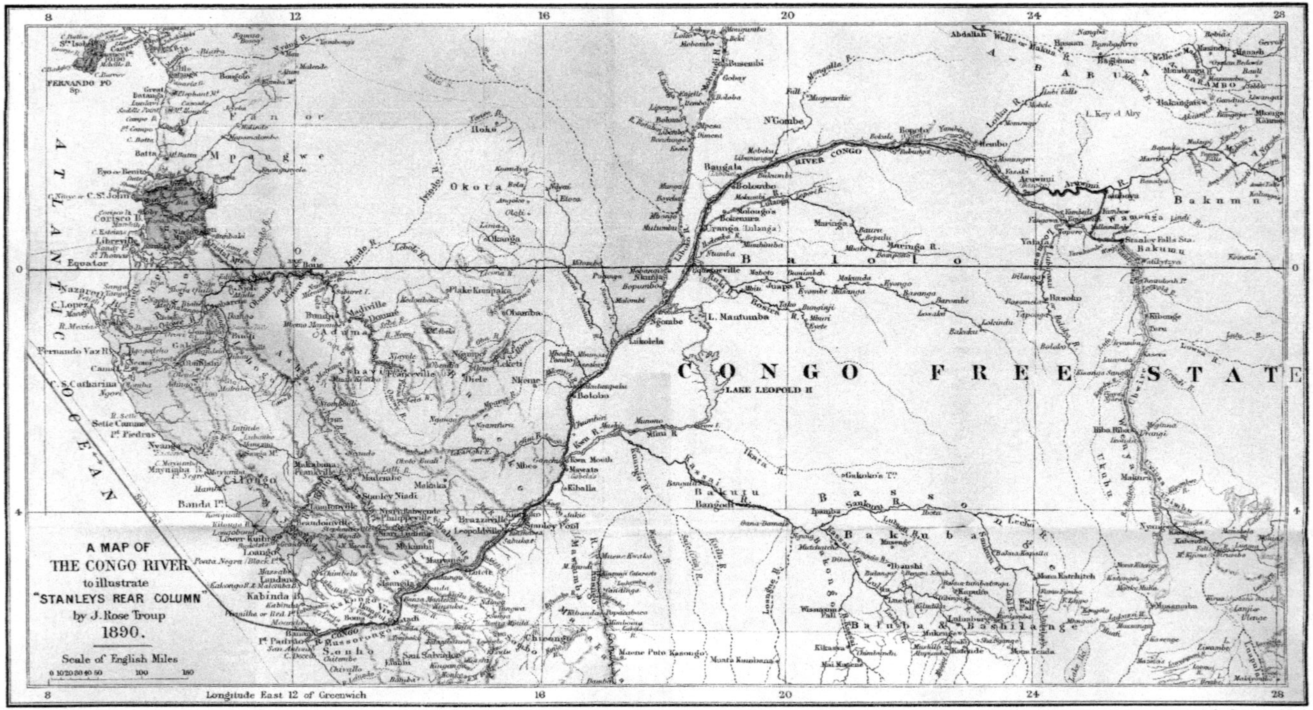

London. Chapman & Hall.

STANLEY'S REAR COLUMN.

CHAPTER I.

INTRODUCTORY REMARKS.

Mr. Stanley has now made public his estimate of the Rear-guard, criticizing and condemning it, and he seems to consider that he has given a sufficiently full account of what happened at Yambuya during his absence. It is necessary to remind the reader that he left that Rear-guard to take care of itself from June 1888 until August 1889, during which period no news of his whereabouts reached those he had left behind. We ought not to object that he touches so lightly upon events of which he could have no personal cognizance, but we cannot admit that it was right for him to condemn his officers unheard, and to write descriptions from preconceived ideas, based on incomplete evidence. The reader of his volume will perceive that Mr. Stanley refers to letters and papers of the two officers who lost their lives in his service, and he gives the accounts written out by two, out of the three, survivors—be it noted, however, that these documents are revised! The third—the senior surviving—of these officers is not allowed to speak for himself in Mr. Stanley's book, the reason thereof being that he was unwilling to revise his story in accordance with Mr. Stanley's wishes. It is given, however, at length in the succeeding pages.

Mr. Stanley began to make public his charges against the Rear-guard long ago, indeed as far back as August 28th, 1888, just after he had met Mr. Bonny at Bonalya with the wreck of the Rear Column. We were inclined to forgive his hasty judgment, based as it was upon what he was able to learn in that short time, and which he could not have thoroughly taken in and weighed carefully before he penned the lines condemning those who were absent. It was hoped that he would say nothing more until he had had an opportunity of hearing fully about the matter. But that hope was vain. Arriving again within range of civilization he immediately recommenced his letter-writing, never asking the other survivors for their reports. Under these circumstances we cannot be blamed for assuming that he had a preconceived idea of what must have been the causes of disaster, and that he rejected as untrue all that did not fit into the line of his argument. It will be admitted by all that he committed a grave error in not asking for reports from all his officers, and not waiting until he had gleaned every scrap of evidence before he condemned those who were in his service. He gives us some short extracts from what he calls a log, the first entry being dated June 11th, 1888, the day the Rear Column left Yambuya, so it contains no information of what happened in camp. He states that he can get no satisfactory information out of Mr. Bonny.[1] He hears nothing from

[1] See *In Darkest Africa*, vol. i. 6 and 478.

Mr. Ward until he receives a letter dated February 13th, 1890, long after his charges are published. He asked no information of me; but I wrote to him, under date of December 4th, 1889, a letter he never acknowledged receiving. Therefore, upon what he based his original charges it is impossible to imagine. The injustice of accusing his officers of grave dereliction of duty without producing a scrap of evidence, and without having full reports, is most serious.

But let us review the situation of affairs and see what he expected, and why his expectations were not realized. Mr. Stanley is on the verge of the unknown country at Yambuya. He has been unable to take with him all his officers and loads owing to the *flotilla* promised by King Leopold being insufficient to transport them. He desires to push on rapidly to Wadelai; he therefore decides to leave two of his officers in a camp at this point, with 127 men and 167 loads; these are to be joined shortly by three more officers, 131 men, and 493 loads. He sets forth with four officers and 400 men, some few of whom carry loads; but as they are to advance with all speed they are in light marching order. They are to make their way quickly to Wadelai, and it was anticipated by the leader that they could accomplish their journey and return to Yambuya in five months. Meanwhile, *if it were possible*, when additional porters had been obtained, the Rear-guard were to follow more leisurely with the bulk of the loads. Having given Major Barttelot a number of orders,[1] he disappears into the forest, and is not heard of for fourteen months.

[1] He speaks of disobeying orders, but now says "the instructions are

What did he expect from the young officer left in command at Yambuya, and his four subordinates? One of the latter Mr. Stanley, in his instructions, associated with the commander, and two he asked should be consulted, and should be allowed to speak freely their opinions. He told Major Barttelot to await the arrival of all the loads and men, and in the event of his not obtaining all the additional porters, *if he preferred marching rather than waiting for the return of the advance column*, he might make marches *twice over*, or else wait patiently at Yambuya. In any event, it was pointed out that the commanding officer's chief duty was to care for the loads and men, for if these were lost "the advance force itself would need to solicit relief in its turn." These were the explicit instructions given to the officer left in command. Is it then marvellous that Major Barttelot used every endeavour to obtain means of transporting the many loads left in his hands?

It should be borne in mind that Mr. Stanley placed a young officer (the youngest in years of all the five, except one), who had had no previous experience with Zanzibari porters, and was ignorant of the peculiarities of life on the Congo, in command of a most trying position. He was expected to treat with a wily Arab, who had been almost more than Mr. Stanley could manage, he was to keep a camp in good condition and the men under control for five months, during which time they would, probably, be forced to remain comparatively inactive—a serious task, for inactivity is apt to breed

not exactly orders. They are suggestions or advices."—*In Darkest Africa*, vol. i. 487.

discontent, and discontent mutiny. Moreover, they were to be in communication with Arabs from Stanley Falls, who could easily contaminate them, and egg them on to mutiny, if it served their purpose.[1] Or if he was not expected to remain in camp, he had a yet more difficult undertaking, especially if he did not receive the porters promised, it is said, by Tippu-Tib, a possible contingency that Mr. Stanley now indicates he believed would be a certainty, for he doubted the Arab. In case the 600 men were not forthcoming, what did Mr. Stanley expect Major Barttelot to do? At the end of his first volume, he gives a pencilled note, said to have been handed to the Major.[2] In this he details 155 men as carriers and 100 as guards, total 255. These men were not to make marches "twice over," as suggested in the original instructions, but four times forward loaded and three times backward empty-handed, that is, traversing the road seven times; added to this the camp would have been insufficiently guarded by the remainder.

We now know what obstacles Mr. Stanley had to overcome with his picked following of 400 men, fairly well disciplined, under the command of a leader

[1] See Stanley's suggestion of this in *In Darkest Africa*, vol. i. 121.

[2] There is no reference to marching without any assistance from Tippu-Tib in the written and signed instructions. This pencilled note, which I never saw, appears to have been an unofficial suggestion. It must be remembered that Mr. Stanley is not one who allows any latitude with written orders—he demands strict obedience. He gave written instructions to Major Barttelot that provided for certain contingencies, and allowed the choice between certain given lines of action, but one contingency he did not provide for in those written instructions, and we had no authority shown to us for acting in the way now outlined at the end of Mr. Stanley's first volume.

of vast experience and ability to control Zanzibaris and natives; his men nearly starved, selling rifles, ammunition, and all they could lay hands on in order to procure food, and deserting on the slightest provocation, so that he lost in fourteen months, or even less, one half of them. Could he have marched even his picked men wearily back and forth? Would he not have required to use discipline of the severest type to prevent mutiny? What, then, would have been the task of a younger and less experienced officer, with half the number of men, and these the refuse of the lot? How could he have coped with the difficulties that had beset Mr. Stanley's path, and by which he was so nearly vanquished? How could his men have obtained food in a district where Mr. Stanley's followers had almost starved while simply passing through it, but where the Rear Column would have been quartered at least four times as long? Mr. Stanley's own words confirm this, for he writes—"We dread your inexperience and your want of influence with your people. If with me people preferred the society of the Manyuema blackguards to me, who am known to them for twenty years, how much more so with you, a stranger to them and their language!"[1]

The men of the Rear Column had been left behind with us because they were feeble, diseased, or otherwise incapable of working. To begin with Mr. Stanley writes (vol. i. 93): "In this Expedition are scores of confirmed bounty jumpers who are only awaiting opportunities. In inspecting the men to-day, I was of the opinion that

[1] Letter to Major Barttelot, February 14th, 1888.

only about 150 were free men, and that all the remainder were either slaves or convicts."

Again (vol. i. 103): "We accordingly selected 125 men who appeared weakest in body, and left them at Bolobo." He refers also (vol. ii. 13), in these words, to our men: "The utter unruliness of the mob of slaves, which had half maddened the officers of the Rear Column." He had picked out the very best, even taking every *headman*, and then demanded that work seven times as difficult should be extracted from this refuse material.

The impossibility of making weary marches back and forth with the miserable men under Major Barttelot's command would have appalled even Mr. Stanley. Can it be wondered at that no one in the camp thought the undertaking feasible, and that the commander shrank from attempting such a "hazardous advance . . . at very great risk to the whole column, and to little purpose"?

The difficulties of the position were almost insurmountable at the best of the times, but they were greatly increased by subsequent events, which led to long delay, and terrible suffering. It will be seen (page 227), as time went on, how impossible it was for our men to carry loads; how, when marching with Mr. Ward from Yambuya to Yangambi, we had to stop to assist the exhausted carriers to their feet, and help to replace their *half* loads upon their heads; for they were too enfeebled, even our best men, to do ordinary porters' work.

The chief cause, then, of Major Barttelot's inability to move was the lack of men to carry his loads. These porters were to have been supplied, Mr. Stanley asserts, by

Tippu-Tib. Concerning the Arab's agreement with Mr. Stanley in this matter, we are possessed of anything but full information. Mr. Stanley simply says (vol. i. 71): "I made another agreement with him about the engagement of carriers to carry ammunition to Lake Albert from the Congo. If there is no ivory I shall be indebted to Tippu-Tib for the sum of £3600." Unfortunately he does not give the wording of this contract, neither is it included in Consul Holmwood's letter (see Appendix), where the other agreement, making Tippu-Tib governor of Stanley Falls district, is given at full length. In fact, we have no reference made to this agreement by the Consul, nor have we anything to confirm Mr. Stanley's *ipse dixit.* If he had given us the text of this contract, we should have known whether the statement I heard, more than once, was true—viz., that each of the 600 men was to receive a certain amount of gunpowder from the stores brought up by me—it might also have informed us whether this ammunition was for philanthropic purposes. However, Mr. Stanley does not refer to this, but indicates that Tippu-Tib was to be paid in ivory, which was supposed to be in Emin's possession, though nothing is said about the formality of obtaining Emin's permission for this disposal of it. There was one very important proviso which I have heard that Tippu-Tib made, but failing the letter of the agreement we cannot prove this is true, viz., that he would provide the men *if he could.*

Mr. Stanley himself says "600 carriers is a large order." They were to be selected from Tippu-Tib's followers at Stanley Falls. Now the greater part of these men were Manyuemas, natives of this vicinity, cannibals, untrained in serving white men, and never accustomed to carry loads. They are half slaves, only having been subjugated comparatively recently by Tippu-Tib's people. When they are employed by the Arabs it is chiefly as fighting men, and it is beneath their dignity to carry loads, such work being relegated to their women and slaves. As it will be seen (page 242), Raschid pointed out to me that "Manyuemas are good soldiers but never do as carriers, while the Zanzibaris are good carriers but cannot fight."

These men are under the control of Tippu-Tib's sheiks, who send them out in marauding parties under the leadership of an Arab to obtain ivory or other valuables; they will attack other natives, loot their villages, kill or make prisoners of whom they please, doing all this in what manner they like. They will be absent on such excursions for weeks or months, but they are obliged, after a time, to return to the Arabs and give up part of their loot. They are provided with guns and ammunition; they are allowed to go and come as they please and do as they like. They are not accustomed to the restraints imposed by the white men, who require their followers to march just so many miles a day, to be in camp at a fixed hour, and to be subject to stringent rules. There were several Manyuemas among Tippu-Tib's people brought round from Zanzibar to the Congo. They would have observed on their way up country the strictness of the white men with the Zanzibari porters, who were obliged to be in every night at a given camp with all their loads, and if they disobeyed rules were severely dealt with.

The Manyuemas would have spread this information among their kindred, and it would tend to make them unwilling to be our servants. We noticed that our Zanzibaris were demoralized by their influence whenever they came in contact with them, just as Mr. Stanley's followers were demoralized at Ugarrowwa's camp. Moreover, it should be remembered that these men had been taught to regard white men as "Christian dogs," and the Arabs about Stanley Falls had lately taken every opportunity to disparage us, on account of the recent fight at that place. Mr. Stanley refers to Tippu-Tib's resentment and desire for retaliation, and then goes on to make very bitter remarks about Mr. Deane.

This gallant gentleman has never received proper recognition for his services. Mr. Stanley says (vol. i. 70): "If King Leopold had sent Deane to fight you, he would not have sent him with 30 men." How then did it happen that, if he was not to fight, he was provided with two Krupp guns, ammunition for the same, rockets, &c., &c.? I was in authority at Leopoldville when he went up to the Falls, and know what he had with him, and the substance of his orders. I also know that the smallness of his force was due to the fact that the time of service of the Houssas had almost expired, and we had great difficulty in obtaining even this number of volunteers. Were these guns, &c., transported all the way from Vivi at great trouble for this occasion, not to be used at all? Indeed, no! The State had been dissatisfied with Mr. Deane's predecessor because he was too friendly with the Arabs, and among other instructions Mr. Deane was told to use his influence to suppress slavery. When the time came, and peaceful methods had failed to effect the settlement of a palaver over a slave, he prepared his station for action, and when the Arabs attacked him, made a gallant defence, until his men deserted and his ammunition was exhausted. He was then obliged to blow up the station and flee for his life. He had been told to suppress slavery as far as lay in his power, but his efforts were not supported by the State, and even when he had done his best, his action was condemned. His great sufferings are now ignored, his gallantry decried, and though he is dead, and therefore unable to defend himself, he is ungenerously attacked by those he served.

The Arabs having gained a victory over the chief of the station, looked with contempt upon all white men. After Deane's brave defence and forced evacuation of the place, the State did nothing to re-establish their authority for nearly a year, and then they *tried* to do so by placing Tippu-Tib, the ruler of the victorious Arabs, in the position of governor of the district, with a salary of £30 a month! In the interim the Arabs, with their Manyuemas, were emboldened, took all the villages on both banks of the Congo from Stanley Falls to the Aruwimi, and laughed at the white men. In this way our prestige in this region received a severe blow, and it was difficult to make these men understand that we were allies of Tippu-Tib. Major Barttelot accompanied the Arab chief to the Falls, and the latter explained to the sheiks that he had made peace with the State.

Here, then, at the outset, Tippu-Tib would find it hard to force the Manyuemas—for this wild tribe was not yet thoroughly subjugated—to become servants to the white men, and he may well have doubted his ability to provide 600 porters at once. However, Mr. Stanley now points out that he did not expect them to be immediately forthcoming, and he even doubted whether they would come at all. He gives his opinion of Tippu-Tib at some length in a conversation with Major Barttelot (vol. i. 126), closing with the remark: "Then, as a last proof of how I regard Tippu-Tib, do not forget that written order to Lieut. Stairs a few days ago, to rake his settlement with the machine-gun upon the least sign of treachery. You have read that letter. You ought to know that the gage of battle is not thrown in the face of a trusted friend." He lauds him in his second volume, and then, speaking to Salim-bin-Mohammed, says: "Neither you nor Tippu-Tib have the least intention of keeping your word" (vol. ii. 19). That he expected Tippu-Tib might fail is indicated by these words: "Assuming that Tippu-Tib really keeps to the letter of his contract," "If Tippu-Tib came, why, well and good; if he did not come, be indifferent."[1] Again (vol. i. 118), he casts doubts on the Arab's faithfulness. "Supposing we had not made agreement with him, how long should we be left to prepare for the march to the Albert, or how long would you be permitted to remain here, before you would have to answer the question why you were on his territory? Could I possibly leave you here—*with my knowledge of what they are capable of*—alone?"

Yet, while knowing the vindictive character, and doubting the faithfulness of Tippu-Tib, who was not without guile, according to Stanley's own showing (vol. i. 113, *et seq.* and elsewhere), he left Major Barttelot at the Arab's mercy. He trusted in an agreement made with one

[1] Letter dated August 31st, 1889.

who, he reminds Major Barttelot, had broken a former contract between them. At one moment, in a letter dated August 31st, 1889, he writes—"At the time he agreed he was sincere in those intentions;" and then adds, "If Tippu-Tib broke his written agreement made with me before the Consul, his promises to you would be still more unreliable."

On the subject of Tippu-Tib's delay in providing the men, Mr. Stanley has much to say: "If Tippu-Tib is here in nine days, it will be the biggest wonder I have met. . . . He will not be here in fifteen days, or even twenty days. We must be reasonable with the man . . . . wait for him patiently until the *Stanley* comes" (vol. i. 121). In his instructions he writes: "It may happen, also, that though Tippu-Tib has sent some men, he has not sent enough to carry the goods with your own force." He points out that the Arab would not unnaturally find it hard to supply so many men at short notice, and we must allow him a fair time. He himself speaks vaguely of twenty days, and even forty-three days, when the *Stanley* would be due. To stretch forty-three days into two months was not unreasonable. But the force of circumstances caused our commander to wait until November, trusting in Tippu-Tib's repeated promises, which were backed by Salim-bin-Mohammed's representation that the men with him were the advance guard of those who were coming presently. Having waited until November, we had arrived at the time when Mr. Stanley had expected to return. As a proof that such was his expectation, we have this postscript to his instructions: "Give one brass rod per week to each man to buy fish, &c., in *five months* these amount to 2580. Give also six cowries per man per week, in *five months* these will number 15,480."[1] If then he did not expect to rejoin us, he failed to give instructions for rationing the men after that date. It is undeniable that he left the impression upon Major Barttelot that he expected to be back in five months, because, in a letter to the Committee, dated August 15th, 1887, seven weeks after his leader departed, he wrote that he would be compelled to stay at Yambuya "till November, when Mr. Stanley expects to return." Added to this written evidence we have the conversation reported by Major Barttelot to have taken place between Mr. Stanley and himself—we may be permitted to rely on a reported conversation, as Mr. Stanley bases so much upon what he remembers saying to Major Barttelot. The conversation at Yambuya was described most graphically by the Major, who said: "As Mr. Stanley passed out of that gate his last words were, 'Good-bye, Major; I shouldn't be a bit surprised to see you here when I return. I shall be away five months, giving myself three months to go up to Kavalli and two to return, so I will be back in November.'" This was said in Mr. Jameson's presence, and he himself repeated it to us almost word for word. It will be well remembered by the other officers in camp. In fact, everything pointed to his return in November; and from this time on we expected daily to see him back or have news of his

[1] This postscript, published originally in the *Times* of November 29th, 1887, in the official copy of the instructions forwarded by Major Barttelot, is for some reason omitted by Mr. Stanley from the copy of the instructions given in his book.

approach. His failure to communicate with us was a grave misfortune; we were left to surmise his movements, basing our conjectures on his own calculations. He anticipated that he would be able to penetrate to Emin in three months; we therefore did not imagine that 296 days would elapse before they met. We allowed a fair margin on Mr. Stanley's own calculations, and were daily looking for his arrival. Tippu-Tib was still making promises, and our commander hoped they would be fulfilled, but as Mr. Stanley might be with us at any moment, why should we abandon our precious loads for the sake of advancing a couple of days' journey into the forest?

If we did not receive all the men promised conditionally by Tippu-Tib, we were to consider marching with such men as he provided, and with such loads as we could carry. At the same time, Mr. Stanley tell us, without these loads and men "the advance force itself would need to solicit relief." To me he writes—"The stores are of great concern to us."[1] He gives a list of loads, some of which must not be discarded. He dwells again and again upon their importance, and only in the last extremity are we to throw anything away. To bring on 493 additional loads, I am left at Leopoldville. To preserve them all, Major Barttlelot, impressed with their value, is left at Yambuya. Mr. Stanley is advancing to relieve Emin by transporting to him loads of ammunition. His object being to accomplish this, he, rather than lose them, leaves them in a camp on the verge of the forest. He forms a Rear-guard to keep them, considering this his base of supplies. Had they

[1] See Letter dated April 26th, 1887, p. 54.

not been so important he could have marched from Yambuya with all his officers and men with such loads as he could carry, or, by waiting six weeks, he could have taken the rest of us and more loads. But no, he advances, leaving us with his last words ringing in our ears—"without these loads . . . . we shall need relief." What more could he have said to urge us to guard them intact?

Now the scene has changed. He had reached Emin and laid at his feet 31 loads!—a small portion of those provided for his relief. This was all that remained of those entrusted to his 400 men. We with our 200 men had not succeeded in transporting the remaining 700. He now says: "If Tippu-Tib came, why, well and good; if he did not come, then be indifferent, adapt your goods to your carriers." "If he comes any time before the arrival of the *Stanley*, all will be well; but if he does not come by that time, it will prove that the man never intended to come. Don't bother your mind about him, but come along with what you can."[1] He now points out that the 700 loads were of no great importance. This was not the tone of his written instructions—"without them we shall need relief." But, he argues, we should have undertaken a march four times as difficult as his, with only 200 men, and these the refuse of the lot. Yet he failed to place in his written instructions that "if Tippu-Tib did not supply any men, we were to march without them." We never received the letters in which he suggests that we may be marching without Tippu-Tib's aid, and in which he writes, "Your column will surely break

[1] Letter written August 31st, 1889.

up if you are alone."[1] On the contrary, our instructions read—"Though Tippu-Tib has sent *some* men;" and we were to wait patiently for him to do so. Can it be wondered at that Major Barttelot, impressed with the importance of his loads, should consider them such a sacred trust that rather than use their contents he preferred to starve at Yambuya?

Again, Mr. Stanley reiterates—"Why did you not march out?" Can he look at what happened when the column did march out and repeat his question? See what occurs. Tippu-Tib supplies 400 men, and Muni Somai 30. We turn to the written instructions. "It may happen, also, that though Tippu-Tib has sent some men, he has not sent enough to carry the goods with your own force. In that event you will, of course, use your discretion as to what goods you can dispense with to enable you to march." Major Barttelot, using his discretion, takes 380 loads, which Tippu-Tib has obliged him to reduce to forty pounds each. Following Mr. Stanley's commands, as soon as he receives Tippu-Tib's men, he marches out; he advances slowly; his path is beset by endless difficulties; the Manyuemas prove unmanageable; they creep on at a snail's pace (but no slower than they would have done had they made marches seven times over); at the end of thirty-eight days they are at Bonalya, about ninety miles from Yambuya. Disaster overtakes them here. Major Barttelot is assassinated; Mr. Jameson seeks assistance and then meets his death, while with difficulty Mr. Bonny saves less than 300 loads; and of carriers Mr. Stanley writes (under date of September 4th, 1888)—

[1] Letter to Major Barttlelot, February 14th, 1888.

"There were only seventy-one, many of them too sick to move, the majority worthless as carriers, and only about ten at all presentable or suitable for the long journey before us." What then would have happened had we marched without Tippu-Tib's assistance? Could any one reasonably expect that we could have succeeded in those weary marches forth and back? Must we then be condemned because, *in obedience to Mr. Stanley's instructions*, we remained at Yambuya rather than throw away our loads, and eventually, still obeying his instructions, when Tippu-Tib's men came, Major Barttelot marched with what he could carry? Can he, after recalling this disaster that occurred when we had the benefit of the assistance of Tippu-Tib's 400 Manyuemas, still hold that we could have done better with our miserable 175 men?

Mr. Stanley, in his *In Darkest Africa* (vol. i. 6), writes—"While I possess positive proofs that both the Major and Mr. Jameson were inspired by loyalty, and burning with desire throughout those long months at Yambuya, I have endeavoured to ascertain why they did not proceed as instructed by letter; or why Messrs. Ward, Troup, and Bonny did not suggest that to move little by little was preferable to rotting at Yambuya, which they were clearly in danger of doing, like the 100 dead followers. To this simple question there is no answer." In return I would ask, what steps did Mr. Stanley take to "endeavour to ascertain"? Not by asking Mr. Ward and myself for reports certainly; and yet he says this before hearing from us, as well as again after he is put in possession of the information given here, showing how utterly impossible it was to march. Was it the business of Messrs. Troup, Ward, and Bonny to suggest marching when they did not, under the circumstances, approve of it? We had been told to place ourselves under Major Barttelot's orders, and he was requested to let us speak freely our opinions. Mr. Stanley does not trouble himself to discover whether we made such suggestions, and fails to point out what good purpose they would have served when it was impossible to move, or how we could have enforced *any* suggestions.

He continues—"I cannot understand why the five officers, having means for moving, confessedly burning with the desire to move, and animated with the highest feelings, did not move along our track as directed." What means had we for moving? One hundred and seventy-five anæmic Zanzibaris, who were to be laden with the bulk of the 700 loads! Was not our moving to be conditional, according to his written instructions, upon the arrival of assistant porters? When that condition was fulfilled did we not march?

He adds—"Or why, believing I was alive, the officers sent my personal baggage down river, and reduced their chief to a state of destitution; or why they should send European tinned provisions, and two dozen bottles of Madeira [1] down river, when there were thirty-three men sick and hungry in camp; or why Mr. Bonny should allow his own rations to be sent down while he was present; or why Mr. Ward should be sent down river with a

[1] These bottles of Madeira did not belong to the original stores of the Expedition ,but were a present to Mr. Stanley brought up through my personal efforts.

despatch and an order be sent after him to prevent his return to the Expedition? These are a few of the problems which puzzle me, and to which I have been unable to obtain satisfactory solutions." Read Mr. Stanley's instructions to Major Barttelot—"Use your discretion as to what loads you can dispense with;" then follows a list of all the loads. They are given in the order of importance—"3rd, private luggage," "5th, European provisions." These were not so important, according to him, as the ammunition, beads, &c. that Major Barttelot took. Mr. Jameson and Mr. Bonny were the only ones left to assist Major Barttelot in selecting the loads. In his instructions to Major Barttelot, Mr. Stanley writes—"There is only one chief, which is yourself." To me he writes—"Place yourself under the Major." In his book (vol. i. 473), in referring to Mr. Ward and myself, we are, he says, "to be associated with the commander and his second in the discussion of every vital step, and no important decision can be taken unless a council of the four has been convened. . . . They are therefore implicated in the consequences of any resolution, and every sequent act;" even, as he states elsewhere, for events that occurred after we had left Yambuya, and were hundreds of miles away from the scene. He charges me (vol. i. 474) with "concurring and assisting" in sending the loads down river, when he knows from the documentary evidence of Major Barttelot's last report, and Mr. Bonny's medical certificate, that I was so seriously ill as to be unable to be present.

He gives (vol. i. 479) the reply of the Committee to Major Barttelot's telegram, and adds this comment—

"A committee 6000 miles away penetrate into the spirit of the instructions instantly, but a committee of five officers at Yambuya do not appear to understand them." What are the true facts of the case, however? The Committee 6000 miles away have in their possession a copy of Mr. Stanley's instructions, which had been forwarded to them by Major Barttelot, enclosed in a letter[1] which contained these words—"Up to the time of writing this Tippu-Tib's men have not arrived, though these marauding Arabs before mentioned may be them. Should it turn out to be so, or should they eventually come, I should march at once in Stanley's track. *If, however, they do not, I shall be compelled to stay here till November, when Mr. Stanley expects to return, or till such time afterwards as he may return.*" With this letter in their hands, knowing that Major Barttelot's intention was to wait for the promised carriers, or failing them to remain at Yambuya till Mr. Stanley returned—knowing this to be his understanding of Mr. Stanley's instructions, they telegraphed endorsing his reading of them, referring him to the instructions, and as if that might not be sufficient, they added—"If you still cannot march in accordance with these orders, then stay where you are awaiting his arrival." *In accordance with these orders* can mean nothing else than *with the loads*, and if he cannot do that he is to stay where he is. In what does this differ from our understanding of the instructions? What new reading do the words suggest? Major

[1] These instructions and accompanying letter were published by the Emin Pasha Relief Committee in the *Times* of November 28th, 1887.

Barttelot did march as soon as he had men to help in carrying the loads.

Again (vol. i. 479), Mr. Stanley transcribes an order written by Major Barttelot, placing Mr. Bonny in command of our camp in the event of his death, or other contingencies, this order bearing date April 22nd, 1888. He then asks—"What remains for the faithful Jameson, 'whose alacrity, capacity, and willingness to work are unbounded,' to do? Where is the promising, intelligent, and capable Ward? What position remains for the methodical, business-like, and zealous Mr. John Rose Troup?" In reply I would say, Mr. Jameson was on that day hundreds of miles from Yambuya Camp at Kasongo, where he might have been made a prisoner by the Arabs. Mr. Ward was nearly as far distant on his journey to the coast. I was seriously ill, having received an injury the effects of which I shall feel all my life. Had any attack been made on the camp at that time, as there was reason to fear there might be, I should have been unable to defend myself even, much less could I have undertaken the responsibility of commanding the defence of the place.

Now, without waiting for the reports of all his officers, Mr. Stanley fulminates accusations against them. He receives my refutation of these and then repeats them in his book, saying that his questions have never been answered. I remonstrate with him about these misstatements, and he then turns upon me, charging me with specified acts, many of which I had already proved I never did, while for others I had no responsibility. He even goes so far as to blame me for events which happened subsequent to my departure from Yambuya. He would have done well not to have made this charge, as he, in truth, is more responsible for them than I am. He, the leader of the Expedition, though absent, is more guilty—if guilt there is—than I, a subordinate, who was also absent.

CHAPTER II.

FROM ENGLAND TO MANYANGA.

In the month of August, 1886, I found myself at Banana, after having completed my term of three years' service with the Congo Free State. For the last nine months I had been chief of Vivi station. Though I had enjoyed comparatively good health during my three years' residence on the Congo, I had during the last few weeks of my time of service succumbed to a severe attack of bilious hemoturic fever, and was still suffering with it when my three years expired. I was in an extremely weak condition when the hour for my departure from Vivi arrived, and was carried on board the little steamer which was to take me down to Boma, where I rested for two days, but at the suggestion of my friend, M. Janssen, the Administrator-General, I went on to Banana to await the homeward-bound steamer which was due on August 17th. Reaching Banana I saw the sea again, a sight which had not greeted my eyes for nearly three years, for my time had been spent on the up-river stations, and I had never visited the coast during that period. I found the place a good deal altered. The chief improvements were a house erected by the State and a hotel! I took up my quarters at the latter, as owing to my shattered state of health I could not have endured the noise and racket of the station-house. Well do I remember that the first day in that hotel seemed to me like being in Paradise; there was a feeling of civilization about it that one missed in the "bush," and before I left I discovered there was not lacking that emblem of civilization, a bill. After three years of rough life on the Congo the hotel seemed so comfortable—almost luxurious—that I took a calm enjoyment in my ten days' stay while I awaited the arrival of the steamer. Day after day, I sat propped up on the veranda feasting my eyes on the breaking waves and listening to their sullen roar, while I breathed in the salt sea air. At length I got on board the Portuguese mail, and took what I believed to be my last look at Banana and the Congo mouth. Little did I then think, as I waved a farewell to my friends and watched the shore fade from my sight, that I should ever again visit that spot. I had no reason to imagine that ere scarcely more than six months had passed I should again put my foot on its shores—and my experience of Congo life had been such that I had no particular desire then to renew it. But I was now homeward bound, and the stoppages at San Thomé and Madeira and other points only increased my impatience to be at home once more. The voyage had done me no perceptible good, and indeed even the good offices of kind friends on board scarcely counterbalanced the many discomforts one must endure who travels in these Portuguese steamers. I was forced to spend a fortnight at Madeira, however, as I was suffering from repeated attacks of fever and a general state of debility, so that it was not until October 1st that I reached Plymouth in one of the Union Co.'s vessels. Not only had I no intention of returning to the Congo, though I naturally took a considerable degree of interest in the scene of my recent labours, but I then believed that had I wished to go back I should not have an opportunity of so doing, as I was well aware that the State desired to have Belgian officers only in its employ, and that having benefited by the energy and industry of officers of other nationalities, they had determined to close the door to other races and reap what benefit they could from the result of their labours; and on this score one can hardly blame them. At all events, I felt at this time that my health must be my first thought before I dreamed of any other work, and with the object of recovering it I buried myself in a country retreat, not beyond easy reach of London; and I soon found the bracing air of Old England was building me up again. With the exception of an occasional return of fever, I found myself getting well after three months of vegetation.

It was, if I remember rightly, about the end of November or early in December of that year (1886) that the newspapers began to mention an Expedition which was to be sent out for the relief of Emin Pasha, who it was believed needed assistance, and they further stated that Stanley was to go to his rescue. Owing to the recent events at Stanley Falls, which I describe elsewhere, the King of the Belgians also called upon Mr. Stanley to return to Europe, in order that he might consult with him concerning the measures to be taken for re-establishing the supremacy of the State at Stanley Falls. Mr. Stanley, in response to these demands, threw up his very remunerative engagements, and returning to Europe, proceeded without delay to Brussels, in obedience to the King's request.

Having concluded his conference he made his way to London, where he arrived on December 23rd. Shortly after his arrival I called upon him, having by this time made up my mind to volunteer my assistance for the relief of Emin Pasha, of whom I had heard enough to make me willing to do all in my power to further his relief. Stanley, I found, was ready to accept my services, and he advised me to send in at once a formal letter of application to the Committee. At that time I was totally ignorant as to what persons constituted that body, and it was a surprise a few days later to receive a letter from my old chief, Colonel Sir Francis de Winton, who had succeeded Mr. Stanley as Administrator-General on the Congo, and with whom I had had considerable personal acquaintance during the time I was in the service of the Congo Free State. He was now the Hon. Sec. of the Emin Pasha Relief Committee, and in his letter he requested me to come up to town to make arrangements for accompanying Mr. Stanley on his Expedition. In compliance with his request I visited Mr. Stanley's rooms in Bond Street, one of which I found had been turned into a suitable office, in which Mr. Stanley, Sir F. de Winton, and Captain Grant Elliott (late Administrator on the Kwillu) were seated at various tables briskly engaged in writing, interviewing, &c., and surrounded by official-looking bundles of papers and all such paraphernalia. I ceased to wonder at this when I heard that they were daily receiving hundreds of applications from "all sorts and conditions of men" who were anxious to follow Mr. Stanley on his bold expedition. Preliminaries were soon got through as regarded myself; a rough draft of a contract was drawn out by Sir Francis de Winton and handed to me for perusal; this was afterwards printed and duly signed in the presence of Mr. Stanley and Sir Francis de Winton as witnesses. After this, in accordance with their urgent request, I consented to give them what assistance I could in their office work, &c.; but not being willing to give up my short remaining time to living in London, and finding that my services were not in constant requisition, I continued to live in the country, and came up to town when necessary.

At the outset it was proposed that I should proceed by the first steamer to Zanzibar, there to help Mr. Mackenzie, Mr. (now Sir) W. Mackinnon's agent, to engage Zanzibaris. It was thought this would be the best way of utilizing the experience with these men which I had gained during my three years' service on the Congo, and it was Mr. Stanley himself who suggested the idea. It should be remembered that I was the only officer engaged in England who had had any experience in this part of Africa, though Ward, who was afterwards a member of the Expedition, had also been in the service of the Congo Free State; indeed, the other officers who composed Mr. Stanley's staff had not even all had the experience of service in the other portions of Africa, such as the Soudan and the Cape. Hence my knowledge of this part of the world, it was thought, would be of use to the Expedition. I was asked if I could get myself ready to start on my journey by the mail steamer which would leave in three days' time. This

I was prepared to do, and I went down into the country that same evening in order to get my baggage together that I might return to town the next day to make all necessary arrangements for my immediate departure. However, this plan was altered, and I received a telegram from Sir Francis de Winton informing me that I was not to proceed to Zanzibar for the present. It must be borne in mind that at this period no public statement had been made as to which route to Wadelai Mr. Stanley would adopt; whether he would proceed to Zanzibar and from thence strike inland, by some of the numerous routes, to Emin Pasha's Province, or would go by the west coast to the mouth of the Congo, then up the mighty river to some point, either Stanley Falls or a place on the banks of the Aruwimi, and from there march across the unexplored region to Lake Albert. It was at the banquet given at the Mansion House, on the occasion of Mr. Stanley being presented with the freedom of the City, that he publicly stated for the first time that he proposed going by the Congo route. His purpose was to go to Zanzibar, there to engage men as porters, and, if he found a steamer there, he would proceed with his followers around the Cape and up to Banana, at the mouth of the Congo. Before he arrived at this conclusion it may be well imagined that he had thoroughly considered the advisability of taking this route, among the advantages of which was one which it may not unreasonably be presumed influenced his decison—that he would thus be enabled to carry out the wish of the King of the Belgians to re-establish the State authority at Stanley Falls. The measures Mr. Stanley adopted to carry out this purpose I have described elsewhere. But to return to the arrangements for starting the Expedition. As far as I was concerned, on the alteration of the plan for my going to Zanzibar, I continued to give my help in whatever way I could in the business affairs of the Expedition, which were carried on at Bond Street, where a thousand and one things kept us busily engaged down to the moment of Mr. Stanley's departure. The officers who were to accompany Mr. Stanley had meanwhile been selected by him; they were Major Barttelot of the Royal Fusiliers, Lieutenant Stairs, R.E., Messrs. Jameson, Jephson, Nelson, Bonny, and myself. Dr. R. Leslie, who is well known to all those who have served on the Congo (having been there on the staff of Sir Francis de Winton, late Administrator-General), not only for his practical knowledge of African fevers, but for his extreme kindness in caring for those who came under his care, had been asked to accompany the Expedition in his professional capacity. Unfortunately for the members of the Expedition, owing to some disagreement with the Committee as to the wording of the contract, this gentleman did not go, and we were thus deprived of his valuable assistance.

Surgeon T. H. Parke, of the Army Medical Corps, obtained permission from the War Office authorities to join the Expedition, and was picked up at Cairo. It should also be mentioned that both Major Barttelot and Lieutenant Stairs had been granted leave of absence from the Commander-in-Chief, that they might accompany the Expedition. It was decided that I should not proceed with the other officers to Zanzibar, but should go direct to the Congo, there to arrange for the transport of the large quantity of relief stores, such as powder, ammunition, provisions, &c., &c., which would arrive at Banana, and which must be carried overland through the cataract region to Leopoldville, on Stanley Pool.

Mr. Stanley left London on January 19th, 1887. I remember well going to see him off at Charing Cross, where a large crowd of his friends and the general public had assembled to wish him God speed in his undertaking. I had to fight my way through the dense throng of privileged spectators on the platform in order to reach him that I might receive his final instructions. His other officers had already departed by steamer from London, and he would overtake them at Alexandria. My turn came next, and I was the last to leave England. It was by the midnight express on the 25th that I left Euston for Liverpool, in company with Mr. Ingham, a member of the American Baptist Mission Union, who was returning to the Congo with his wife. He was an old friend of mine, for I had known him when he was in the service of the Congo Free State. It was owing to his energies that the transport service between Lukungu and Matadi, in the cataract region, had been established and worked successfully. The Emin Relief Committee were most anxious to secure his services, but as he was going out to resume his missionary labours, they could not prevail upon him to join the Expedition, but he willingly agreed to help in getting porters from Lukungu, his old district, where he had great influence among the natives, and he was very ready to aid me in my task in whatever way he could. At Liverpool we caught the Pacific mail steamer *Sorata*, for Lisbon, embarking with us some 100 bales of cloth, &c., to serve as money wherewith to engage carriers on the Congo. At Lisbon we were obliged to stay a day or two, and on February 6th we left in the Portuguese mail steamer *Angola*, for Banana, at the mouth of the Congo. This was my second experience of this line of steamers, and I cannot miss this opportunity of saying that the whole of their arrangements for accommodation of passengers and everything else are the worst on record. At Liverpool we had been joined by another old friend, Dr. Sims, also of the American Baptist Mission Union, who, after a short visit home, was returning to the Congo to resume the labours which he had carried on for many years in the same field. We reached Banana, at the mouth of the Congo, on February 28th. Here I was back again at the same hotel after an absence of only a few months. I felt quite at home in these parts, meeting many old friends who were all as much surprised to see me as I was to find myself back again. When the news got abroad, that Mr. Stanley with hundreds of men was coming round from Zanzibar, and was going up the Congo, the excitement was great. I was soon besieged on all sides by all sorts of questions concerning the Expedition, which I answered to the best of my ability. Our 100 bales of cloth had been landed at the Dutch Factory, and we were very anxious to make arrangements for getting them up to Matadi without delay. I had called upon the Commissaire de District at Banana, who informed me that it would be impossible for the small steamer belonging to the State—the only one that had been sent down to meet the mail—to carry a heavy cargo in addition to what was already on board. But an old friend came to my assistance, Mr. Donald Fraser, chief of the British

Congo Trading Company, who offered to transport me and all my loads as far as Boma, in his steamer, the *Albuquerque*. So the morning of March 2nd found me on board, and we were soon steaming merrily up the Congo. Fraser and his colleague, Dennett, were on their way up on business to their branch house at Boma. They did their best to make the journey an agreeable one to me, placing every comfort at my disposal. Pleasant were the hours spent under their sheltering awning as we made our way along the familiar river, many the pleasant chats over old times as we leaned back smoking comfortably. I little thought then that my sturdy comrade, Fraser, after his fifteen years or more service in this part of Africa, would succumb to the deadly fever ere I came down the river again; no thought of such a possibility saddened our cheerful day together. It was a year and a half later that I heard the full details of his illness and death from his companion, Dennett.

Ingham, his wife, and Dr. Sims stayed behind at Banana, purposing to come up in the Dutch House steamer a few days later. The little *Albuquerque* steamed up to Boma, where we arrived at four o'clock that afternoon. Here are the head-quarters of the Congo Free State—the residence of the Governor—as well as the various offices of administration and a large depôt of stores. Arrangements had been courteously made to take me up in comfort to the Governor's house. M. Janssen, at that time Governor of the Congo Free State, was absent, on leave, in Europe. The administration of affairs had been left in the hands of an executive council, consisting of Lieutenant Valcke, Major Parminter, and the Judge. It was to the first-named of these three gentlemen that I presented my letters. These letters had been obtained by the Emin Relief Committee from the State officials at Brussels, and contained requests to the various officers of the State to do all in their power to assist me in forwarding the interests of the Emin Relief Expedition. I was courteously received by my old friend, M. Valcke, who assured me that he would pass on the instructions to the other officers, and promised me that I should receive whatever assistance I desired.

It might be as well, for the benefit of my readers, for me to state clearly here with what object I had been sent out in advance of the rest of the Expedition. Mr. Stanley, on his arrival at Banana, would have with him several hundred Zanzibaris, permission having been obtained from the Sultan to take 600 men round with him to the Congo. These would not be sufficient to transport all the loads over the land route from Matadi to Leopoldville. It therefore became my duty to engage native carriers from this region to supplement Mr. Stanley's men; but Ingham had undertaken to obtain men from his old district, Lukungu, who would carry the loads over the first portion of the road; that is, from Matadi to Manyanga. It is the custom for natives to carry loads through certain districts, usually their own and the neighbouring country. These men would not go beyond Manyanga. It was hence necessary that I should find carriers to complete the transport between that point and Leopoldville. My intention therefore was to make my way as speedily as possible to Manyanga to accomplish this portion of my task; but at the outset I was detained at Boma owing to the illness of M. Valcke, from whom I was to obtain the necessary letters; he was to provide me with one to M. Franqui, Commissaire de District at Lukungu, and another to M. Lieutenant Dannfeld, Commissaire de District at Manyanga. M. Valcke had already complied with my request for a tent, camp-bed, and other necessaries of life, for I had been obliged to leave England without these supplies, as I was told that mine had been shipped with the general lot to Zanzibar, and I was further informed that I would no doubt find the State authorities on the Congo able to supply me. Although it was well known that there are never any spare articles to be had out there, yet fortune favoured me, and I was able to borrow articles for immediate use until the Expedition arrived from Zanzibar. A day or two before my departure Ingham had proceeded in the Dutch House steamer to Matadi and thence to Lukungu, where he found his task simplified, as M. Franqui had engaged all the obtainable carriers of this district for the service of the State, and they were now placed at the disposal of the Emin Relief Expedition. As soon as M. Valcke recovered and was able to complete the necessary arrangements on my behalf, I left Boma on the s.s. *Heron* for Matadi. This was not until March 11th. I had been somewhat troubled by these few days' unavoidable delay at Boma, but though anxious to start as early as possible, I could not leave by the previous steamer because M. Valcke was not at that time well enough to attend to business, and I was obliged to await his recovery. When that occurred I took the first opportunity to proceed to Matadi. I would here mention that I received the most courteous attention from

M. Valcke, Major Parminter, and all the State officials at Boma, who did what they could to expedite matters.

The bales of cloth which had been brought up by my friend Fraser, he had kindly carried on as far as Ango-Ango, a point a little short of Matadi. On my arrival at the latter place I rowed down in the whale-boat to Ango-Ango and brought up the bales, which I sent off to Lukungu by caravans furnished by the State. I took advantage of the arrival of the next caravan to start on my journey up country. I left Matadi on March 14th, late in the afternoon, being accompanied as far as Palaballa by my old friend Baron von Rothkirch, of the Sandford Exploring Association, who, by means of a large caravan, was taking up some machinery for one of their steamers. I had to leave him at this point, as he was travelling slowly with his heavy loads, whilst I was desirous of pushing on more rapidly. There is no use in describing the well-known beaten track between Matadi and Lukungu; mine was an uneventful journey, except that I met Ingham, accompanied by the carriers he had immediately got through M. Franqui, at the Kwilu river, on his return journey from Lukungu. We lunched together, and he informed me that he had seen our mutual friend Ward, who had been for some time in the service of the State, and afterwards in the employ of the Sandford Exploring Association. He had just left the latter service and had been on the point of returning to Europe, when, hearing of Stanley's proposed Expedition, he determined to await his arrival, and offer him his services. Ward had accompanied Ingham from Matadi to Lukungu, and had remained at the latter place for the purpose of bringing down some more carriers when they came in from their villages to report themselves at the station as ready to take service. About three days after leaving Matadi we came to Masamba Wood, the only piece of forest in this region. At the edge of the wood I saw a small encampment, which proved to be in charge of two Houssas. On coming up to it I recognized one of these men as Danganna, who had served under me when I was with the State. He with the others were stationed here to look after some heavy pieces of machinery which had reached this point, and were waiting for a fresh relay of carriers. He was undoubtedly delighted to see me, and I fancy he drew the long bow when he undertook to describe to his Houssa companion my importance. Having picked up by this time a few words of English, Danganna proceeded to tell me all he could about his former comrades who had belonged to my old Houssa force at Leopoldville. When his glee at seeing me again and finding I had not forgotten him had somewhat abated, he went to his hut, from which he produced a large bugle, and made the woods re-echo with its notes. I remembered in the old days his one ambition in life was to become a bugler, and he had employed all his odd moments in trying to learn this accomplishment—for, be it known, the position of bugler is much coveted, because the duties are rather light, and it gives the holder a large degree of importance among his comrades.

I soon found that it was the intention of Danganna and his fellow Houssa to do escort duty for me, and with this addition to our party we continued our journey through the wood. Not satisfied with his own discordant attempts to make a noise, he persuaded his companion to play upon a concertina he possessed; thus accompanied by merry music our march was by no means silent, the notes reverberating among the trees. Danganna's progress in the study of bugling had not brought him to such an advanced stage as would please a fastidious London audience, and at times I rather wished his wind would fail, but in *this* direction he had developed wonderful power, and his anxiety to show me due honour led him to make the most of his ability. As it was I had great difficulty in restraining my risibility, but knowing it was his method of showing his admiration for me, I did my best to let him know I appreciated it. During the time I was at Leopoldville, where my old friend Captain Saulez was Chief of Division, I had command of the Houssas, who were soldiers of the State. I had taken considerable interest in Danganna, who was something of a character in his way. I remember on one occasion, while he was on sentry duty I came upon him near his post, and in the dusk noticed that he was carrying a queerly-shaped bundle on his back and was walking with a more swinging gait than usual. As I drew near I discovered he was uttering peculiar crooning notes, which he doubtless considered a song. On asking him what he was carrying he showed me a tiny bit of ebony, which being disturbed gave unmistakable signs that it was a black baby of the smallest dimensions; he had skilfully attached it to his back by means of a scarf, somewhat in the style adopted by Indian squaws with their papooses. I knew he was not a married man, so I called up my Houssa servant, Tonko, who acted as interpreter, to inquire how Danganna happened to be carrying about this morsel of humanity while on duty. His long explanation revealed the fact that this was the latest addition to the family of Mrs. Abu Bukare, the wife of one of the buglers; she was ill, and the man had undertaken to play nurse. Tonko added, "Danganna he very fond of small boy. Oh! he like him plenty too much." Though laughing at the ludicrous disregard of the eternal fitness of things in combining sentry duty with nursing, I did not find it in my heart to order him to dispense with his queer knapsack, especially as I knew that the child would act as an incentive to wakefulness. I never found out whether this care of the baby acted on its father, the bugler, as an inducement to give Danganna free lessons on his favourite instrument; in fact, I did not pursue the matter far enough to discover if any contract had been made between them. The memory of this amusing incident came back with great force when I saw him on this occasion produce the bugle and put it to his lips in a way that reminded me of old times. I had after all enjoyed his efforts to do me honour, so that I experienced a feeling of regret when we were out of the woods, and he looked upon it as necessary to return to his post. It was a pleasure to see his beaming smile, which brought out in strong contrast the whiteness of his teeth and his dark skin, when, with a correct military salute, he left me to pursue my journey to Lukungu, while he retraced his steps.

On the morning of the day I arrived at Lukungu I met another old friend, Casement, in charge of a caravan with loads for the Sandford Exploring Association. We lunched together, and he accompanied me some distance towards Lukungu, which I reached late in the afternoon. Here I found MM. Franqui, St. Marcq, and Ward. After talking over the situation with the former, I found matters were as Ingham had stated, that all the carriers in the district had been engaged for State service, but had been placed at the disposal of the Expedition, and would be sent down as they came in on the days fixed in their agreements. I was again delayed, this time, as before, through no fault of my own, so that it was not until the 26th that I was able to start for Manyanga, in company with M. Franqui. Nearly half-way between Lukungu and Manyanga the Dutch House have a trading station, at a place called 'Ndunga, afterwards the scene of Mr. Stanley's encampment. Here we were hospitably entertained at lunch by my old friend M. Greshoff, manager of the Dutch House at Kinchassa. It was six o'clock that same evening when we arrived at Manyanga, where I was heartily greeted by Lieutenant Dannfeld, whom I had long known in the service of the State. After I had made known to him my requirements, he entered body and soul into the spirit of the affair, and knowing my friend well, I felt sure that my task would be lightened by his able and varied experience in the transport service.

CHAPTER III.

STANLEY'S VOYAGE TO ZANZIBAR AND ON TO LEOPOLDVILLE.

MR. STANLEY after leaving England made his way at once to Cairo, where he remained a few days, and consulted the Egyptian Government about his proposed plan of action. Although he had already made up his mind to go *viâ* the Congo to Wadelai, he had some difficulty in convincing the Cairo authorities that this was the most expeditious route. I use the word consulted, for it should be remembered that it was necessary to show some deference to the Egyptian Government, as it had promised to give £10,000 towards the expenses of the Expedition. It is well known that they favoured one of the inland routes from Zanzibar, and were supported in this by the advice of Nubar Pasha and Dr. Junker. The latter had just returned from Wadelai, *viâ* Uganda to the East Coast, after having spent some time with Emin Pasha.

Sir Evelyn Baring, on the other hand, if we can rely on the newspaper reports, was convinced that the Congo route must be adopted, so he loyally supported Mr. Stanley in his project. Eventually Mr. Stanley had his own way, and persuaded the Egyptian Government to give in to his views. He left Cairo on February 3rd with about 100

Soudanese soldiers, a few Somalis, and two interpreters. There also went with him Dr. Parke, of the Army Medical Corps, who had been granted permission to accompany the expedition as medical officer. These shortly after joined the other officers who had left England previously. Mr. Stanley and his seven officers all arrived at Zanzibar, where Mr. Mackenzie, agent for Mr. (now Sir) W. Mackinnon, chairman of the Emin Pasha Relief Committee, had engaged for him 600 porters. The pay of these men was to be at the rate of four dollars per month, and the niamparas, or headmen, were to receive six dollars per month. At Zanzibar also a numerous string of donkeys were purchased for the Expedition, as well as hundreds of sacks of rice, bales of Zanzibar cloth, sacks of cowries and assorted beads, coils of brass and iron wire, and such goods. The ammunition provided consisted of 100 cases of powder, amounting to sixty pounds in weight each (all the loads were of the same weight, this making the standard load carried by the Zanzibaris and natives), some hundred cases of Remington and Winchester cartridges and percussion caps, and cases of ammunition for the Maxim machine-gun. These made up the relief stores. Added to this was a small supply of tinned European provisions, supplied in cases by Fortnum and Mason, for the use of the officers of the Expedition; a supply of medicines and also camp equipages, such as tents, beds, &c., pack-saddles, saddlery, spades, axes, billhooks, &c. Each officer was allowed, besides his tent and bed, personal baggage amounting to two loads of sixty pounds each. All these above-mentioned loads were taken round from Zanzibar in the steamer *viâ* the Cape to Banana, there transhipped into the various small river steamers obtainable, to be carried to Matadi. Thence 600 of them were to be taken up by the 600 Zanzibaris to Leopoldville, the remainder were to be transported to the same place by natives engaged for that service by Ingham and myself. At Leopoldville they were to be loaded into whatever steamers Mr. Stanley could procure, to be taken up the Congo some 900 miles, either to Stanley Falls or to the Aruwimi, and from there carried some 400 miles to Wadelai! All the Zanzibaris were armed with Remington rifles, and a water-bottle was served out to each man. Each officer then with Mr. Stanley had been provided with a Winchester rifle and a Colt's army revolver.

There was also the Maxim machine-gun, which was in charge of Lieutenant Stairs. It was a marvellous triumph of mechanical skill, and worthy of the greatest admiration, yet its usefulness might become impaired by difficulty in obtaining water for the water-jacket, or the whole thing might be entirely thrown out of gear by being choked with sand.

At Zanzibar, also, Mr. Stanley met again Tippu-Tib, the great Arab chief, whose head-quarters were at Stanley Falls on the Congo, and whose influence is felt far beyond that point at Kasongo. He had come over from the latter place by the well-known caravan route to Zanzibar, *viâ* Ujiji and Tabora, by which route he also sends all his ivory to Zanzibar, and in return he has hitherto received, by means of it, his supplies of powder, cloth, &c. He is a large landowner in Zanzibar, having invested considerable sums of money in property, and no doubt he intends to settle there with his numerous wives when he is too old for, or when he has had enough of, slave-raiding in the regions of the Congo.

Remembering that Stanley had two objects in view, viz., the relief of Emin Pasha, and the re-establishment of the State authority at Stanley Falls, it strikes one as very opportune that Tippu-Tib was at Zanzibar at this time. No one knew this man better than did the great explorer. Yet knowing so much about his history Stanley approached him, perhaps, with the idea that without his assistance he could scarcely accomplish either of his objects.

I have reason to believe that Stanley had considerable difficulty in persuading Tippu-Tib to fall in with his plans; I have been told by one of my comrades that when the Arab chief was informed that Stanley was going up the Congo to Stanley Falls, and intended to march into the interior, he was very suspicious, and in the first heat of his wrath declared he would not allow Stanley to use any of his caravan routes. Yet I cannot for the life of me see what he could fear, considering that the bulk of Stanley's men would be Zanzibaris. But, on the other hand, Tippu-Tib was quite shrewd enough to suspect that there must be something behind all this. His followers at Stanley Falls had just a few months previously overcome Mr. Deane's resistance and had taken Stanley Falls Station, and it was but natural that he should fancy that the white men intended to have their revenge, and he evidently could not see how that could be accomplished by placing him in control of the station he had already conquered, unless some sinister design lay hidden under this offer. However, as I have said, Stanley knew Tippu-

Tib well, and had great influence with him. He succeeded, to all appearances, in allaying the Arab's fears, and persuaded him to accompany the Expedition on its voyage to and up the Congo. Stanley proposed, in the name of the King of the Belgians, to make him Governor of Stanley Falls, which he had in reality been for some time back, although heretofore unrecognized by the State. It has never been publicly mentioned how highly Tippu-Tib prized this recognition.

Stanley entered into an agreement with the Arab, which I copy verbatim from the correspondence of Mr. Holmwood, then Acting Consul-General at Zanzibar, as it appears in the parliamentary paper issued by the Foreign Office. [His letter of March 3rd, 1887, is of so much interest, and so fully confirms much that I have said, that I give it elsewhere.] The agreement is as follows:—

"Mr. Henry Morton Stanley, on behalf of his Majesty the King of the Belgians, and Sovereign of the Congo State, appoints Hamed-bin-Mohammed al Marjebi, Tippu-Tib, to be Wali of the Independent State of the Congo, at Stanley Falls district, at a salary of £30 per month, payable to his agent at Zanzibar, on the following conditions:—

"1. Tippu-Tib is to hoist the flag of the Congo State at its station near Stanley Falls, and to maintain the authority of the State on the Congo and all its affluents at the said stations downwards to the Bujine or Aruwimi river, and to prevent the tribes thereon, as well as Arabs and others, from engaging in the slave trade.

"2. Tippu-Tib is to receive a resident officer of the Congo State, who will act as his secretary in all his communications with the Administrator-General.

"3. Tippu-Tib is to be at full liberty to carry on his legitimate private trade in any direction; and to send his caravans to and from any places he may desire.

"4. Tippu-Tib shall nominate a *locum tenens* to whom in case of his temporary absence his powers shall be delegated, and who, in the event of his death, shall become his successor in the Waliship, but his Majesty the King of the Belgians shall have the power of veto should there be any serious objection to Tippu-Tib's nominee.

"5. This arrangement shall only be binding so long as Tippu-Tib or his representative fulfils the conditions embodied in this agreement.

(Signed) "HENRY M. STANLEY.
"TIPPU-TIB (in Arabic).

(Signed) "FREDC. HOLMWOOD.
"KANJI RAJPAR (in Hindi).

"ZANZIBAR, *February 24th*, 1887."

At the same time, we are told, Stanley entered into an arrangement with Tippu-Tib, whereby the Arab was to furnish him, on his arrival at Stanley Falls, if he could, with 600 porters to carry loads from his camp on the Aruwimi to Lake Albert. It was quite apparent that the 600 Zanzibaris, who were going to sea with him, would not be sufficient to carry all the loads of the Expedition, and Stanley knew he would require additional men on this overland journey, just as he would need native carriers in the cataract region, below Stanley Pool. All these preliminary arrangements having been satisfactorily settled, Stanley was ready to set out from Zanzibar. A steamer belonging to the British Indian Steam Navigation Company had been placed at the service of the Expedition, by Mr. (now Sir) W. Mackinnon, chairman of that company, to take them round to the Congo *viâ* the Cape. The start was made from Zanzibar on February 25th. At the Cape they stayed for a few days, for coaling, they also picked up a few additional stores in the way of biscuits, &c., and took on board Mr. Walker, who had been engaged as engineer-in-chief to the Expedition, but who, having been unable to catch the steamer to Zanzibar, had proceeded direct to the Cape. This gentleman was to be employed in overhauling whatever steamers Mr. Stanley might be fortunate enough to find at Leopoldville. It is hardly worth while to enter here into the details of the voyage of this vessel, which was laden with Europeans, Arabs, and Zanzibaris; suffice it to say that the Expedition arrived at Banana, at the mouth of the Congo, on March 20th. All the river steamers which were at this port were most promptly placed at the service of Mr. Stanley, to carry his men and stores to Matadi, some miles up the Congo. The British Congo Trading Company, the Dutch House, the Portuguese House, and the State, all joined in lending their aid for this task, and even a Portuguese gunboat rendered assistance. On the arrival of the Expedition at Matadi, the men and stores were disembarked, and arrangements were immediately made for the despatch, by native carriers, of the more than 1000 loads over and above what the 600

Expedition Zanzibaris would be able to transport to Leopoldville. It was on March 22nd that the first caravan with some of the loads started along the overland route to Leopoldville, and it was not much later that the officers of the Expedition began their journey. Stanley had made systematic arrangements for their procedure.

The Zanzibaris, some 600 in number, were told off into companies of eighty, under the command of a white officer. My company was temporarily under the command of Dr. Parke, until I joined the main column. Major Barttelot had the command of the 100 Soudanese. Stanley himself had charge of number one company of Zanzibaris, composed of a picked lot of men. Jephson's company carried the sectional boat, while Bonny's men looked after the donkeys. Stairs and Nelson both had companies allotted to them. Each officer was responsible for the number of loads his men carried, and at the end of each day's march was obliged to check them off, and to report upon them. By all accounts the Zanzibaris gave a considerable amount of trouble.

CHAPTER IV.

TIPPU-TIB'S HISTORY.

TIPPU-TIB, the Arab chief and slave-dealer, had already played such a prominent part in the events that took place on the Congo in 1886 and previously, and was destined to be associated so closely with the history of the Emin Relief Expedition, that it seems right to insert here a few notes about him and his doings, that the reader may clearly understand the sequence of events and their relation to what is subsequently described.

Mr. Stanley first met Hamed-bin-Mohammed, *alias* Tippu-Tib, in the year 1876, near Nyangwé. In his volume, *Through the Dark Continent*, after giving a full description of his personal appearance, he adds :—"I came to the conclusion that this Arab was a remarkable man—the most remarkable man I had met among Arabs, Waswahili, and half-castes in Africa." Such was his first impression of Tippu-Tib. It was about this time that Stanley first saw the mighty Lululaba (Congo), "the river that flows north for ever," and which he soon after descended in that memorable feat of exploration. Tippu-Tib was not at that time the big chief he is now, but he was even then of considerable importance, as Stanley discovered. He entered into a contract with Tippu-Tib that he might obtain that chief's assistance in starting to follow the big river to the sea. But the two great men did not continue long in company, because the Arab was discouraged by the density of the jungle they encountered, and turned back, leaving the explorer to continue his journey without him.

At this period the Arab chief's head-quarters were at Nyangwé, some miles above the cataracts of Stanley Falls. It was not until several years after Mr. Stanley's first trip down the Congo that the Arab slave-raiders succeeded in making their way down the river, and in establishing their influence over the natives on both banks. They had conquered the Manyuemas, a native tribe which inhabited the country around Nyangwé and Kasongo, to the west of Lake Tanganyika, and were able to train them for their (the Arabs') service, and employed them to assist in subduing other natives further down the river.

In 1879 Mr. Stanley, under the auspices of the King of the Belgians, as President of the African International Association, undertook to form stations and open up the country on both banks of the Congo. He returned to Europe in 1882, having founded a series of stations on the river as far as Leopoldville on Stanley Pool. But his absence from Africa was of short duration, as we find him again on the Congo at the end of the same year. He proceeded to transport the small steamers he had brought out, overland to Leopoldville, and in April 1883 started on a voyage up the Congo from that point. In December of that year, after steaming up a clear waterway of 900 miles, he approached the first barrier to navigation, the cataracts, which he had previously named Stanley Falls. Some little time before reaching this point he found a band of Arabs encamped at Yomburri—they had been raiding and pillaging the country between the Falls and the Aruwimi river, but had as yet formed no permanent settlement here—therefore, when he decided to establish a station at Stanley Falls, it was with the natives that he made negotiations, thus recognizing their authority as superior to that of the Arabs. It was in 1884 that Tippu-Tib made his way down from Nyangwé to Stanley Falls, and there established himself, not far distant from the State station. From this date the Arab influence began to predominate in that region.

After two officers had occupied the position of chief of Stanley Falls station, during the years 1883, 1884, and 1885, the authorities, under the administration of Sir Francis de Winton, began to take steps to place this remote outlying station in some state of defence. To carry out their intentions the late Mr. Deane was chosen by the Administrator-General to succeed Mr. Wester, who had been chief of Stanley Falls; he was to take up with him an extra force of Houssas, two Krupp guns, rockets, and such defensive stores.

I remember well when I was acting chief of Leopoldville the preparations made for Mr. Deane's departure for his up-river post with his small armed force, and the instructions I received from the Administrator-General to assist him by providing men, guns, and ammunition. However, Mr. Deane on his way up was wounded, and returned to Leopoldville; but after a time new arrangements were made for his second departure to Stanley Falls, where he took over the charge of the station. It was not very long after his arrival that he was obliged to use all the weapons in his possession to defend the place against the Arabs who attacked it. Their enmity had been roused by his having carried out his instructions regarding the suppression of slavery. As his men deserted and the ammunition, originally very defective, was almost entirely exhausted, he was obliged to terminate his gallant defence of the station, but before leaving he destroyed, as best he could, the guns, &c., and blew up the buildings. He and his second in command, Dubois, had to flee for their lives, but the latter was drowned as they were making their way to the mainland. Mr. Deane, after a month's wandering and suffering in the bush, was at last rescued. The story of his escape has already been written.

But the Arabs had now gained possession of the Stanley Falls station, as well as all the surrounding country. Tippu-Tib himself was not there at the time, but it was his nephew Raschid who led the attack and took the station. Sefu, Tippu-Tib's son, ruled at Kasongo, and other chiefs controlled Nyangwé, Riba Riba, Kibonge, and many places in this district. It is almost a misnomer to speak of them as Arabs, for very few if any are pure-blooded. All those I have seen are half Arab and half Zanzibari; in some the latter blood predominates, and they can be recognized by the blackness of their complexion, while others are fairer, more like real Arabs. There is not a large number of these half-castes in these parts, but almost all there are act as sheiks over the native Manyuemas, whom they have trained, to a certain degree, to fight in their service. Each sheik has his head-quarters in some village, and rules the surrounding district, but every sheik is tributary to Tippu-Tib, who may be considered as a sort of king, ruling dependent princelings. The Manyuemas are employed chiefly to obtain ivory for their masters, either by taking it by force from the natives, or by capturing the native women and children and demanding ivory for their ransom. These Manyuemas are not kept under strict surveillance by the Arabs, but are sent forth on their missions often enough simply under the control of their own headmen, and having attacked the natives in whatever way they deem best, they deliver up to the Arabs the ivory when their raid is ended. This network of sheiks ruling over the Manyuemas extends throughout the region about Stanley Falls and down the caravan routes towards Zanzibar, but it should not be forgotten that they are dependents of Tippu-Tib. Recognizing his superior power, Stanley made his agreement with Tippu-Tib at Zanzibar. Such was the system, the Arab chief, through his dependent sheiks, being in command of the district at and around Stanley Falls, when Stanley at Zanzibar agreed, under certain conditions, that the Congo Free State should recognize Tippu-Tib as Governor of Stanley Falls station, and it was with the purpose of placing him in possession that he took the Arab chief from Zanzibar, round by the Cape, and up the Congo.

CHAPTER V.

AT MANYANGA.

AFTER this somewhat lengthy digression, to pick up the connecting links in the history of events I must return to my personal narrative. The station of Manyanga at which I had arrived was a place with a history, and such as it was this history did not speak well for the salubrity of the place. As I spent nearly a month here, some description of it will not be out of place. Manyanga is situated 140 miles from Vivi, or 250 from Banana, and is 95 miles short of Stanley Pool. In Mr. Stanley's book, *The Congo and the Founding of its Free State* (vol. i. p. 266 *et seq.*), he gives the following description of this region:—"I ascended in the whale-boat as far as Manyanga, even up to the foot of the cataract. I knew the cataract was impassable, but we were now nearing the conclusion of our river journey, and it was necessary to decide upon the site of the station to be established. A more lifeless, cheerless, unlovely scene than that around Manyanga it is scarcely possible to conceive. The slopes of the upland, which rises on either side 500 feet above the river, are extremely steep, in some places even precipitous. All the soil in which vegetation might thrive seems to be washed clean off the red clay down to a narrow terrace, or into the depths of the narrow ravines, where we see dark lines of trees. Where these slopes run sheer down to the Congo we have only masses of grit-stone piled one above another in admirable disorder. A few projecting points of these rocks have permitted a broad deposit of white sand to fill up indentations in the shore, which, enriched by soil from above, have become fertile terraces. It was at one of these nearest the cataracts that I proposed to settle my camp until we could decide where we should build the station—whether on the terrace or upon some available hill near. Since Vivi and Isangila were on the north bank, of course I wished to continue the chain of stations on that bank, so that, in case of accidents to boats, communication might be continued by land. . . . We commenced . . . . to remove our *personnel* and *matériel* to the highest terrace nearest the cataract of Manyanga. The landing-place was all we could wish it to be at this season—a calm haven, shielded from the currents by a sedge-covered bank. The terrace sloped smoothly to the river. . . . The fertile ledge where we proposed to lodge ourselves temporarily, the more I looked at it the more promising it appeared, though it had the disadvantage of being cultivated. In length it might be over a mile, while its width varied from 80 to 300 yards. A small perennial brooklet close by appeared to promise drinkable water."

While he was at Manyanga he suffered from a severe

attack of fever, so severe that he gathered his companions about him to hear his farewell messages. But it was not to be the end; the famous explorer was to be spared for many years to come. He was not the only one to suffer in health from a stay at this spot; many a friend of mine has been down with fever after being here, and too often has it proved a fatal illness. Stanley writes elsewhere (vol. ii. p. 287): "Yet old Vivi, excepting Manyanga, is the most sickly spot in all our possessions, if the sick list is at all reliable." It was found that there was great danger in the situation of Manyanga North, as it was called; it was exposed to the intense heat of the sun, and as soon as night fell the chill winds suddenly lowered the temperature, so that more than one case of illness was due to this great change. A man heated by his day's work would perhaps be insufficiently wrapped up at night, and the sudden chill would bring on an attack of fever. The theory that a low-lying station is the most unhealthy must be abandoned, or at least modified by experience in this region. If there is danger arising from malarial influences at such a place, yet is there greater danger in some stations built on hills or high land. Referring to Stanley's book (vol. ii. p. 320) we read: "At Equator station, with a river only five feet below its foundations, creeks sable as ink surrounding it, the ground unctuous with black fat alluvium, Europeans enjoy better health than at Manyanga, 240 feet above the river and 1100 feet above the sea. Fourteen miles away from Manyanga, and eight miles from the river, we have a station on the plain of Ngombi, 1500 feet above the sea, where our people have enjoyed better health than at Manyanga Hill."

This, he holds, is largely due to the strong currents of cold air which rush through the deep gorge cut by the Congo. It is true enough that danger lurks in this circumstance.

In view of this unhealthy condition of affairs it was decided to abandon the station on the north bank and to use only the small post on the south bank, which had been established as a point of departure for caravans passing through the cataract region. But this place was not deserving of the name of *station*, for the only buildings were a small dwelling-house and a little store; the former was built of mats and thatched with grass, while the latter was made of clay. They were both erected on a sandbank close to the river, with little or no shelter from the broiling sun. By this it will be seen that the accommodation was of the scantiest, and there was nothing for me to do but pitch my tent on the sand near M. Dannfeld's quarters, for he preferred living in a tent to being mewed up in a stuffy hut, and I considered his decision so wise that I followed his example. Making myself as comfortable as circumstances would permit, I began my work here, which was to continue during the next four weeks. M. Dannfeld assisted me in obtaining the men necessary for the transport service.

The method of collecting carriers is accomplished after this fashion:—The Commissaire de District keeps at his station a record of the "capitas" [headmen] of the surrounding country; these capitas, according to their influence, are able to promise the service of a number of men belonging to their village. When it is necessary to obtain porters for a caravan the Commissaire sends out his interpreter to make known the fact among the leading men, who forthwith come to the station. As each man arrives he is admitted to the presence of the Commissaire, who inquires his name. "Capita Lesa," "Capita Mwanga," "Capita Kupidi," as the case may be, is the reply. "How many men can you supply?" "Thirteen, "twenty-five," "thirty," answers the native, or mentions even a larger number if he is a great man. "When can you bring them?" "After Nkandu market,"—perhaps four or six days hence—for it is at the market that they make arrangements and obtain provisions for the beginning of their term of service. "Very well," assents the Commissaire, "here are so many pieces of cloth," and he gives out to the capita an amount in accordance with the number of men promised, the regulation sum "for rations," as it is called, a sort of engagement money, being half a piece of handkerchief—that is six squares of bright-coloured handkerchiefs—per man, while the capita receives for his own pay a whole piece. The Commissaire notes in his book some such entry as this: "Capita Kupidi of Nzonsa, twenty-five men on March 18th." Sometimes a half-dozen capitas will come to the station on one day, and perhaps only one the next day.

On this occasion the men poured in with considerable rapidity, and it will be seen that we lost no time, for on the 28th, two days after my arrival, I despatched a special messenger to Matadi, bearing a letter addressed to Mr. Stanley, and in it I was able to inform him that I had already engaged 500 native porters, who were to come in during the next month, prepared to carry on the loads to Leopoldville as soon as these arrived at Matadi. I was well pleased that I could report thus favourably to my commanding officer, as it had been impressed upon me by Mr. Stanley that it was most important to forward the loads with the utmost despatch, and that I must do all in my power to hasten their transport. So far so good, but I foresaw there would be considerable difficulty in the matter, owing to some elements of uncertainty that must be allowed for. In the first place it was impossible to foretell positively the day the loads, under Mr. Stanley's charge, would arrive at Matadi. This I presumed would be about the middle of the month, and I based my calculations on this presumption. Then there was another item to be considered, the length of time required to transport them from Matadi to Manyanga. This was varied by circumstances, but the usual time allowed was from eight to ten days. Hence I arranged that the native carriers should begin to come in about the end of the month of March. But my anxiety was by no means at an end, for when the men arrived I questioned whether I should have loads on hand for them to transport. I could not be sure that delay might not occur at some point, and I well knew that it would not be possible for me to keep the carriers waiting idly about the station, for they would return to their villages at once if I had no employment for them, and my outlay of handkerchiefs as earnest money would be wasted, unless I was willing to go to the expense of feeding them while they lingered about. This would be almost impossible too, because the surrounding country was suffering from a famine. So great was the scarcity of food that I had immense difficulty in obtaining provisions for my own personal consumption.

The consequences of all this would be that, rather than see the men return to their villages, M. Dannfeld would employ them, as he had a perfect right to do, in carrying the State loads, which would be delayed because all the porters of the district were placed at the disposal of the Emin Relief Expedition.

Major Parminter, an old comrade of mine, had promised to send me up a special courier immediately upon the arrival of Mr. Stanley at Boma. I knew that Ingham would have sufficient carriers at his command to start the first few caravans of Expedition stores whenever they arrived, and he would have relays of men to continue the work. As far as I was concerned they could not come too quickly, and having made all possible arrangements for their transport, I awaited the arrival of the loads at Manyanga. I had not long to wait, for on March 30th the first two caravans with the Expedition loads reached Manyanga, accompanied by a note from Mr. Stanley. By the way-bill I noted that these two caravans had left Matadi on the 22nd, having taken eight days to accomplish the journey, a very good beginning as far as time record goes. This was, in fact, my first intimation that the great Expedition had arrived on the Congo. The caravans brought up fifty-six sacks of rice and beads, and these were all started the next day to Leopoldville by my first detachment of carriers. It was not until April 1st that the promised messenger arrived from Boma, bringing a letter from Major Parminter, informing me that Mr. Stanley had arrived at Banana on the 20th of March. He also confirmed the truth of the vague rumour, which had already reached me, that Stanley was bringing with him

Tippu-Tib, the famous slave-dealer. The same courier brought me a letter from Ingham enclosing, according to Stanley's instructions, for the commander was too fully occupied to write himself, a detailed list of the stores, &c., belonging to the Expedition, which should pass through my hands on their way to Leopoldville. The total amounted to considerably over 1000 loads, so that it was incumbent upon us to immediately set about engaging more carriers. M. Dannfeld sent out his interpreter and even went himself to arrange for the number required. My work had now fairly begun, and I found my hands most fully occupied. I had no assistant, except the small boy who had been given me by the State as a body servant, and though M. Dannfeld did all in his power, he had no men to place at my disposal. The staff of men at the station was very small, and the number of useful ones had been further reduced by sickness, incident on the scarcity of food. In consequence of this all the work fell upon me, and I was fortunate enough to keep well and able to attend to my many onerous duties. My task was no easy one. As the carriers came in from Lukungu my first duty was to check off each load as it arrived; I had then to examine it to see that it came up to full weight, as there was a chance that some of its contents had been stolen. As it was the rainy season the sacks would often come in soaking wet, so that, in order to save the contents from destruction, I was obliged to give them my personal attention, and see to their drying and repacking. At times I would have a number of men collected at the station ready to take on the loads, but there would be none on hand. I would persuade them to wait a couple of hours, while I would at once make my way down to the 'Npioka river, which ran at the back of the station. This bit of water the caravans from Lukungu must cross; there I would watch in the hope of seeing the *avant garde* approaching. As soon as the first members of the caravan came up to me I would hasten them on to the station, and display them to the impatient porters to prove that loads were coming. When the loads had been properly attended to I placed them in the hands of my men and started them on to Leopoldville. Here is an entry from my notebook, showing one of the causes of delay I had to contend with: "N.B. Capita Wamba, bringing way-bill for eleven loads of brass wire, only nine arrived; one case, No. 73, with half the mitakos taken out and one case missing." Below this I find added: "This palaver settled, the mitakos turn up all right."

I was able, however, to forward the loads either the very day of their arrival or the day after, with the exception of the damaged goods; these, I am sorry to say, were somewhat numerous. I found that the rice, beads, and cowries brought from Zanzibar, as well as the mitakos bought on the Congo, suffered much from having been badly packed in the first instance; but I did my best, with the scanty materials at my disposal at Manyanga, to patch them up and make them more taut for their journey to Leopoldville. But I had a good many difficulties to overcome, among them the lack of twine, needles, or sacking, and I could not easily obtain these necessaries. With my experience of the rough handling such things must undergo in the course of their transportation, it was marvellous to me that they should have been allowed to be started on their way in such bad condition. I foresaw that they would lose greatly in value, owing to the gross carelessness of the persons originally responsible. No London merchant would send goods so badly packed as far as Brighton, and how was it possible to expect these loads to endure without damage the many knocks and blows incident to their passage through so many hands? First, there would be the loading at Zanzibar; then at Banana they would be re-shipped to Boma; again they would be roughly handled in unloading at Matadi; in their transit from that point to Manyanga their carriers would deposit them with more or less force twenty or even fifty times. Hence it is little to be wondered at that by the time they reached me the sacking was more or less worn through by the rubbing of the angular contents. When I had repaired them as best I could with the materials obtainable and started them onward, it was not to be my last experience with them, for until the very end, after their arrival in Yambuya camp, they were a source of trouble to me.

I found that some of the caravans were taking a very long time to come up from Matadi, some even consuming seventeen days in the journey. I knew by my former experience that the men were only too apt to linger when they were passing through or near their villages in Lukungu. It was a great temptation to meet their families or friends here, and they would sometimes spend several days loafing about when they should be on the road. This caused so much detention that I was obliged to write to M. Franqui, chief of Lukungu, asking him to be good enough to look into the matter, suggesting he should send men to see that no caravans lingered in these intervening villages on their way to Manyanga. I got no satisfactory answer from him about this, and not having a single man of my own, hence none to act as a sort of police, I decided reluctantly that I must devote a day to going to Lukungu to see about the matter. On April 7th I started, taking with me my servant. We had just crossed the 'Npioka river and climbed the high ground on the further side, from whence a good view down the Congo could be obtained, when I noticed a sail rounding the point. I knew there were no State boats running in these parts at this time, so I came to the conclusion that it must be some of Stanley's party, for I was aware that he was carrying a sectional steel boat, and would most likely put it in the water when he came to the first clear reach. As I looked I saw they were heading for a large sandbank, and to save them from getting aground I fired my Martini, and believing I had attracted their attention, I snatched a handkerchief or scarf from my boy, and waved it in lieu of a white flag to warn them off. My signals were understood, and I had the satisfaction of seeing them clear of the bank. Through my glass I could see the white shirts of the crew, which proved to me that the boat was manned by Zanzibaris; and I saw, too, that they were in the charge of two white officers. I felt that under these circumstances I must give up, at least for that day, my proposed excursion to Lukungu, so I re-crossed the river, and waited on the sandy beach for their arrival. In a few moments I caught the sound of the unmistakable Scotch accent of a cheery-voiced man, who hailed me with "How are you, Mr. Troup?" When he sprang ashore, I had no difficulty in recognizing my old friend Mr. Walker, who had been for two years superintending-engineer in the service of the State. Having completed his term, he had returned to Europe, where he had been engaged by the Emin Relief Committee to exercise his talents in preparing the steamers at the Pool for their voyage up river.

The other white man I soon found was Mr. Mounteney Jephson, another officer of the Expedition, whom I now met for the first time. In fact I had not previously had the opportunity of meeting any of my fellow-officers, with the exception of Lieut. Stairs, who had been introduced to me in Stanley's rooms in Bond Street.

With my two companions I made my way to the station. From them I learned that Stanley with his officers, some 600 Zanzibaris and 100 Soudanese, had begun his march from Matadi to Leopoldville. They had left him at Banza Manteka, a village on the well-known caravan route, about three days' march from Lukungu; from there they had been sent off the ordinary inland route to the nearest point on the river, where they had launched the boat and sailed up to Manyanga. That night our little encampment on the sandbank was swelled by the addition of the tents of Jephson and Walker, which were pitched near ours, while their Zanzibaris made themselves comfortable not far off.

That very same afternoon a special messenger arrived at Manyanga, bearing two letters from Mr. Stanley addressed to me. I give them *in extenso.*

[LETTER No. 1.]

"KWILU FERRY, *April 4th*, 1887.

"DEAR SIR,

"What rice you may have sent to Lutete's care of Baptist Mission only, ask them to let it remain with them until I appear only. I have sent thirty-two sacks (canvas) with Mr. Jephson, by boat from near Banza Manteka. He will probably be with you as soon as this arrives. I wish you to remain at Manyanga until Ward or Ingham appears with the last of the goods sent from Matadi.

"Your information respecting boats received, but it is of no use to this Expedition. We only need the boats already at the Pool.

"If you could send 200 men from Manyanga to Matadi it would assist me very much. Manyanga is tolerably near the Pool, but Matadi is distant, and my anxiety is great respecting the arrival of those goods within a reasonable period.

"Ward's engagement with me depends on the early transport of the Matadi loads.

"I shall be three days hence at Lukungu, Friday, April 8th. On the third day after I hope to be at Lutete, where I hope to meet all the rice with you, and in the boat there to ration my people.

"Yours truly,

(Signed) "HENRY M. STANLEY.

"J. ROSE TROUP, ESQ."

It was apparent to me that it was his intention not to pass through Manyanga, but to take the upper road—the direct inland route—from Lukungu, and from thence to Leopoldville. As regards the rice he asked for I had been on the point of forwarding the sixty sacks on hand to Leopoldville, but that very morning I had received a despatch from M. Franqui, stating that Mr. Stanley had given orders that all the rice should be returned to Lukungu, and it was now too late to recall it. Although this counter order business perplexed me much, in the end the affair turned out all right, as Stanley was able to ration his men on their arrival at Lukungu instead of waiting until they reached Lutete.

I had been much puzzled as to the means Mr. Stanley would adopt to find steamers for the transport of all his men and loads up the river from Leopoldville. It had been stated in the newspapers at home that the King of the Belgians had generously placed the *flotilla* of the State at the service of the Emin Relief Expedition. This read well enough on paper, but *I* was only too well aware of what that *flotilla* consisted. The only steamer of any size was the *Stanley*, which I remembered had arrived at Leopoldville in August 1885, when it was badly put together in a great hurry. It had been constantly running up to date, being in continual demand, and there were no facilities at Leopoldville to allow of an overhauling; so matters had been allowed to run on till she was in a very dilapidated condition. There had been three other small steamers belonging to the State at Leopoldville, the *En Avant*, the *A.I.A.*, and the *Royal*, but these were little more than steam-launches. I had heard that the *Royal* had been broken up after her engines had been taken out; these were to be placed in the *A.I.A.*, as hers were played out. However this had not been done, but the *A.I.A.* was lying at Leopoldville without engines, and could, consequently, only be used as a barge. Moreover, the *En Avant*, according to information I received, was cruising about between the up-river stations, and no one knew at what date she would return. Therefore the available *flotilla* consisted of the *Stanley*—the only steamer—the *A.I.A.* as a barge, and two whale-boats. These combined I knew were incapable of carrying half of the men and loads of the Expedition. Of course I was aware that there were two missionary steamers, the *Peace* and the *Henry Reed*, but it was doubtful whether they would be at Leopoldville, as they are constantly, like the State steamers, plying between that place and the up-river stations. I was also aware that there had been some correspondence at home with the secretaries of the Missionary Societies regarding the use of these steamers, but I had reason to believe that nothing definite had been settled. I knew from my experience while at Leopoldville, that these steamers had been continually requisitioned by the State, and though the missionaries had usually complied with the requests without demur, they had been troubled by the inconvenience entailed. They also felt that their property, transported with so much trouble, was looked upon as at the service of whoever chose to ask for it. I therefore had my suspicions that there would be some difficulty in obtaining the use of these vessels. I had likewise been informed that the Sandford Exploring Association had a steamer in course of construction at Kinchassa, the sections of which had been put together, but they were awaiting the arrival of the engines from Matadi, and these would not be at hand before Mr. Stanley was ready to start from Leopoldville. While knowing all this I was aware that it was no affair of mine, especially as Mr. Stanley knew well the difficulties that lay in his path as far as river transport was concerned. Yet, in the interests of the Expedition, having been fully informed of the conditions of things at Leopoldville by Mr. Davey, the engineer in the employ of the Sandford Exploring Association, who arrived at

Manyanga about this time, I felt it incumbent upon me to ask him to examine three large whale-boats, State property, which I found were near Manyanga. I forwarded to Mr. Stanley his report of their condition, as well as his description of the state of affairs in connection with the steamers at Leopoldville. In a letter sent with the report I pointed out to Mr. Stanley, that if he thought it necessary these whale-boats could be transported to Leopoldville to help, in a way, to obviate some of the difficulties which seemed likely to hinder the progress of the Expedition up the river. It was in reply to my communication that he makes mention of the boats in his above letter, numbered 1.

The other letter which I received from Mr. Stanley was in these words:

[LETTER No. 2.]

"*Tuesday, April 5th*, 1887.

"DEAR SIR,

"Mr. Walker, engineer, is to be started off to the Pool as soon as possible to render immediate help to patch up the *Stanley* for the transport of the Expedition. Pray let there be no delay about this, and write to the Commandant at Leopoldville per Walker, stating why he is sent up, that he may turn his talents to good account quickly.

"Yours,

"H. M. STANLEY.

"MR. TROUP."

In compliance with the above Mr. Walker was started off to Leopoldville the next day (April 8th). Carrying out my instructions, I provided him with a letter to my old comrade, M. Liebrechts, who was in command of the station at Leopoldville, and in it I embodied Mr. Stanley's orders. Mr. Jephson remained with me at Manyanga, waiting for the fresh detachment of Zanzibaris, which Mr. Stanley had promised to send him for the purpose of carrying the sections of the boat to Lutete, where Mr. Jephson would again fall in with the main column of the Expedition.

About midnight of the 9th another messenger arrived from Mr. Stanley with a third letter for me. It was dated from Lukungu that same day, and in it he asked me to send him a statement of the loads which had reached me up to date at Manyanga, and how many I had sent on to Leopoldville. He sent me also some provisions, of which I was greatly in need. These supplies consisted of a box of biscuits, sugar, tea, dates, and hulwa, and he told me to detain a bag of rice for my own use. He adds: "I have no meats of any kind with me, except the live stock, goats, and these are few enough for the officers with me. You have the north bank, whence to purchase a few fowls occasionally. The one hundred or so bags of rice have come in very handy to me. I am able to provision my men as far as the Pool. People will leave here to-morrow for the boat. Tell Jephson that Barttelot leaves here this afternoon, while I follow on to-morrow, and that on the 13th I hope to be at Lutete."

On Easter day (April 10th) I heard that Stanley, with the main column of his Expedition, had camped at 'Ndunga, a few hours distant from us. Whilst I should have very much liked to have gone over to see him and my fellow officers, I felt that I could not leave my post, for I was obliged to be on the alert for the reception of caravans from Matadi, and to attend to the despatch of the same to Leopoldville; there was no one I could leave in my place for the few hours of absence.

M. Dannfeld, however, went out to the camp and kindly took a letter from me to Mr. Stanley, giving him all the information he had asked for in the letter he had sent the day before. On Monday Mr. Jephson, having received the Zanzibaris he had been waiting for, got off. He was so disgusted with our sandbank and all things connected with Manyanga post that he left it joyfully, breathing forth anathemas upon the spot, where he considered he had suffered the extreme of discomfort. He was filled with pitying wonder how I could survive life in this forsaken place. Later in the day M. Dannfeld returned from 'Ndunga; he had spent the night with the officers, and he gave me a detailed description of his experience. He had been graciously received by the *great man*, and had been invited to drink tea in his tent. He had also been introduced to the famous slave-trader, Tippu-Tib. It was a striking sight, he said, to see this large encampment of some 800 men. Stanley had a most imposing tent; it alone required nearly forty men to transport it. It consisted of one large covered space, with an inner compartment as a sleeping-chamber; the spacious reception-room was approached by a species of vestibule that ran parallel with the bedroom. The entire affair was constructed of green waterproof material, which did credit to Messrs. Edgington. I afterwards slept in the edifice one night at Kinchassa, and right snug I found it.

Tippu-Tib also had a tent of his own, which he had in all probability picked up at Zanzibar, and in which he installed himself and his wives. He had moreover what might be called a large back-yard, enclosed with some light canvas or cloth, "to cheat surprising, prying eyes" from obtaining a glimpse of his numerous spouses whilst engaged in their domestic occupations. Some of the officers of the Expedition had small single tents, each with a fly, and there was also a big tent with the mess-room in the centre and sleeping compartments on each side to accommodate the other officers. All of this tent, too, was made of the capital green waterproof material above mentioned.

The Zanzibaris, as usual, when arriving at the camping-ground, soon made themselves as comfortable as circumstances would permit. They ran up small huts of sticks, roughly thatched with grass, just large enough for a man to crawl into; but the more fortunate ones, who were able to get possession of large pieces of thin sheeting-cloth, constructed impromptu tents. The Soudanese did not take the trouble to obtain even such comfortable quarters, but generally made the best of what shelter they might find under the trees and bushes.

The Zanzibaris caused some trouble by not getting their loads in to the appointed encampment at a reasonable time. Dannfeld's description of the stragglers coming long after dark was not encouraging when one thought that marching in the open caravan roads was mere child's play to what they would be obliged to go through in the dense jungle and swamps between the Aruwimi and Wadelai!

M. Dannfeld had a long interview with Mr. Stanley, for he was anxious, like so many others, to join the Expedition, but he could not prevail upon the leader to accept his services. The fact was, Stanley was not willing to increase the staff of officers he had engaged while at home, where he had many to choose from; he made but the single exception of taking Ward, who happened to be of immediate practical service to him. Dannfeld's disappointment at Stanley's refusal was not even alleviated by a good dinner. He certainly messed with the officers in the aforesaid tent, and after the short commons of Manyanga he had every reason to expect something better in the way of *food* provided for the great Expedition, which had so recently arrived from Europe. Alas! the poor food obtainable could not even be eaten in comfort, for the rain soon percolated through the badly-pitched mess-tent. By his description it seems to have been one of those awful nights so common during the rainy season, when the rain fell and the wind blew and beat upon that tent in a truly scriptural manner. When he retired for the night to one of the *sleeping* compartments of this ponderous structure, he could obtain no rest, and was inclined to think that "bathing compartment" would have been a better description of his quarters. He was not alone in discomfort, as the air was soon warmed by the exclamations of his companions in misery. This tent proved a failure for more than one reason. In the first place, it was a great deal too large, and it needed nearly as many carriers as Mr. Stanley's tent required. It took a long time to put it up, and, unless it was pitched with the greatest care, it was sure to let in water, and the constant tornadoes blew it about much to its damage. The idea was excellent when applied to a marquee on the Thames, or such kindred places, but it was of no practical use in Africa. When I saw it afterwards at Kinchassa it had been split up, the end compartments being made into separate tents to accommodate two officers, and even this makeshift did not answer nearly as well as a small single tent with fly. It would have been much better if each officer had been supplied with a tent of this latter description, as they only require two porters, and are speedily pitched and struck, and, moreover, the officers would often have to be separated, and would then absolutely each require a tent of his own. But M. Dannfeld had survived his night at 'Ndunga, and had returned to Manyanga to tell the tale. The letter he brought with him from Mr. Stanley was as follows :—

[LETTER No. 4.]

"KINDANDUNGA, *April 10th*, 1887.

"DEAR MR. TROUP,

"Your letter is perfectly intelligible, and I have a clear idea of what has been done so far. I pray you write as soon as possible to Ward and Ingham to obtain from them somewhat similar conclusive and authentic data respecting number of goods left and goods sent, if any, in order that at any cost neither you nor Ward may be detained too long below Stanley Pool. As I suggested before, I believe that 200 [carriers] sent from Manyanga direct to Matadi would finish everything, and beyond that could bring us quite a valuable stock of rice to Stanley Pool.

"Natives report Ward on the road, but that may be all fudge. Even if they brought—supposing all goods sent up—nothing but rice, it would be quite worth the expense, if done quickly.

"I have written already about the bugle, that I need it badly.

"I left at Lukungu for immediate transport to Stanley Pool:

6 boxes biscuits.
27 sacks rice.
3 sacks of empty sacks.
—
Total 36

"Do not be easy in your mind until this stock has gone up to the Pool. It is so small and yet so near to you that it might be forgotten, but the biscuits and rice we shall need badly at the Pool.

"Urgency now is everything.

"Yours very truly,
(Signed) HENRY M. STANLEY.

"J. ROSE TROUP, ESQ."

In accordance with his request I sent 200 carriers down to Matadi, and made all haste to transport the articles mentioned to the Pool as soon as they arrived at Manyanga from Lukungu.

Mr. Stanley spent but one night in the camp at 'Ndunga, and the following day continued his march. After his large Expedition had passed on, leaving the road from Matadi clear, I found the native caravans transporting loads to Manyanga came in more rapidly. All Stanley's 600 Zanzibaris were carrying loads; such a large caravan moved slowly, and completely blocked up the road. The native carriers, also bringing up Expedition loads, were delayed in consequence; moreover, the sight of such an imposing number of Zanzibaris frightened them a good deal. A few of them had tried to pass Stanley's main column, but they had been roughly treated by the Zanzibaris, so the other native porters were by no means encouraged to attempt to come on until the road was quite clear, this accounting for a good deal of the delay.

On April 14th I heard that Ward was at Lukungu, and from this I inferred that all the loads had been started from Matadi on their road to Manyanga, as Stanley had given strict orders that he was to remain at Matadi until every load on the list given him had been despatched from that place. However, at that date I had not received more than 603 loads out of the 1000 odd that I was told were to pass through my hands. One principal cause of delay I have already stated, *i. e.* that Stanley's caravan occupied the road; another reason was the loafing of the men at Lukungu on their way to Manyanga. These matters I had brought to the notice of Mr. Stanley, as they caused me great anxiety, because I further anticipated that unless reprimanded, the Zanzibaris would repeat the intimidation when the parties again met on the road between Manyanga and Leopoldville. This would cause additional delay in the arrival of the stores, for which I would be held responsible. However, Mr. Stanley took the matter up, and ordered that the "capitas" should be provided with slips of paper—no matter what writing was upon them, as neither party could read—and should be instructed to shout "Boula Matadi!" to give notice that they were engaged in his service. I have reason to believe that Mr. Stanley warned the Zanzibaris of the Expedition that if they in any way interfered with the native caravans, which were then as important factors of the Expedition as were the others, they would be severely punished.

Ward arrived at Manyanga on April 16th. On comparing notes we came to the conclusion that there must be several native caravans with loads from Matadi loafing at Lukungu. I again drew the attention of the State officer at Lukungu to this fact, and urged him to send some of his Houssas into the village to stir up these men. Ward on his arrival was a great assistance to me. I had about this time 100 carriers awaiting the arrival of loads, which they were ready to transport to Leopoldville. It was very tantalizing to think of losing the services of these men for the Expedition, and I was at my wits' end to devise some means of keeping them for my employment. Ward and I were constantly going down 'Npioka river, playing the game of "Are they coming, sister Anne, sister Anne?" but all in vain on the day in question. As the afternoon closed in it seemed as if I should be obliged to let the men carry State loads, as I had none of the Expedition loads left to forward to Leopoldville. However, M. Dannfeld, as usual, came to the rescue. He presented us with a perfect treasure in the shape of a fine large "squeaker," "porker," "snorker," or any other endearing term that can be applied to the *genus* pig. The carriers consented to stop that night with us, when they received this *ration*-al supply, and when orders were given to end the porker's existence, our present anxiety was ended. In triumph they bore the prize squeaking from our sight, and, having divided the spoil, settled themselves down to a thorough enjoyment of the feast at some little distance from our camp. On the next day some loads arrived, and we had the satisfaction of knowing that our efforts of last night had not been useless. But on the twenty-third 201 carriers arrived from Matadi, all bearing Expedition loads. All these men had taken from 18 to 21 days to accomplish the distance between the two points; these were no doubt the carriers about whom we had been so anxious, who had been loafing at Lukungu.

By my list I could see that these were nearly the last of

the Expedition loads. But on that day my supply of native carriers for forwarding to Leopoldville failed. It must be borne in mind that in this period of three weeks I had received and forwarded to Leopoldville over 800 loads by a like number of carriers. I had had more than enough men to complete the transport of the loads, but by the delay in their arrival I had been forced to lose their services. I had still on my list of engaged men a large caravan due from chief Makito, but as urgency was everything I decided to go over to a large village on the north bank of the Congo, with the purpose of having two strings to my bow. Ward remained behind in charge of the loads, so that if Makito's men should come there would be no delay in forwarding the stores. But I received only vague promises from the chief of this village, and learning that there was to be a local market here next day, I knew I could persuade him to do nothing until that was over.

The markets are an important feature of Congo life. In this district they are chiefly held on the north bank, and are of daily occurrence. To distinguish one from another they are designated by the name of the day on which they are held. The Congo week consists of four days, viz. Wukaunzo, Nkengi, Nsona, Nkându. The Wukaunzo market may be held near a village at one end of the district, while the Nkengi market may be quite at the opposite end. The places where the markets are held are well known to the natives, who will go a great distance out of their way to attend them. I myself have been to many of these markets, and was always much struck by the animated scenes I witnessed. It is not wholly to be regretted that very few are held in the direct caravan

VIEW IN A NATIVE VILLAGE NEAR MANYANGA.

route between Matadi and Leopoldville, for it is the most difficult job in the world to persuade a native caravan to pass one by without stopping; even if they have no need of additional rations, or have no money to spend, they must at least be allowed to pause for a chat with their friends, or they will try to induce the *Mundelli* (white man) to buy them some malafu (palm wine). If they only had a little malafu, they will say, all would go well, they could carry their loads easier, they could go so much more rapidly, and many kindred reasons would be advanced; but unless you know your men well and consider it wise to yield to their requests, in order to save time you usually turn a deaf ear, but if you do buy malafu for them you take care to distribute it cautiously. The markets are held in a cleared space within easy reach of most of the villages of the district. The particular one I visited on this occasion was some four hours' march back of the old station of Manyanga North, in a district well populated by natives. About noon of the day in question we wended our way along a path already thronged with native men, women, and children, all intent on reaching the same destination. Nearly every woman of the company was carrying some article she was prepared to sell; the heavier burdens they bore on their backs in baskets suspended by a strap or band across the forehead. The scene of action was a large clearing with a few shrubs and straggling trees scattered about it, with numerous paths converging at this point in a manner suggestive of the Seven Dials. When I arrived the market was fairly on, and business seemed brisk. There was a perpetual and almost deafening roar of voices; some of the "sporting gents" who shout the odds at race meetings, even the fraternity who hail from the north, Yorkshire to boot, would find themselves "not in it" with some of these dusky Manyanga natives, who not only can make a noise, but can keep it up at high pressure all the time. Above the almost sullen roar of the men one hears the shrill, penetrating voices of the women raised as they chaffer over their wares, which they display in what they consider a tempting manner. Here you may buy with strings of blue glass beads or coloured handkerchiefs—the currency in this part of the country—such delicacies as sun-dried or smoked fish, smoked rats, palm-oil, bananas, plantains, chiquanga, maize, or even a pig, goat, fowl, or some eggs, but to-day I notice that the supply is far from being as plentiful as usual. Among themselves they either buy or barter, and the amount of haggling over bargains would do credit to an Italian Jew. The women do most of the business, the men condescending to attend to such matters as the sale of goats, or an occasional "snorker," which is led by a string, and invariably lends its voice to the general uproar. Among those assembled at the market are a large proportion of loafers, who have no other business than gossiping or hanging about in the hope of getting a stray drink of malafu from a friend. There is generally a good supply of malafu, or palm wine, at these markets, but if one wishes to obtain it one must do considerable dodging about in shady corners to find it. It is a delicious drink when fresh tapped from the palm, a light, white beverage, in colour somewhat resembling the milk of the cocoa-nut, though it tastes much sweeter. In this condition it is not likely to inebriate any one, but after it has been kept a few days it ferments and becomes a powerful intoxicant, and it is in this latter form that it is best liked by the natives. Here is a new and almost untrodden field for Sir Wilfrid Lawson, if he wishes to extend the scope of his proposed Local Option Bill.

At this market were present a number of petty chiefs, who could easily be distinguished from the common herd by their dress. Many of them had got themselves up as great swells, for in addition to the usual native loin-cloth they had donned black coats and black felt hats, while a few even swaggered about under cover of huge red umbrellas. They were all old, seasoned topers, and patronized the malafu vendors *early and often;* but all the malafu drinking is done, so to speak, on the sly. A little coterie, intent on a palaver of some kind, adjourn to a sheltered spot in the bush, and there ensconce themselves, with a huge gourd full of malafu in the centre; one of their number is appointed "*ladler in chief*," and there is no doubt that he takes the greatest care that he is not left out of the drinking part of the ceremony. He pours out the wine into a discarded mustard tin or similar "loving cup," which, having completed the circle of thirsty mouths is refilled, to be started again on its journey. There is a general smacking of lips as it circulates, while between the draughts some knotty point is settled or a bargain is clinched. I would by no means accuse these worthy chiefs of being ashamed of being seen drinking; the secrecy of their conclaves is probably largely due to a desire to keep out of the way of the many *bar loafers* who are anxious to have a pull at the malafu jar. There are many such gentlemen lingering about, whose mouths would accommodate themselves to anything, from a pickle bottle to a bath tub. Occasionally after one of these bouts the men grow quarrelsome; but taking them all round their markets are as well conducted as many a one in this country. For, be it known, they have a strict code of rules, which are stringently enforced: any infraction of them is summarily dealt with, and, in some cases, the death penalty is inflicted in anything but an attractive form.

On this occasion I sent the State interpreter, who had kindly been lent to me for this excursion, among these clusters of chiefs to apply for native carriers, but all he could succeed in obtaining for me were vague promises that they *might* send men at some future time. This was not at all what I wanted, and was by no means satisfactory, so I was obliged to give up my attempts. Rather disgusted, I returned to Manyanga in the afternoon, wondering whether the men promised by Makito had arrived during my absence. I was annoyed to find from Ward that no such good luck had happened. This was the first check I had experienced in my work. I had always been able heretofore to start the loads onward as soon as they arrived, and this delay of two days was a source of vexation to me. However, next day Makito's men arrived late in the afternoon. I was glad to see at last that this task of mine was drawing to a close, and I was more than rejoiced to find that we now had sufficient men to carry all the remaining loads, and even our own private baggage, so that we were able to make this a *personally conducted* caravan. It cannot be denied that I breathed a sigh of relief when I knew that we were to leave Manyanga sandbank, with its many discomforts.

We started that very afternoon for Leopoldville, at which point we were confident the main column of the Expedition had already arrived. We could not go far that night, indeed, we were only able to reach the neighbouring village, but this was not wholly unsatisfactory, as by so doing we could get the carriers under way. The next day we made a very rapid march to Lutete, in spite of the discomfort caused by the rain, which came down in torrents. It was late in the afternoon when we reached there, where we were hospitably entertained by the missionaries of this station.

After leaving Lutete we came upon von Reichland encamped at the Inkissi river, and a day or so from Leo we met Mr. Billington, of the American Baptist Mission Union, who was going down to meet his *fiancée*, a lady who had been engaged in missionary work for some years on the Congo, and whom he had met in the course of his labours. From them we gathered that there had been exciting palavers concerning the missionary steamers, which Mr. Stanley wished to use for the forwarding of his expedition. Mr. Billington also told us that the matter was settled, and that Mr. Stanley was at Kinchassa (two hours' march beyond Leopoldville) preparing for an immediate start on his voyage up river. This made Ward and me increase our speed to a terrific pace, leaving our carriers to follow as quickly as they could. Our last march, from Mbees village, was, in spite of the rain, made at such a rattling rate that we reached Leopoldville on April 30, having accomplished the entire distance between Manyanga and Leopoldville, though it was the wet season, in six days' march.

CHAPTER VI.

STANLEY'S DEPARTURE FROM KINCHASSA.

WE got into Leopoldville at 2.30 P.M. of April 30th. Here I was, back at my old station, where I had spent eighteen months in the service of the State during the years 1884-1885. It did not strike me at the first go off that there had been any improvement in its general condition since I left it. The long broad avenue acting as approach, which in the time I was there was always kept scrupulously clean, was now overgrown with long grass, &c. It was even difficult to trace it, as the luxuriant growth made the road look like the surrounding bush. But one could not wonder at this state of affairs when it was known that there had been a regular famine in these parts, so that the chief of the station, M. le Marinel, had been obliged to send the bulk of his men away, as he was unable to find sufficient supplies to feed them; hence he could employ no more men than were absolutely necessary. Some improvement had been made in the way of building, but even this did not help to take away from the neglected look such as one often finds in a tenantless estate where the grounds are allowed to go to rack and ruin. Here we found my old comrade, Lieut. Liebrechts, who had only a few weeks previously taken over the duties of Commissaire de District. He had served three years in the State, chiefly on the Upper River, and after a few months' holiday in Brussels, he had just returned to the Congo. We learned from him that Mr. Stanley would be starting from Kinchassa on the morrow early, so that there was no time to be lost. After a hearty but hasty lunch, which friend Liebrechts had quickly prepared for us, we pushed on with all speed to Kinchassa to report ourselves to Mr. Stanley. Whilst we partook of our repast we got all the latest news out of Liebrechts. Of course the most engrossing topic of conversation was the palaver which had taken place between Mr. Stanley and the missionaries, Dr. Sims and Mr. Billington of the American Baptist Mission Union, when the former demanded the steamer *Henry Reed* for the purpose of helping to take the Expedition to the Aruwimi. The difficulty occurred exactly as I had anticipated. Mr. Stanley arrived with his large following and found the *flotilla* of the State such as I have described elsewhere. Of course it was quite inadequate to transport his men, let alone his stores. Mr. Stanley wrote home a somewhat lengthy description of this unfortunate palaver, which was published in the newspapers at the time. It must be remembered, as I have before mentioned, that there had been a kind of famine throughout the region, and there was great difficulty in obtaining food for the few men of the State station. For example, when I was at Leopoldville in 1884 we could get as much chiquanga as we wanted. We had a regular fixed price for it, the native women of the neighbouring villages used to make this staple article of food from the manioc plant or cassava root; in the Leopoldville district it is made up into large round balls the size of a Dutch cheese. The contract price used to be, in my time, 3 mitakos (brass rods) for 2 chiquanga, and often we have been obliged to refuse the too plentiful supply offered. This plenty was due largely to the protection that had been extended to the natives hereabouts, but during the time of the transport of the sections of a steamer, which necessitated encampments of large bodies of Zanzibaris at different points on the caravan route, these had, owing to a lack of strict surveillance on the part of the officers in charge, obtained their food by ill-treating the natives. Consequently now the villages on this route were all deserted, the inhabitants having gone further back in the bush. This circumstance, which happened some time previously, had been followed now by a bad season, and the natives who had chiquanga demanded 6, 7, or 8 mitakos for a single ball of it! Even this was of a smaller size. This complicated matters seriously. Mr. Stanley could not get sufficient steamers to take his men away, yet it was impossible to obtain food for his over 600 men even for a few days' stay. It should not be forgotten that before his arrival at Leopoldville Mr. Stanley knew the exact situation of affairs, and after his long experience we must credit him with the ability to foresee exactly what the consequences would be.

After rather excited palavers between Mr. Stanley and the missionaries of the American Baptist Mission Union, regarding the serious check to the progress of the Expedition, the incident was happily terminated by the intervention of M. Liebrechts, who grasped the situation, and forthwith requisitioned the *Henry Reed* for urgent State service. As soon as he obtained possession of it he handed it over to Mr. Stanley. The s.s. *Peace*, belonging to the English Baptist Mission, had already been placed at the service of the Expedition, and the unfinished *Florida*, of the Sandford Exploring Expedition, had been also requisitioned and would be towed up. So Mr. Stanley had managed to get all three steamers, viz. the *Stanley* belonging to the State, and the two steamers *Henry Reed* and *Peace*, the property of the missionaries, added to these were two large whale-boats, two barges (the *A.I.A.* and the *Florida*), a missionary whale-boat and his own sectional boat. Such was the substance of the state of the case that was unfolded to us while we sat at lunch. After that meal M. Liebrechts placed a canoe and men at our disposal to take us to Kinchassa. Our carriers had not yet come in, so we could not change our clothes, which had been reeking wet but had now somewhat dried, for a scorching sun had dispelled the rain.

We were soon under way across the old familiar Pool, hugging the shore to avoid the strong current which here rushes down with a terrible force to reach the rapids below the Leopoldville station. We safely pass the bad water at Kalina Point, and then strike the calmer reach beyond, from whence we could see through a cleared space on the banks the new quarters of the English Baptist Mission, who had transferred their station from the hill at the back of the State station at Leopoldville to this more suitable spot. This and the Dutch House further up towards the old State station of Kinchassa were quite new to me. The south bank of the river from Leopoldville to Kinchassa in my time was one unbroken stretch of scrub and bush. As we came in sight of the cleared beach in front of the Dutch House, a busy scene presented itself to our view. Here were collected the steamers, barges, and boats which were to transport the men and stores of the Expedition to the Aruwimi some 1000 miles away. Instead of going on to the State station of Kinchassa (now rented by the Sandford Co.) a little further up, Mr. Stanley had formed an encampment on the cleared ground at the side and back of the Dutch quarters; here were all the men of his Expedition, whilst his craft were close at hand on the beach. Our canoe was soon alongside of the s.s. *Stanley*, and we clambered up her sides. The first man to greet us was our old friend Capt. Shagerström, the skipper; standing on the beach not far off and easy to be recognized was Mr. Stanley, giving instructions as to the loading of the steamers and barges. Everything around showed that business was going on in real earnest. But we had no time to loiter with friend Shagerström after we had returned his warm greetings, so we made our way across the deck and down to the beach to report our arrival to Mr. Stanley. I was much struck by the great improvement in his appearance; he had put on flesh, and was looking much better than when I last saw him in London three months previously. In fact I never saw him looking so fit before in my life. He was arrayed in his usual African *war-paint*, viz. a Norfolk jacket, with continuations in the shape of knickerbockers, which displayed most

MR. J. ROSE TROUP.

decidedly well-developed calves, his head-gear was a large flat-topped cap with a peak in front such as is worn by officers in the German army. What a contrast he presented to the last time I saw him thus arrayed! That was at Vivi in June 1884; he was on the eve of returning to Europe after having been engaged for eighteen months in forming stations on the Upper Congo. He was then very light, to say the least, gaunt and yellow, reduced to a mere shadow by repeated attacks of fever, from which he had suffered also while at Vivi. I had my Houssas drawn up on the beach to give him the customary salute. Besides these there had assembled all the officers of the station, as well as Sir Francis de Winton, who was to be his successor in the administration of affairs. Here were also Col. Pollok, Major Vetch, Dr. R. Leslie, Dr. Allart, Massey Shaw, Valcke, Deane, Posse, Monet, Spenser Burns, Bathurst, Tugman, and a host of others. How few of these comrades, personal friends of mine, are now left! Among those who have gone are Vetch, Spenser Burns, Posse and Deane.

On the present occasion Mr. Stanley seemed very pleased to see us; he had been anxious because we had not come in before, but he knew well that we had been detained by the work intrusted to us. He informed me that he had sent me a letter that very day, but perhaps it had not gone off yet; he told me to make inquiries, and if possible to get it. As it proved, it had not been despatched, and it was delivered to me. Another which had crossed on the road, but had been returned, reached me some days later. These letters are as follows:—

[Letter No. 6.]

"Dutch Agency, Kinchassa,
"*April* 30*th*, 1887.

"Dear Mr. Troup,

"Neither you nor Mr. Ward have arrived. I am truly sorry. But never mind, we shall have work enough for you both. Wait at Leopoldville until *Stanley* steamer comes down, then you and Mr. Ward embark with all goods and ascend river to Bolobo, there you will find Major Barttelot and his detachment of about 150 men.

"The *Stanley* can carry 400 (four hundred) loads and 180 to 200 men. A man weighs 130 lbs., a load 65 lbs. Make your calculations accordingly, because if excessive you will have to add another vessel or two, lighter or something. I have taken with me 119 boxes brass rods, 186 should arrive at Leopoldville from Matadi. All Dutch brass rods are longer than ours, and packed in bundles. If I find I have too many turn the surplus over to Greshoff, add a pound sterling for freight for each box or load. Get a due bill from Greshoff, and send it to William Mackinnon, care of Gray, Dawes and Co., 23 Austin Friars, London. Letter, telling you how many you will find, on steamer *Stanley* on her return from up river.

"Yours very truly,
(Signed) "Henry M. Stanley.

"J. Rose Troup, Esq."

(The other letter which reached me some time afterwards was in the following words.)

"Camp, near Leopoldville,
"*April* 26*th*, 1887.

"Dear Mr. Troup,

"After all the goods you mention in your first column, namely, 864, have left Manyanga, then you and Mr. Ward hasten along to Leopoldville to check the goods received here, so as to correspond with that number. It may be that we shall have left Leopoldville before your arrival, in which case you will remain here, both you and Mr. Ward, until the steamer *Stanley* comes down for you. You will take all the goods remaining for us at

Leo', and accompanied by Mr. Ward, ascend to Bolobo and report yourself to Major Barttelot, who commands the second column. From the Major you will learn where you shall meet us.

"Yours truly,

(Signed) "HENRY M. STANLEY.

"JOHN ROSE TROUP, ESQ.

"P.S.—I need not tell you how much we miss both you and Mr. Ward. Your companies are waiting to be placed under your command, and they are sadly in need of you. I wish you were here to-day. Still our stores are a great concern to us.

"Yours sincerely,

(Signed) "H. M. STANLEY.

"J. ROSE TROUP, ESQ."

After reading the letter of April 30th, we had another chat with Mr. Stanley on the beach, and he then said to me, "Well, as you have got up, I will take Ward up with me, but I wish you to remain behind until the steamers return, as we cannot take up all our loads on one trip." Ward therefore hurried back to Leopoldville to get his personal baggage, which should have arrived by this time, whilst I remained on the beach, and had time to have a chat with my comrades, some of whom I now saw for the first time. I had a cheery welcome from Stairs, who was somewhat difficult to recognize with his well-developed beard and rough Karki suit, as the same man in his neat and trim town dress I had met in Stanley's Bond Street rooms. I dare say he found it equally hard to place me in my picturesque get-up, which I had reduced to decidedly light marching order—loose flannel trousers or pyjamas, tucked into my long boots, flannel shirt and broad *sombrero*. All the other officers were scattered about attending to the loading of the different boats, but they were just knocking off work, as it was getting on towards sunset. I rejoined Mr. Stanley and walked up with him to the Dutch House, where he was the guest of M. Greshoff.

I had a wash and brush up in his room and swallowed a good big dose of quinine; he lent me a cashmere coat as it was growing chilly and I had no change of garments with me. After I was thus made more comfortable, we sat down together on the verandah talking for the half hour that remained before dinner was announced. He told me then his plans and arrangements, explained why he left me behind, and gave me his instructions. When he first told me that he wanted me to stay behind until the steamers returned and then bring up the loads, I cannot help honestly admitting that a feeling of considerable disappointment came over me, as it seemed I was to be left out of the hunt altogether, and I expressed as much as that to him. However, it was only momentary, for he continued his remarks, pointing out to me how much depended on the safe transportation of the remaining stores, adding how pleased he was with what I had already accomplished, and it was this that made him select me to undertake this task. After he had shown me that it was in the interests of the Expedition I must remain behind, all personal feeling of disappointment was cast in the shade. He had already sent Major Barttelot and Dr. Parke with some 150 men overland from Leopoldville to Kwamouth where they would be able to get more food, and also give him more room on his steamers. He purposed sending back for them after he had formed his camp at Bolobo, thus making room for them. His plan at that time was to leave Major Barttelot at Bolobo with these men (as there was plenty of food in that district), and in the second trip of the steamers I was to pick him up and his men to carry them on to the Aruwimi. By the time he had made all this known to me, dinner was ready. We were all the guests of M. Greshoff that night —Mr. Stanley, Stairs, Jephson, Jameson, Nelson, and I, sat down to dine with our host. Bonny was seedy, Ward had gone to Leopoldville for his traps, while Barttelot and Parke had started before for Kwamouth. Ours was a very enjoyable evening. Mr. Stanley was in his most entertaining mood, full of anecdotes, and inclined to chaff one and all of us. No wonder the great explorer was feeling at his best. On the morrow he would have started from Kinchassa, and the main body of his Expedition would be on their way to the Aruwimi; he had gained his object when he seemed likely to be thwarted, his tiresome overland journey would for a time be ended, and after all his petty annoyances he seemed likely to have smooth sailing. His great Expedition was fairly started and all was, so far, going well. He could doubtless see in his mind's eye his ultimate triumph; he had conquered preliminary obstacles, what was to prevent him from being equally successful when contending with his future difficulties? Would he not before many moons had waxed and waned approach Wadelai bearing to Emin the relief stores? Would he not issue from the heart of Africa, having triumphantly accomplished his task? No wonder then that he was cheered if such thoughts flitted through his mind. We all did ample justice to the excellent dinner supplied by our host. While we were sitting over our coffee and pipes, the great Tippu-Tib came in, and having greeted Mr. Stanley and the other officers his quick eye rested on me, in whom he recognized a new arrival, and he promptly asked Mr. Stanley who I was. Our leader called me forward to introduce me to the Arab chief, adding by way of explanation that I had been three years on the Congo, and had had considerable experience with the Zanzibaris. It was with a good deal of curiosity that I approached the Arab of whom I had heard so much but had never before seen; to his greeting of "Yambo?" [Is it well with you?] I replied, "Yambo sana" [It is very well]. After dinner we strolled through the encampment, which looked very picturesque, lighted up by the blaze of the camp fires; around them were gathered the dusky-hued men standing out prominently against the darker shadows behind them—a picture that Rembrandt alone could have done justice to. In these groups I came across several old friends among the Zanzibari niamparas (head-men) who had served under me in the days gone by when I was in the State service, and many of the men recognized me; so I did not come among them as a stranger. I finished the evening by sitting with Stairs and Nelson in their tent, which was one of the sections of the big mess-tent, where we yarned away right merrily. They lamented with me that I was not going up with them, but we made the best of the situation, hoping to meet soon again, and remembering we were all working towards one common goal.

Mr. Stanley was to sleep at the Dutch House, and kindly placed his big tent at my disposal for that night. It was pitched alongside of the Zanzibari encampment, and I was only too glad to accept its hospitable shelter, for the tent I had borrowed from the State was no doubt at Leopoldville, to which point my carriers should by this time have brought it with my other things. The other occupant of these roomy quarters was William, Mr. Stanley's servant, whom I had not seen since I had helped him out of a difficulty at Charing Cross Station on the day of Mr. Stanley's departure, where the man had been greatly excited because he had not sufficient money to register his master's luggage. To-night I was glad to turn in. I had been marching from early morning until I reached Leopoldville, and after that had had the canoe voyage to Kinchassa. The excitement and bustle had caused me to forget my fatigue, but now I felt dead tired; not even the chatter of the Zanzibaris round their camp fires, nor the troublesome annoyance of the mosquitos, could keep me from soon getting off to the land of nod. At daybreak all was astir again. Stairs, who was in immediate charge of the camp, had his buglers early at work, and the music of their instruments was soon taken up by the "tom-toms" of the Zanzibaris. All the arrangements for departure had been made the night before; everything in the shape of stores had been placed on board the various boats, there only remained the embarkation of the men, donkeys, and the personal luggage of the officers. After completing my toilet, which was of necessity brief, I wended my way down to the beach, and was soon lending a helping hand in the troublesome task of getting the donkeys on board the *Florida*, which was prepared for their reception, and was lashed on the further side of the *Stanley*. We had some difficulty in getting them on to the *Stanley* even, as there were no proper appliances at hand. They were walked or driven up the plank to the steamer, but generally slipped off and floundered about in the water. However, we managed this job after a deal of trouble, and then got them safely on board the *Florida*, which was laid with dried grass to make the little *Neddies* as comfortable as circumstances would permit. By this time the beach presented quite an animated appearance; the Zanzibaris were all drawn up ready to take their places in the various crafts, and there was also assembled a fair gathering of white people who had come from the Mission station and the Sandford Association quarters, and to these were added the employés of the Dutch House. I now felt that my assistance might be dispensed with for a time, so I made my way up to the Dutch House to see if there was any breakfast going on. As I walked up I was hailed by numerous old friends who had not been over the night before. Here were Swinburne and Casement of the Sandford Association, Mr. and Mrs. Bentley of the English Baptist Mission, and others. Swinburne was one of Mr. Stanley's oldest followers, having come out to the Congo with him in 1879. He had served many years in Africa, first in the Association and State, and now in the Sandford Expedition, of which he is chief; his head-quarters were at the old State station of Kinchassa close by. He had, at considerable risk and expense to his Company, generously placed the *Florida* at the disposal of his old chief, in order to help him out of his difficulties. As I have already stated, the *Florida* was on the slip at Kinchassa, all but completed, almost ready for launching, only awaiting the arrival of her machinery that she might start on her first trip up river to purchase ivory, &c. But instead of doing so, she was launched as she was while Mr. Stanley was at Kinchassa, and placed at his service as a barge. Although detained by these old friends, I did reach the Dutch House in time to find Stairs and some others still at breakfast. By this time I had a healthy appetite and was ready to join them; but while thus pleasantly engaged that little imp Baruti, Mr. Stanley's black boy, came in to say that "Boula Matadi (Stanley) he want to speak to you;" so I had to rush down again to the beach. Here I found Mr. Stanley, who asked me to make a note of his final instructions, which he forthwith proceeded to give me. He also handed me a list of stores which he was able to take up out of the lot I had transported by native carriers. The rest of the loads with him were the bulk of those brought up by his Zanzibaris. Besides the stores he left three donkeys and two donkey boys.

Everything now seemed ready for the start, but a delay was caused by the *Henry Reed*, which was to be the first to get away, not being prepared. She was to take up Tippu-Tib and his following, who were slow in taking their places. It was nearly eight o'clock before he and his women were stowed away, and Mr. Stanley, from the beach, was able to give the word to cast off. The *Henry Reed* was towing the *A.I.A.* on one side, and a large whale-boat on the other, so with her heavy freight she had all her work cut out. Mr. Bonny, an officer of the Expedition, was on board, and she was piloted by Captain Martini, one of the State skippers, while Mr. Walker, of the Emin Relief Expedition, was acting as engineer, holding himself in readiness to look after the machinery of the other steamers also. By this time the bulk of the Zanzibaris who were to be transported in the s.s. *Stanley* had taken their places, being packed on board like so many sardines, and the *Florida*, lashed alongside, was equally crowded, and so was the whale-boat which was on the other side of the steamer. We all bid farewell to Stairs, Jephson, Jameson, and Nelson, who were on board, and the *Stanley*, amidst the cheers of the company on the beach, which were re-echoed by Stairs and his men, steamed out, her old stern wheel slowly revolving. She was skippered by Captain Shagerström, a Swede, one of the most experienced and painstaking officers the State ever had, and her engineer was M. Deman. All the men on board were under the immediate command of Lieutenant Stairs. When they were off there only remained the missionary steamer the *Peace*, which was to carry "Boula Matadi" and his picked company of Zanzibaris. The little vessel looked quite gay, decked with the mission flag as well as the Stars and Stripes, with some device in the corner which turned out to be the "bourgi" of the New York Yacht Club; this had been presented to Mr. Stanley some time ago by Mr. James Gordon Bennett, of the *New York Herald*. Mr. Stanley took with him, besides William his servant, Ward, who was in charge of the picked Zanzibaris. Mr. Whitcley was pilot of the *Peace*, and Mr. Charters engineer; both of them belonged to the English Baptist Mission.

The commander of the Expedition, after impressive farewells to all assembled on the beach, and wishing me, in particular, health and a speedy meeting, took his place. As the *Peace* steamed out, with two whale-boats in tow, I called for "Three cheers, and one cheer more for 'Boula Matadi'" which were given with a will. Away went the great Expedition, having entered on its second phase. The place that had been the scene of so much bustle and excitement had now resumed its accustomed stillness. I turned away from it and started with Swinburne and Casement on the half-hour's walk to their quarters further up the river. On our arrival there we were much disturbed by being greeted with the news that the *Peace* had put in here, as she had met with disaster: the head of her rudder had been snapped off while passing through a bad bit of water just above. It was found on examination that this could not be repaired at Kinchassa, but must be sent to Leopoldville, so Mr. Whiteley and Charters went over with it. As the job could not be finished until late that day, Mr. Stanley decided to remain with Swinburne and make another start on the morrow. The other steamers had gone on, and no doubt wondered what had become of their commander. Swinburne readily placed all his accommodation at the service of his old chief, and Ward, Casement, and I joined them at lunch, an agreeable, partly because unexpected, event. Mr. Stanley was anxious to get some rice over from Leopoldville with which to feed his 117 men, as this extra day's delay in this district would encroach on what they had. I undertook to attend to this, as I bethought me that I should be far more comfortable if I could get out of the garments I had not been able to change for the last forty-eight hours, and to do this I must get hold of my personal baggage, which was at Leopoldville, so my journey would accomplish both objects. It was late in the afternoon when I started in a canoe down river, but I arrived in time to dine with friend Liebrechts. I waited until rather late when the moon came up before starting on my return journey, taking with me the rice and my own personal baggage. It was eleven o'clock at night when I got back to Kinchassa, and found all had turned in except Casement, who gave me a "nightcap," and showed me to the quarters Swinburne had assigned to me. Next morning, as the rudder was mended, all was quickly in readiness for a fresh start, and the rice, which Mr. Stanley was pleased to find I had been able to bring, had proved of service. Again the men all filed into their places, and steam being up, they started off again, followed by cheery "God speeds."

CHAPTER VII.

AT KINCHASSA AFTER DEPARTURE OF MAIN COLUMN, AND MY JOURNEY UP RIVER.

THE perpetual strain of the last six weeks and the terribly high pressure of the last two days was now ended, and I was able for once to sit down quietly to take a survey of the situation of affairs—past, present, and future—as regarded the Expedition.

I had been invited by M. Liebrechts to be his guest at Leopoldville until the steamers returned, but as I knew well he had a full complement of Europeans to provide for at his station, I did not feel justified in upsetting any of his arrangements, which I knew I should do if I took up my quarters with him. So I declined on these grounds, and decided to remain with my old friend Mr. Swinburne, of the Sandford Exploring Association, at Kinchassa, who had also pressed me to be his guest. Casement, also in the Sandford service, had gone down country again, a few days after the departure of Stanley's Expedition, in order to look after the transport of the machinery of the *Florida;* so Swinburne and I were left by ourselves in his comfortably situated station, only two hours' march from Leopoldville.

I made arrangements with M. Cloetens, the commercial agent at the latter place, that he should keep me informed of the arrival of some of the caravans which I had sent off from Manyanga, but which had not yet appeared at Leopoldville, and I was constantly going to and fro between my new quarters and Leopoldville, having selected one of the donkeys of the Expedition, which were left in my charge, to convey me from one point to the other. These journeys became necessary, for I found I was not to remain idle, because the troublesome loads again required a deal of personal attention. A great number had suffered considerably since I had sent them off from Manyanga; the same old loads that had been my bane before again cried out for assistance; the loads of rice, cowries, and beads packed at Zanzibar, and the mitakos and handkerchiefs purchased on the Congo, were the chief victims.

As I had plenty of time at my disposal, I endeavoured to put them straight for the long journey that still remained to be accomplished, and in which they must endure plenty of rough handling. The rice had all got a pretty good soaking again between Manyanga and Leopoldville, so it required to be dried and repacked; in addition to this, the original miserable material was by this time perfectly rotten. Luckily Mr. Stanley had brought out with him a number of stout canvas sacks well sewn together, and some of the rice had been transferred into them at Matadi, consequently these loads had suffered very little; there were also a good many of these sacks sent up empty, and I repacked all the rice I could in them, but there still remained a great quantity that had been damaged, some to such an extent as to be unfit for human consumption. The smell arising from some of these sacks was almost unendurable, and they had to be condemned and thrown away. The beads and much of the rice had been done up in a most ridiculous manner. They were in bags made of the material used for pillow-slips, and were tied round the mouth with string; over all this was fastened a rough kind of straw matting. In nearly every case the sheeting bags were split lengthwise by the force engendered by repeated "dumping" in transit,—not that I think it required more than one well-conducted "dump" to do the trick; the consequence was, the rice and beads were rattling about quite loose in the matting, which in its turn leaked. The cowries were not packed quite so badly; they had been put in coarse inferior sacking, which had rotted more or less in consequence of the many soakings received on the way up.

I found great difficulty even here in obtaining the proper materials required to make good loads of the rice and beads. The only stuff procurable was bits of sacking regions bales of cloth. Out of these I manufactured bags for them, and improved their condition somewhat, though they were still very far from being what I should have liked to have seen them. With the same material (old sacking) I patched up the cowrie sacks, as well as the bales of handkerchiefs and mitakos, which had been purchased from the Dutch House at Ango-Ango. The latter gave me the most trouble of all. The mitako, or brass rod, is the currency among the natives at Leopoldville and most of the regions of the Upper Congo. It is in general imported to the Congo by the State in large rolls or coils of 60 lbs. in weight. After its arrival at

Leopoldville it is cut up into the regulation lengths (about 2 feet), making it resemble in size and appearance an ordinary stair-rod; the value of each of these pieces at Leopoldville is reckoned at 1½*d.* But the mitakos which Mr. Stanley had bought on the lower river were already cut in lengths and put in 60 lb. loads done up in something of a waterproof material, and then sacking was sewn outside of it. With the constant dumping of the loads the sharp edges of the brass rods quickly cut their way through, and in most cases either one or both ends were thus laid open, and the carriers had every opportunity to help themselves, which they had done in many cases. They arrived in this condition whilst I was at Manyanga, and though I then put *nose-bags* on each end of these loads, I knew that before they reached Leopoldville they would be reduced to the same bad condition, and I found them now just as I expected. I had bronght some mitakos out from Europe, for the payment of carriers, but they were not required for that purpose. These had been cut in England, but were all packed in narrow wooden cases, 60 lb. loads each; they stood the transport all right, and Mr. Stanley took them all, some 116 boxes, up with him in the first trip of the steamers.

It was impossible to make good loads of the mitakos in sacking unless they were put into boxes, but it was impossible for me to get any boxes at Leopoldville, so I was forced to do the best I could, which was to put on new *nose-bags* of two or three thicknesses of the strongest material obtainable. The bales of handkerchiefs were also badly put up. Those sent to the Congo from Europe are excellently packed, for it is well known by this time to the manufacturers of these cloths that packing is most important, considering the amount of reshipping and rough native transportation these bales must undergo. They are put under hydraulic pressure and reduced to a compact mass, and are packed in stout brown paper, then wrapped in strong waterproof material, and over all is put good sacking strongly sewn up and hooped. But the few bales purchased on the lower river had been repacked loosely, sewn up in inferior material, and formed most ungainly loads, with too much slack room; in some cases the sacking had burst out, while in others it had rotted with the wet. It was a case of patch up all round, and the worst of it was, that, do what I could, I was totally unable to make any of them up to my standard of a good load.

All these above-mentioned stores were never properly packed in the first instance, and to all parties concerned in putting up these loads in such a slipshod way great blame must be attached for the damage and consequent loss to the Expedition. Of course the rainy season is not exactly the time to choose for transporting perishable articles like rice from Matadi to Leopoldville, and one must expect to have it somewhat damaged; but in all my varied experience at Leopoldville I had never seen loads arrive in such a wretched condition as did these rice, beads, cowries, mitakos in sacking, and handkerchiefs. I am thankful to say the other loads were in a more satisfactory state. The 100 cases of powder, the 200 cases of ammunition, percussion caps, coils of brass and iron wire, turned up in good condition, having withstood the rough usage fairly well. With a hammer and a few nails they were easily put right. There were 90 bales of Zanzibar cloth which were packed tightly, being wound round and round with thin sheeting, tied up with thick cord, then sewn up in straw matting. This was all right, as it formed a solid load; but it was absurd to send rice and beads in such bad packages as they were. Beside these, there were 30 large cases (60 lbs. weight) of ship's biscuits in bulk, purchased at Zanzibar and the Cape, the latter not in tin-lined boxes, so that long before they had reached Leopoldville they had become rotten and totally unfit for food. I was obliged to make constant visits to Leopoldville, where I was fully occupied in attempting to set these loads straight, and I was much worried by the bad way in which they had been packed. Yet with all my trouble and worry I could only make the best of a bad job. I had every assistance from Swinburne at Kinchassa and Liebrechts at Leopoldville in giving me what material they had to patch up the miserable loads.

At Kinchassa I knew I should have to wait at least forty days until the return of the steamer from the Aruwimi. I considered well the subject that cropped up in my mind, of making a trip down country to Matadi, or even as far as Banana. There were several things which I thought I might accomplish by such a journey, such as obtaining some good stout canvas for repacking the loads properly, and also I should endeavour to get some fixings for myself. But although I considered both these matters very important, I could not shut my eyes to the fact that my place was here; and though I had accomplished all that I could do at present in the interests of the Expedition, and could only wait patiently for the return of the steamers, yet I had to bear in mind that there might be some mishap to one or other of the steamers which might cause their return, or that Mr. Stanley might change his programme and form a large camp somewhere short of Stanley Falls, where he could get food, and in this event might send down for me and the loads earlier than I anticipated. As I was the only officer of the Expedition remaining here, if I left my post, on even such an important mission, there would be no one on hand to carry out my instructions should any of the contingencies I suggest have happened. Hence I came to the conclusion, much as I regretted it, that it would be inadvisable to leave my post to go down river. I was thus obliged to sacrifice, for the chance of furthering the good of the Expedition, the opportunity to get material for the loads as well as articles for my own future personal comfort. I was entirely devoid of the necessary paraphernalia for travel in Central Africa. Goodness alone knows that we in these parts have little enough of anything; but there are a few articles which are really indispensable, such as a tent, camp-bed, rifle, canteen, and some provisions. It might strike most people as ludicrous that an old traveller like myself should be forced to admit that he was without any of the above-mentioned articles. But, gentle reader, I had not, I am sorry to say, sufficiently taken to heart the admonition of the Psalmist, "Put not your trust in princes, nor in any child of man, for there is no help in them."

When Mr. Stanley and Sir Francis de Winton at last made up their minds in what way my former African experiences could be best turned to account in the interests of the Expedition, and gave me my sailing orders, I applied for the tent, bed, canteen, rifle, and revolver with which in the contracts the Committee had agreed, among other things, to supply each officer. In reply, I was informed by Sir Francis de Winton that all these had been sent down to the docks and were on the eve of being shipped to Zanzibar with all the other stores of the Expedition, and he added it would be a great trouble to get out these articles for me. While admitting the truth of this last suggestion, I could not help pressing hard for them, as I had not forgotten that a bird in the hand is worth two in the *bush* of Africa as well as elsewhere. If it was too much trouble to get those already packed and ready to start, ought I not to be supplied with others? For the inconvenience or economy of the Committee, was I to be the sufferer in comfort as well as, more important, health? My late experience on the Congo had shown me how very difficult it was to get these same necessary articles. I have seen numbers of men arrive fresh from Europe at Vivi, where they received orders to go up country, and when they applied for tents and beds they were told that there were none at the station; on their remonstrating, the only satisfaction they got was that they must clear out all the same. They too, in their contracts, had been promised all kinds of things, but it would be a curious and enlightening table of statistics if one could find out how many out of the hundreds I met have ever had a tent or bed!

On this occasion I was assured that orders would be given that these necessary articles would be left for me, and I should have them when I joined the Expedition. "Don't, like a good fellow, give us any trouble over this small affair," added the Colonel in comfortable Bond Street; "you know this is a different affair from the Belgian State, you can rely on us as English gentlemen." This was in answer to my urging that I ought to have these things *with me*. I had already said to him, "You know what trouble there was on the Congo about tents and beds, and you cannot be surprised that I wish to have them with me, as you cannot buy these things out there." Well, the end of the matter was, that sooner than give *any* trouble I had to be reluctantly satisfied with his assurances and go without, though I surmised what the result would be. It may be remarked that I might have bought these things, but the £40 allowed for our outfit was far from sufficient to cover the other necessaries, and I had already been forced to add £30 to that amount to get anything like a satisfactory kit, and I could not very well go to the expense and trouble at the last moment to buy articles which were mentioned by name in my contract as to be furnished me. On my arrival at Boma, M. Valcke lent me from the State stores a tent, bed, and canteen, all of which I agreed to return when I reached Leopoldville to the store of that station.

Although I should have in any event fulfilled this promise, yet I was reminded of it by Mr. Stanley during the few hours I was with him at Kinchassa. I took the opportunity to ask him when I was to receive my own tent, &c. He replied, "Man alive! you won't require these things on the steamer, you will get them when you reach the camp."

Just so, but I had already made inquiries among the officers, and could not discover that any of these things had been saved for me—yet now I must trust to Mr. Stanley's promise. The Committee dealt more reasonably with Mr. Walker, who was sent out direct to the Cape, there to be picked up, and he was supplied with a tent, &c., to take with him. Even Mr. Ward, who was engaged out on the Congo, and for whom of course the Committee had made no provision, was more fortunate than I was, as he found a tent and bed among the Expedition stores lying at Matadi after the departure of Mr. Stanley's column, and like a sensible man promptly annexed them. It was the same old joke, "Every man for himself and the devil take the hindmost." It was not that I was embittered by the personal discomfort I endured that I enter into these details, but it is only right that some comment should be made. It ought not to have been difficult to have properly supplied the officers of the Expedition, who did not exceed the vast number of nine. However, having accepted the situation and given up to the State my borrowed supplies, the only thing for me to do was to try and get together on my own account what I could in the way of bed, canteen, and other things, for I knew it was quite impossible to obtain a tent. I had many old friends at Kinchassa and Leopoldville, and found them quite ready and willing to place at my service whatever they had to spare. I got an excellent camp-bed from Dr. Sims of the American Baptist Mission Union. It was quite new, and, what was more to the point, proved to be most serviceable. He had got it at Silvers; it was made of good stout canvas mounted on strong, seasoned wooden folding trestles; it could be put up and taken down speedily, and, what was a great consideration, weighed very little. Tents made by Edgington have been brought to a state of perfection, but there is ample scope for some inventive genius to use his talents to devise a camp-bed that will combine strength and simplicity, and ability to be made ready for use quickly, and withal be light to carry. Edgington had supplied the Expedition with camp-beds: these consisted of a piece of canvas, through hems on the side of which jointed poles were put, then two cross-bars were fixed at the ends, and to this were added three sets of trestles that had to be carefully adjusted. All this required the superintendence of the owner. The drawback to this construction is that it takes time, and is hard to put together, while in addition it is heavy as a load. Some of the officers discarded these beds, preferring to have the weight thus disposed of made up of some other personal comforts! trusting to make impromptu beds of some kind as they went along.

I might have got a Martini-Henri rifle, as more than one was offered me, but I should have had to take with it ammunition (for the Expedition had nothing but Winchester and Remington); this would have encroached much on the weight of the two sixty-pound loads allowed me, so I was obliged to give that up. I had learnt from my comrades during the short time I saw them at Kinchassa that they had seen mighty little in the way of provisions on the march up from Matadi,—in fact I gathered from them that food was exceedingly scarce. As my numerous friends at Kinchassa and Leopoldville offered me all kinds of things, I saw my way to take up a few provisions. Dr. Sims kindly furnished me with tea, coffee, and a small supply of necessary medicines; Swinburne, Liebrechts, and Greshoff made up three small cases between them of most acceptable articles, such as tinned milk, tea, sugar, biscuits, butter, Brand's essence of beef, &c. Thus was I overwhelmed with kindness in every way by these good fellows, and I knew that in most cases they were depriving themselves in letting me have these things.

But I had to limit my supplies for two reasons: I knew that the *Stanley* would be well filled up with Expedition stores on her second trip, and that even if I found room for a large number of private stores I should not be able to get carriers to transport them onward to Lake Albert, though, as the end will show, this latter difficulty need not have been counted, as long before our end of the Expedition could start these supplies might have been well used to mitigate our sufferings. However, I was promised enough, and could take some to divide among my comrades without adding too much to their personal baggage. I managed also to get two extra beds, as I had heard that Barttelot and Bonny were among those who had dispensed with theirs, and thought by this time they no doubt regretted having done so. The time passed away at Kinchassa rather drearily while I was longing for the return of the steamers that I might be again up and doing. My companions at this place changed from time to time. Davey the engineer returned from his trip down country to supervise the transport of the machinery of the *Florida*, Von Rothkirch turned up with some of the pieces, while Casement came in his wake a few days later. Meantime Swinburne had gone down country. Thus, with few events to mark its flight, time went on till July 3rd, when sixty-five days after her departure the s.s. *Stanley* returned. I had calculated that forty days would be consumed in her trip, so I had ever since daily visited the beach anxiously on the look-out for her. On this day I was in my room dressing quite early in the morning, when some men employed in the outskirts of the station came rushing in shouting "Sail ho!" I hastened down with all speed to the beach, followed by Casement and Davy, and there I saw the *Stanley* coming round the bend at full steam, and in a few minutes she was alongside. She had a large number of men on board, who turned out to be Bangalas brought down for service in the State at Leopoldville and down-river stations. The *Stanley* had brought back the *Florida*, but neither of the other steamers, the *Peace* or *Henry Reed*, turned up. We lost no time in getting on board, and, as soon as we had greeted Shagerström and Deman, eagerly asked for news of the Expedition. Where had they been left? Was Mr. Stanley all right? How were the other officers? Had they had any trouble in taking Tippu-Tib back to Stanley Falls? and scores of similar questions were showered upon Shagerström. We received satisfactory answers to all, and he handed me my letters—one from Mr. Stanley and two from Ward. Shagerström and Deman breakfasted with us before going on to Leopoldville. I took the earliest opportunity of reading Mr. Stanley's letter, which is as follows:—

[Letter No. 7.]

"Camp, Rapids of the Aruwimi,
"*June* 18*th*, 1887.

"Dear Mr. Troup,

"I extract from list of goods which were to be sent to me from Matadi, and which I need here.

"You will then ship for me on board the following articles or loads per *Stanley*—

	104	cases Remingtons (ammunition, old and new).
	100	,, powder.
	10	,, caps.
	90	bales Zanzibar cloth.
	63	loads beads, cowries, and wire.
	60	,, brass wire (mitakos).
	22	bales handkerchiefs.
	8	boxes biscuits.
	12	bags rice.
	8	,, salt (purchase it).
	2	boxes soap (ditto).
	6	bales cloth for presents.
	2	bags empty sacks.
Total	487	

"The remaining articles you will kindly dispose of to M. A. Greshoff and the State, and my friend Swinburne. From each receive a proper due bill, payable where and by whom? We don't wish to make money on them, but simply to obtain cash spent and charges on the goods. These goods will consist of brass rods, rice and biscuits. For the oil please ask a simple receipt from M. Liebrechts.

"Out of the 60 boxes brass rods you bring with you from Leopoldville, 31 boxes and 100 rods are to be delivered to Bangala station. Demand receipt. 270 brass rods are to be paid Glave at Equator Station.

"A few, I believe, are due at Lukolela at the English mission.

"Your expenses in brass rods from Bolobo to this camp will roughly amount to as follows—

"Bolobo to Bangala, 15 days—125 men at 17 mitakos (1 mitako per man per day)	=	$4\frac{1}{4}$ boxes.
Bangala to Camp, 12 days—at 12 mitakos per man ($\frac{3}{4}$ mitako per man per day)	=	3 ,,
Three officers (Troup, Ward, and Bonny) 30 mitakos per day, 1050	=	2 ,,
		$9\frac{1}{4}$
Boxes to be paid out *en route*		32
Total		$41\frac{1}{4}$

"You thus ought to have 18 or 19 boxes left on reaching this camp. You will halt *en route* thus—

"One day at Bolobo, one day at Lukolela, one day at Equator, one day at Bangala, one day at Upoto to purchase provisions.

"Bolobo to Lukolela . .	3 days.
Lukolela to Equator . .	4 ,,
Equator to Bangala . .	4 ,,
Bangala to Upoto . . .	5 ,,
Upoto to Aruwimi . .	5 ,,
Mouth of Aruwimi to Camp .	2 ,,
	23

"The *Stanley*, being without the *Florida* this time, may do it much quicker on your trip.

"This is all about business I believe. And now, remember, do not let there be any stupid delay at Leopoldville. On this first trip the *Stanley* had on board 180 men and 456 loads, on yours she will have 125 men and 483 loads.

"If the captain thinks necessary, demand the large whale-boat to assist.

"Officers on returning from up river generally fall sick from the subsequent inaction, but do you urge departure for the sake of health and time. The *Stanley* was lent to us for 90 days, beginning of course the day she sailed from Leopoldville. The captain has lost us three days out of that. I presume the boat will be down to you 1st of July. You therefore have no time to lose.

"Should there be, by any freak or mischance, an impossibility to obtain the *Stanley*—take note of all the facts in connection with it—then turn your attention to other means at once in the most energetic manner.

"When you join us we shall be all heartily glad, and you need have no fear of separation afterwards. Wishing you a good trip, in health and speed.

"I am,

"Yours very sincerely,

(Signed) "HENRY M. STANLEY.

"JOHN ROSE TROUP, ESQ.
etc. etc.

"P.S.—On arriving here place yourself under the Major, for he will no doubt, when you join, push on after me with you all.—H. M. S.

"P.S.—It is a capital camp here—impregnable to all Arabs and natives, with a good garrison and plenty of chop (food) and most healthy.—H. M. S."

It will be seen that he had somewhat altered his arrangements as to the distribution of his officers, for instead of leaving Major Barttelot at the camp at Bolobo, as he originally intended, he had told off Ward and Bonny for that duty.

I had now to set to work at once to get away on the second trip with as little delay as possible. I had already made every preparation for departure that I could make, and among other things I had got sufficient wood ready at Leopoldville and Kinchassa to start the steamer without delay on this score. The s.s. *Stanley* went on to Leopoldville, while I made my way there overland, and accompanied by friend Casement. On our arrival we found that the *Stanley* had not come in, for she had met M. Liebrechts going over to Brazzaville on the opposite shore to call on the French Commissary, and had in consequence been delayed. It was late in the afternoon before she reached Leopoldville, so that it was impossible to do anything that day; but before returning to Kinchassa, I made arrangements to have her loaded up next day (Monday), so that she could start on Tuesday morning. I was at Leopoldville early on Monday, and soon had all the stores required by Mr. Stanley on board the *Stanley*. She had had another hole knocked in her on her way up with the first detachment of the Expedition, which had been patched up; taking her all round, she was in an exceedingly shaky condition. However, there was no time to be lost, and we must make the best of her, as it was impossible under existing circumstances to do anything to her. My next task was to try to dispose of the surplus stores of rice and mitakos, in accordance with Mr. Stanley's instructions. I sold to the State 65 sacks of rice and handed over to them 41 drums of machine oil, in lieu of what they had supplied for the use of the Expedition. There were also 23 large cases of the aforesaid biscuits in bulk, but I could get no purchaser for these, as they were all damaged and totally unfit for human food! The Dutch House purchased 18 sacks of rice and 28 loads of brass rods; the Sandford Exploring Association, 7 sacks of rice. This accounted altogether for 90 sacks of rice, 28 bundles of brass rods, and 23 cases of biscuits, or 141 loads not wanted by Mr. Stanley. Besides all this there was also some rice which had been left at Matadi, and which was not included in the total of loads that were to be brought up by my native transport, but Mr. Stanley had ordered this to be brought afterwards, and I had sent men down to fetch it. I had been now informed by letter that some of this had arrived at Manyanga, so I arranged with M. Liebrechts that he should purchase all this as well. Among the sacks of rice which I sold to M. Liebrechts were a number of those that I had put into good stout canvas bags when I was repacking these loads. I was very anxious to get these back, but as he had no other sacks to put the rice into, I was obliged reluctantly to part with them, for he would not take it in bulk because he had no stores sufficiently dry to keep rice in. In the long list given in Mr. Stanley's letter were mentioned 8 bags of salt and 2 boxes of soap, against which he had noted "purchase it," as he knew these were not among the Expedition stores that were in my hands. I was fortunately able to carry out his instructions as far as the salt was concerned, but the soap required I could not possibly obtain.

M. Liebrechts had added to his many personal attentions and kindness to me, by doing everything in his power to aid me in my efforts to get the steamer started without delay. Consequently, if I could do it consistently with my duties to the Expedition, I could not refuse the request he now made of me. He had forty time-expired Bangalas who should be returned to their homes, and also some 60 loads that were to be carried to the same station. This would be his only opportunity of sending them back for some time to come, as the *Stanley* would be away on this trip for at least another fifty days. This delay in sending up the men so long after their term of service had expired might lead to trouble, and would create a bad impression, which might cause difficulty at some future date when it was desirable to engage more men from the Bangala district, that station being 500 miles further up the Congo from Leopoldville. Under these circumstances I agreed to take the men and loads on the condition that I could have one of the State whale-boats should they arrive before I started, or if not, I must be authorized to take one when I met them coming down. I had made my calculations, and had found that I should be crowded right up if I did not have the addition of a whale-boat. I could not have the *Florida* as a barge again, as it was distinctly arranged that she should be lent for only one trip. I found that on leaving Leopoldville I should have 546 loads of the Expedition, including what I had to pay out *en route*, and I was to pick up 125 men and 2 officers at Bolobo, four days farther up the river.

I completed all my arrangements at Leopoldville that day, and returned to Kinchassa late the same night, Casement and I having a most pleasant ride by the glorious light of the moon. I had bidden farewell to Liebrechts and the other State officials, for I had arranged that the *Stanley* should call for me at Kinchassa the next morning, and she would now be able to leave Leopoldville in good season.

The morning of July 5th, 1887, saw me up early, getting my few personal things together. There were in all four Zanzibaris in my party. I had to leave one of the two donkey boys behind because he was very ill. Dr. Sims kindly undertook to look after him, and I left rice and mitakos in the doctor's charge for his rations. During my stay at Kinchassa these Zanzibaris who had belonged to Stairs' company had made their way back here; they had been left behind at Kimpoko, further up the river, having missed the steamer. So I had these four men as well as the Cabinda boy, lent to me by the Sandford Association as a servant, who was to accompany me as far as the Aruwimi and then return by the steamer. The s.s. *Stanley* turned up all right, and I got my small party, as well as two of the donkeys, on board. The third donkey I was obliged to leave behind; he was in such a wretched state that he was only fit for the knacker's yard.

I had hoped that one of the other steamers with the whale-boats would have returned before I started, but we had seen nothing of them.

After farewells with Casement and Davey, we were soon off, steaming up the Pool, at last under weigh, once more moving on towards our goal. This was the last detachment of the Emin Expedition, and I have no doubt that our friends hereabouts were all glad to see the last of it, for although I had received every assistance and kindness from them, one and all, yet there had been a great strain upon the State as well as every one else during the stay and passage of the Expedition on the Lower River. Even Shagerström and Deman were glad to get away. My own feelings on the subject may well be imagined. At present the steamer was fairly clear, but we had great difficulty in finding suitable quarters for the two donkeys. The only available place where they would be out of the way was right forward, but this was close to the boilers; still here they had to be put up and make the best of it. We had twenty Zanzibaris on board who were sent down from the camp at Bolobo in order to cut wood for us from day to day on our way up.

We pulled up at Kimpoko (where Bishop Taylor's people had established a mission station in the old State quarters) in the afternoon, as we were on the look-out for the steamers and whale-boats which we were afraid might pass us in the large expanse of water of the Pool; but it was not until the next day when we had really got out of Stanley Pool and were steaming up the river that we caught sight of one, which we at first thought was the Mission steamer *Henry Reed.* She turned out to be the other Mission steamer *Peace*, and it was not long before we met and got alongside of her. On board her were Messrs. Whiteley and Charters of the mission and Mr. Walker, engineer, of our Expedition. We learnt from them that the *Henry Reed*, with the *A. I. A.* and whale-boats had been stopped at Bangala on their way down from the Aruwimi by Capt. Vangèle, of the State, who demanded them on urgent public business. Mr. Stanley had evidently expected that something of this kind might happen after he saw Capt. Vangèle at the Equator station on his way up. Consequently Walker had been warned when he started down again not to put in with the *Henry Reed* and boats at Equator Station, and following his instructions did not intend to do so. But when he stopped at Bangala, much higher up than the Equator, he found Capt. Vangèle had come on to this point, and immediately demanded the *Henry Reed, A. I. A.*, and the whale boats; the latter were State property, but the s.s. *Henry Reed* belonged to the missionaries, and it was the very steamer which it will be remembered Mr. Stanley had had so much difficulty in getting! Mr. Walker did all he could in the way of remonstrance but all to no purpose, so he was obliged to come down on the *Peace*.

This news was very serious to me, for now it was apparent that I could get no whale-boats. I had then to consider whether I ought not to send back to Leopoldville the time-expired men and the loads destined for Bangala in the *Peace*, as we should be overcrowded if we had not the whale-boats, after picking up the men at the camp at Bolobo. I consulted with the captain of the *Stanley*, who assured me that we could manage to take them all, so it was decided to go on as we were, taking Mr. Walker with us in his capacity of engineer, in case anything should go wrong with the engines of the *Stanley*.

On July 10 we reached Bolobo, where we found Ward and Bonny in their encampment, with everything looking very neat and trim; we had to stop overnight at this place. We found a large supply of wood had been got ready for the steamer. We were all in high spirits at the idea of starting on the final stage of our river journey the next morning. Ward and Bonny were both looking very well, and I was right glad to meet them again.

Early the next morning we got all the men on board, as well as another donkey. We were just crowded right up, the men had all to be fitted into their places on the lower deck like so many sardines, leaving no room to pass round the deck houses; the upper deck was full up also, while nearly all the compartments were packed tightly with the stores, not leaving much space for the necessary fuel for the steamer. It was lucky that no Government Inspector was within hail, for if we had had a license he would most certainly have revoked it when he knew of our overcrowded condition. That the public may judge how much we had on board, I append here a list of officers, men and stores, on the s.s. *Stanley*.

Captain Shagerström . . Captain s.s. *Stanley*.
Deman . . . . Engineer ditto.
Troup . . . . .
Ward . . . . .
Bonny . . . . .
Walker . . . .
} Officers of the Emin Relief Expedition.

Total 6 officers.

14 Bangalas, crew of s.s. *Stanley*.
3 Zanzibaris, left at Kimpoko.
1 ,, donkey-boy.
40 time-expired Bangalas.
125 Zanzibaris, from camp at Bolobo.
4 officers' servants.

Total 187

STORES.

493 loads of Emin Relief stores for transport to the Aruwimi.
50 ,, for repaying what Stanley borrowed on his way up and for rations of officers and men.
60 ,, of the State for Bangala, missions, &c.
25 ,, officers' private baggage.
40 ,, belonging to returning Bangalas.

Total 668

3 donkeys.

It must be remembered that besides the above we had to carry a large enough supply of wood to permit of our steaming for about ten hours. As it was we were quite unable to get on board anything like all the wood that Ward and Bonny had had cut for us. The men had all been rationed for three days, their bananas and chiquanga helping to increase the weight on board. This was renewed when we reached places where they could get food, that they might have supplies in hand.

The reader may be curious to know how we passed the days on our way up. We generally started about 5.30 A.M., and steamed on until between 4 and 5 o'clock in the afternoon, when we would pull up at some place chosen by the captain as being a suitable spot to obtain a fresh supply of wood for the next day's steaming. As soon as we were securely moored, the men were all put ashore with their bundles of food. They were divided into parties, and provided with axes and cross-cut saws, the general order been given "*leete choka*" (bring axes), while *choka, choka* was echoed down the line. Then under the general supervision of the white men they began wood cutting. Trees that were dry enough for firewood were selected, and the task of cutting down and drawing them to the water's edge was entered upon. At dusk huge fires were lit, and by the blaze of these the men cut up the logs into small chunks, three or four feet in length; thus they would be occupied all night in preparing fuel for the next morning. It must be remembered that the men had nothing to do all day but to sit stock still on the steamer, for this wood cutting was the only work they had. It was a picturesque enough sight, attended with its noise and bustle, the thud, thud of the axes, the crash of the falling trees, then the firelight scene, with the scraping of the saw, and chop, chop, followed always by a dull thud as the blocks were thrown on the piles, and were then tossed from hand to hand till they were all loaded on to the steamer; added to all this were the weird cries and monotonous songs of the men at their work.

The next day steam would be got up early, we were generally off at 5.30 A.M. The shrill whistle of the steamer was the signal for all the men to get on board; then would come the task of *fitting* them properly into their places and seeing that the headmen had collected all the axes and saws. Shagerström, the captain, would be at his post on the bridge, shouting his orders, and we would soon be again on our way. M. Liebrechts, before we left Leopoldville, had kindly put on board a sufficient supply of provisions such as tea, sugar, milk and biscuits for us officers of the Expedition, and we had a ration of mitakos to buy animal food when we could get it at the villages we came across. All that we had to eat was partaken of in the little "saloon" of the steamer, followed by many a jolly yarn. After breakfast we would go up on the upper deck, the fore portion of which was covered, with seats on each side, making a comfortable wheel house. Here Shagerström used to take up his quarters from the time we started till we pulled up for the night, some ten hours, and would have his breakfast and lunch sent up to him while he was engaged in piloting us through the intricate channels of the mighty Congo! There was no one to share this task with him, to relieve him of his anxieties. To us it was a painfully monotonous existence; the only thing that kept us going was the knowledge that each turn of the stern wheel was bringing us nearer our destination. In this way time went on until the 14th, when, after having started as usual at 5.30 A.M., we reached Lukolela at 8.30 A.M. At this point there had been a State station, but it had been abandoned; there was now only a mission station left. We put in here, as we had the mail and a few cases for the missionaries, with the less hesitation as we knew it was a very good place to get wood, for we had not been able to get a large enough supply the night before. We quickly set our men to work, so that we were able to get away again a little after midday with a capital lot of wood. All had gone well with us so far, but about 4 o'clock in the afternoon, while we were going at full speed, we suddenly felt a shock. We had struck something, had rebounded, and then struck again and stuck fast. At the time I was sitting with Walker in the little saloon, and we rushed out immediately to learn what had happened. Walker with the captain and Deman soon had opened the hatches over the fore-compartments, and found that one of these was filled with water. Luckily there was no panic among the men; the order was passed around that they were to sit still. The bulk of the wood was at once thrown overboard to lighten her; we then tried to steam off, but there was no moving ahead or astern, we only swung round. Soundings were taken, and we found we were in six fathoms of water. A sail was drawn underneath from the bows, and we discovered we were fast about midships. On examining the compartments beneath the captain's cabin we found the plates were indented, but fortunately she was making no water here. The conclusion arrived at was that we had struck on a big snag, tearing a large hole in the fore compartment, and had then stuck fast on to it further aft. Had the compartments not been water-tight we should have gone down at once. We were a good distance from the shore, with a very strong current running, so that the outlook was discouraging to say the least of it. Presently we caught sight of some canoes, and after shouting and beckoning to their occupants to come to our assistance, after some time we managed to make them understand us, and they came alongside. We now decided that our best plan was to transfer by this means all the men, if possible, to a sandbank as near as the shore, as we were afraid they might cause trouble if they got into the native villages. I am thankful to say that still there was no panic among the men, in fact, I don't think they knew their danger; we said very little, but watched them closely. After striking a bargain with the natives, we persuaded them to take some of the men off, part to the sandbank and others to the shore, where they told us they had a village. Time was getting on, it would soon be sundown, and the canoes were very slow in returning. We had kept one canoe alongside, and in this Ward went ashore to hasten them. He soon sent back a lot more canoes, and we were able to get off all the men before dark. We then made another attempt to steam off, but without success, so there was nothing to be done but to wait till morning, when we could remove some of the cargo. Ward arranged with the natives to send their canoes out at dawn for this purpose. Steam was kept up all night, while we were keenly on the alert for further developments. We were afraid that the steamer might bump a hole through the midship compartment, the point where she was stuck. The fore compartment was baled out and the hole stuffed up with tarpaulins, &c., which helped to keep the water somewhat under. It was an anxious night for us all. Speaking for myself, I can say I did not sleep much; the chance of losing the precious stores after all the care I had bestowed on them was too horrible to contemplate. I lay awake counting the moments to daybreak. At dawn of July 15th some of the canoes were visible, but it was blowing very hard, so they took some time to reach us, and when they came alongside were pretty full of water. Ward went ashore to receive the cargo on its arrival there, and afterwards sent word that he wanted assistance, as it was too much for one person to watch all the canoes as they landed at different points, and there was danger that some of the loads might be made off with. So Bonny went to his aid, while I stayed on board to despatch the cargo.

As we thus by degrees lightened the steamer, Captain Shagerström had steam up, and was trying hard to get off the snag, whilst Walker and Deman were watching the efforts, ready for instant assistance if any fresh damage was done to the midship compartment. It was not until we had taken out more than 100 loads that at last we got off, without any additional injury to the steamer. This result was greeted with loud shouts, and we immediately steamed to the shore, where we picked up a portion of the men and loads there, and then made for the sandbank, which we considered a more suitable place for beaching the steamer for repairs, as well as safer for our men and loads. On arrival here we immediately commenced to take out all the loads, stacking them carefully and covering them with tarpaulins. It was then decided by the skipper to fill all the aft compartments with water in order to lift the bows of the steamer, that Mr. Walker and Deman might get at the hole in the fore compartment to repair it. It was quite clear that we would be obliged to stay here two or three days before the necessary repairs could be completed. Hence we made the Zanzibaris camp on the sandbank, giving them their usual rations of mitakos for buying food; this the natives brought over, and a brisk trade was done. We told off a guard of men to take care of the stores under one of the officers. Ward, Bonny, and I took watch and watch for three hours at a time throughout the night. In the meantime Walker and Deman were hard at it working over the hole; they found it difficult to get at, though they had prized up the steamer in the way mentioned above. To accomplish their job they had cut up an old oil drum, flattening it out to serve as a plate to patch up the hole, riveting it on the best way they could, and *framing* it all with red lead. Mr. Walker and his fellow-engineer deserve the highest praise for having so successfully achieved their task under such great difficulties, standing in the water most of the time, and sticking faithfully to their work.

We remained at "Snag Point" three days, the repairs being finished on July 17th, and all was then in readiness for the reception of the stores again, which had in no way been improved by their wet canoe journeys. We worked hard, so that by nightfall we had all prepared for a morning start. The next day, after getting all our men on board, we bade a cheerful good-bye to the unfortunate "Snag Point," and made our way to the other side of the river to get wood. We were all in good spirits, feeling rejoiced at our escape, for we had had a very narrow squeak; had the accident occurred in some place beyond reach of the villages, or had the natives been unfriendly or afraid to come to our assistance, we should have been obliged to throw some of our stores overboard! However, we were off again, and right thankful we were that the natives had aided us. We did not forget them, for they were well paid for their trouble, and had, moreover, done a good trade with our men during our enforced stay in their district. No blame could be attached to Captain Shagerström for this unforeseen disaster. The Congo river, with its intricate channels, numerous rocks, strong currents, &c., &c., is enough to try the skill of the most experienced navigator; but in addition to all these dangers there are also found, to our cost, huge trees or snags sunk in deep water, with branches, like supple prongs, sticking up nearer the surface, and upon which it is a serious matter to run. We had lost three days when time was valuable, but we consoled ourselves by thinking it might have been worse.

Nothing further of importance happened until the 21st, when we arrived at Equator station. Here there had been an important State station, but it had been abandoned, and now it was in the possession of the Sandford Exploring Expedition. Here we met Mr. Glave, an old friend of Ward's and mine, who had been in the service of the State for three years, and after a short sojourn in England had come out again in the service of the other expedition. A mission station is also here. Glave took dinner with us, and afterwards we adjourned to his quarters, where, as he was a banjoist, he amused us with some very quaint songs. Up to this date we had all enjoyed good health, but Bonny was down with a slight attack of fever since yesterday. We had arrived at the Equator too late to cut wood that night, so we had to remain the next day for this purpose and to allow the men to buy food. Hence we could not get off again until the morning of the 23rd, by which time Bonny was much better.

We arrived at Bangala about 8.30 A.M. of the 27th July, all well. We had now accomplished half the distance —that is, 500 miles. We found here several State officers, all old comrades: Capt. Vangèle, MM. Lieut. Vankerckhoven, Baert, Danis, Liénart, &c., and Mr. Werner, engineer of the s.s. *En Avant*. We had a very friendly reception from them all, and M. Vankerckhoven, Commissaire de District, entertained the whole party at dinner that night.

We had noticed that the s.s. *Henry Reed* with the *A.I.A.* and the whale-boats were moored at the landing-place. We got rid here of the time-expired Bangalas and their loads, as well as the stores for Bangala station. I also repaid some thirty-eight loads of mitakos that Mr. Stanley had borrowed at this station, so we were somewhat lightened. However, I deemed it advisable to ask for one of the whale-boats; for in case of any further accident, knowing the shaky state of the *Stanley*, it was prudent to have some kind of boat for communicating with the shore in case of any fresh disaster. But after all we did not get one, as I was informed by M.

Vankerckhoven, as well as by Capt. Vangèle, that all the craft were required for the service of the State.

Next day, that is, July 28th, we were off again, saying farewells to the State officials. This was the last station on the up river. We had still 500 miles more to make before we reached Yambuya, our destination, and we were now on the eve of leaving this last outpost of civilization on the river. From here onwards our days were more uneventful and monotonous than heretofore, steaming by day, cutting wood by night, only relieved by the excitement and worry of half a day's delay owing to the donkey pump breaking down. So time passed, and everything went quietly until August 11th, when we reached the mouth of the Aruwimi river. Here matters looked for a moment threatening enough. A native village called Basoko is on the river bank here; it was burnt down some time ago by Mr. Deane when he was in charge of Stanley Falls, five days' journey higher up the Congo, as a punishment because the inhabitants had taken some Houssas prisoners and had killed and eaten them. For this act of cannibalism they were made to suffer, but as the people hereabouts are all given to this horrible practice, there is little doubt that the burning of the village had not improved the moral tone of the district to any great extent. Mr. Stanley on his way up went ashore here, and the frightened natives fled from their village. But he did not, as it happened, escape from all trouble thereby, as his boy, Baruté, who had been in England with Mr. Stanley, and who was a native of this village, deserted, in company with another Aruwimi boy, taking with

MR. J. S. JAMESON.

them several things of Mr. Stanley's, including a presentation revolver and a Winchester rifle. On the occasion of our visit to this part of the world, the natives rushed out, lining the bank, brandishing their shields and spears, and howling like all possessed in a most warlike manner, with drums beating and a general row that was beyond description. No doubt had we turned towards the shore their bold defiance would have been turned into undignified attempts to get out of our way. This diversion was not repeated by all the inhabitants on the river, as, farther up, when we stopped to cut wood that night, the natives came out unarmed, and were rather afraid of us than otherwise. They were a wild-looking lot, splendidly made, strong-limbed devils, but we showed them our friendly disposition, and after a time they came alongside and sold our men plantains, palm-oil, and fish. We had still three days' steaming to make before reaching our destination, for it was on August 14th that we reached Yambuya. Shagerström and Walker, who had come up with the advance column, were acting as our guides, but when they pointed out the spot as Mr. Stanley's camp, we began to think they must be dreaming, or that if this was the place it was now deserted, for we saw no sign of life on the cliff above us; but at last, in response to the whistle of the steamer, we saw some men coming down the winding path in the face of the cliff, and were thus assured that no mistake had been made.

It was not long before we greeted our fellow-officers, Major Barttelot and Mr. Jameson, whose figures we had seen descending the path. We were well satisfied to have reached our present destination. In truth, we had made the trip up river in remarkably good time.

Mr. Stanley had fixed August 10th as the day we should be at the camp, and it was now only four days later, but he had based his calculation on the presumption that the *Stanley* would reach Leopoldville on July 1st; as it was, she was two days late. Moreover, we had no assistance from barges or whale-boats in carrying our loads; if we had had these, our delay at "Snag Point," which accounts for the other two days, might not have been so long. So anxious had Mr. Stanley been that our detachment should arrive promptly, that he left two £50 cheques in Mr. Jameson's hands, to be paid to Captain Shagerström and M. Deman should they make the trip within a given time. They succeeded in earning this reward, which was handed to them.

CHAPTER VIII.

AT YAMBUYA.

AFTER greetings with Major Barttelot and Mr. Jameson, who were soon on board the *Stanley*, they were asked to stay to luncheon with us, as that meal was about ready. Shagerström could give them a better meal than we were likely to get in camp. I handed Barttelot and Jameson the bundle of letters which I had brought up for them, and also made over to the former the mails for Stanley and the other officers who had gone on. Barttelot soon told us all the news; the substance of it was about as follows:—When the Expedition with Stanley came up the Aruwimi, Barttelot did not come with them, but was sent in the s.s. *Henry Reed*, with Mr. Walker, to take Tippu-Tib to Stanley Falls. From what I could gather, there seems to have been considerable commotion at the sight of a steamer in this region. It must be remembered that no steamer had been up to the district since just after Mr. Deane had had to evacuate the station after his gallant defence, more than a year ago, and the Arabs were suspicious, fearing that the State would come up to have their revenge. But when it became known that their great chief Tippu-Tib was on board, matters assumed quite another aspect.

Barttelot told me that one of the questions asked by Tippu-Tib was if the powder promised to his men had arrived, and he seemed much put out to find that it had not, though Barttelot added that it would be brought up by me.

After seeing Tippu-Tib safely landed, Barttelot came down the Congo and up the Aruwimi to the camp, where he rejoined Stanley. In the meantime Mr. Stanley had formed his camp at Yambuya, and in a letter written to Mr. (now Sir) W. Mackinnon he relates the story of his arrival and the capture of the place. It is dated Yambuya Rapids, Aruwimi river, June 19th, 1887. After describing his approach to Yambuya village, which is the furthermost point one can go up the river, owing to the rapids here barring navigation, he decided to land, and began a palaver with the natives, who signified their unwillingness to have him come ashore. He writes that they told him, through their interpreter, "'Camp on the opposite bank if you will. We will bring to you whatever we have to sell, but if we permit you to land here our village will become the common resort of the Watamba-Tamba, the Arab slavers.' We argued that if we were with them, they need have no fear of the marauders, but we wasted our breath. We had been nearly two hours at this work of negotiating, and the natives, being addicted to palavers, would not have minded very much had the palavers lasted a week. We therefore signalled to the *Stanley* to appear with the troops. A few minutes later, at a second signal, both steamers set up a hideous steam-whistling, under the protection of which the troops disembarked, and in a few seconds we were in possession of an empty village. There was no occasion to fire a shot, for the natives had disappeared as completely as the vapour of the steam-whistling had dissolved. We have commenced the construction of a proper encampment. The dense bush which is in the background of every African village has been cut down to a distance of 100 yards. To-morrow the axe-men will cut trees down to form a palisade. . . . We have to build a storehouse for our goods; we have to select what we need for the forced march to Lake Albert. We have to send out scouting parties in every direction to spy out the land, to discover the fields for future forage if the natives should absolutely refuse to return and resume their own villages and possessions."

Mr. Stanley left this camp on June 28th with the advance column, taking with him Lieutenant Stairs, R.E.; Dr. Parke, Army Medical Corps; Mr. Mounteney Jephson, and Mr. Nelson, and his white servant William. The column consisted of 414 men—360 rifles and 54 supernumeraries—the bulk of them Zanzibaris, with a few Soudanese. He had his steel boat carried in sections, as well as the Maxim machine-gun. Part of his men carried loads, chiefly of ammunition; the Soudanese—about forty or fifty—being the only soldiers in the column.

He left written instructions with Major Barttelot, and a copy of these was forwarded by the latter to the Emin Pasha Relief Committee in London, as he had been instructed, accompanying it by a letter. Both these documents were published by them in the *Times* of November 28th, 1887, and were as follows:—

FROM MAJOR BARTTELOT, YAMBUYA VILLAGE, TO MR. W. MACKINNON.

"*August* 15*th*, 1887.

"SIR,

"I have the honour to inform you that Mr. H. M. Stanley went eastward from here on the 20th of June. A lagger on the march came in on August 12th, and reported he had left Mr. Stanley at a river flowing north into the Aruwimi, eighteen days' march out from here. He reported all as well. Since Mr. Stanley's departure, Mr. Jameson and myself have been employed in fully carrying out his instructions, a copy of a letter of which is attached for your information. The trench has been dug, the platforms put up, the bank of river cleared, and the clearing to the east side kept clear.

"I have ascertained the position of all villages within a four mile radius on the southern, eastern, and western sides. Our relations to the natives are most friendly. They trade with the men in a small way, and the chiefs come in constantly to see us. Lately they have come to claim our protection against some marauding Arabs, who may or may not be Tippu-Tib's people.

"I have tried to ascertain, but cannot obtain any information. The steamer *Stanley* arrived yesterday, August 14th, with Messrs. Troup, Ward, and Bonny, three donkeys, and 479 loads.

"Our present strength is—Soudanese 44, Somalis 2, Zanzibaris 200; total, 246. The sentry duty is entirely furnished by the Soudanese, the Zanzibaris only finding picquets. Up to the time of writing this Tippu-Tib's men have not arrived, though these marauding Arabs before mentioned may be them. Should it turn out to be so, or should they eventually come, I should march at once in Mr. Stanley's track. If, however, they do not, I shall be compelled to stay here till November, when Mr. Stanley expects to return, or till such time afterwards as he may return.

"Since Mr. Stanley's departure we have lost—Soudanese 1, Somalis 2, Zanzibaris 4; total 7. Attached also is a copy of the camp orders for your information, which will show you that all has been done to preserve discipline, alertness, and also cleanliness.

(Signed) "EDMUND M. BARTTELOT, Major,
"Commanding Yambuya Camp."

"Camp, Yambuya Village,
"Aruwimi Falls, Africa,
"*June 24th*, 1887.

"To Major Barttelot, etc.

"Sir,

"As the senior officer of those accompanying me on the Emin Pasha Relief Expedition, the command of this important post naturally devolves on you. It is also for the interest of this Expedition that you accept this command, from the fact that your Soudanese company, being only soldiers and more capable of garrison duty than the Zanzibaris, will be better utilized here than on the march (road).

"The steamer *Stanley* left Yambuya on the 22nd of this month for Stanley Pool. If she meets with no mischance she ought to be at Leopoldville on the 2nd July. In two days more she will be loaded with about 500 loads of our goods, which were left in charge of Mr. J. R. Troup. This gentleman will embark, and on the 4th July I assume that the *Stanley* will commence the ascent of the river, and arrive at Bolobo the 9th July. Fuel being ready, the 125 men in charge of Messrs. Ward and Bonny, now at Bolobo, will embark, and the steamer will continue her journey. She will be at Bangala 19th July, and arrive 31st July. Of course the lowness of the river in that month may delay her a few days; but having great confidence in her captain, you may certainly expect her before the 10th of August.

"Though the camp is favourably situated and naturally strong, a brave enemy would find it no difficult task to capture if the commander is lax in discipline, vigour, or energy. Therefore, I feel sure that I have made a wise choice in selecting you to guard our interests here during our absence. The interests now entrusted to you are of vital importance to this Expedition. The men you will eventually have under your command consist of more than an entire third of the whole Expedition. The goods that will be brought up are the currency needed for transit through the regions beyond the Lakes. There will be a vast store of ammunition and provisions, which are of equal importance to us. The loss of these goods and men then would be certain ruin to us, and the advance force would in its turn need to solicit relief. Therefore, weighing all these matters well, I hope you will spare no pains to maintain order and discipline in your camp, and make your defences complete, and keep them in such a condition that, however brave an enemy may be, he can make no impression on them. For this latter purpose I would recommend you to make an artificial ditch of six feet wide, three feet deep, leading from the natural ditch, where the spring is, round the stockade. A platform like that on the southern side of the camp, constructed near the eastern as well as the western gate, would be of advantage to the strength of the camp. For, remember, it is not the natives alone you have to fear, or who may wish to assail you, but the Arabs and their followers may, through some cause or other, quarrel with you and assail your camp.

"Our course from here will be true east, or, by magnetic compass, east by south, as near as possible. Certain marches that we may take may not exactly lead in the direction aimed at; nevertheless, it is the south-west corner of Lake Albert, near or at Kavalli, that is our destination. When we arrive there, we will form a strong camp in the neighbourhood, launch our boat, and steer for Kibero, in Unyoro, to hear from Signor Casati, if there, of the condition of Emin Pasha. If the latter is alive and in the neighbourhood of the lake, we shall communicate with him, and our after conduct must be guided by what we learn of the intentions of Emin Pasha. We may assume that we shall not be longer than a fortnight with him, before deciding on our return to this camp, along the same route traversed by us when going east.

"We will endeavour, by blazing trees and cutting saplings along our road, to leave sufficient traces of the route taken by us. We shall always take by preference tracks leading eastward. At all crossways where paths intersect, we shall hoe up and make a hole a few inches deep across all paths not used by us, besides blazing trees when possible. It may happen, should Tippu-Tib have sent the full number of adults promised by him to me—viz., 600 men (able to carry loads)—and the *Stanley* has arrived in safety with the 125 men left by me at Bolobo, that you will feel yourself sufficiently competent to march the column, with all the goods brought by the *Stanley*, and those left by me at Yambuya, along the road pursued by me. In that event, which would be desirable, you will follow closely our route, and before many days we should most assuredly meet. No doubt you would find our bomas intact and standing, and you should endeavour to make your marches so that you could utilize these as you marched. Better guides than these bomas of our routes could not be made. If you do not meet them in the course of two days' march, you may rest assured that you are not in our route.

"It may happen also that, though Tippu-Tib has sent some men, he has not sent enough to carry the goods with your own force. In that case you will, of course, use your discretion as to what goods you can dispense with to enable you to march. For this purpose you should study your list attentively, viz.:—

"1. Ammunition, especially fixed, is important.

"2. Beads, brass wire, and cowries rank next.

"3. Private baggage.

"4. Powder caps.

"5. European provisions.

"6. Brass rods as used in the Congo.

"7. Provisions (rice, beans, peas, Matama, biscuit).

"Therefore you must consider after those, sacking tools, such as shovels (never discard an axe or a billhook), how many sacks of provisions you can distribute among your men to enable you to march, whether half the brass rods in your boxes could not go also, and there stop. If you still cannot march, then it would be better to make marches of six miles twice over (if you prefer marching to staying for our arrival) than throw too many things away.

"With the *Stanley's* final departure from Yambuya you should not fail to send a report to Mr. Wm. Mackinnon, of Gray, Dawes, and Co., 13 Austin Friars, London, of what has happened at your camp in my absence, of when I started away eastward, whether you have heard of or from me at all, when you expect to hear, and what you purpose doing. You should also send him a true copy of this Order, that the Relief Committee may judge for themselves whether you have acted or propose to act judiciously. Your present garrison shall consist of eighty rifles and from forty to fifty supernumeraries (the *Stanley* is to bring you within a few weeks fifty more rifles and seventy-five supernumeraries under Messrs. Troup, Ward, and Bonny).

"I associate Mr. J. S. Jameson with you at present. Messrs. Troup, Ward, and Bonny will submit to your authority in the ordinary duties of the defence and the conduct of the camp or of the march. There is only one chief, which is yourself; but should any vital steps be proposed to be taken, I beg of you to take the voice of Mr. Jameson, and when Messrs. Troup and Ward are here, pray admit them to your confidence, and let them speak freely their opinions.

"I think I have written very clearly upon everything that strikes me as necessary. Your treatment of the natives, I suggest, should depend entirely on their conduct to you. If they do not molest you, suffer them to return to the neighbouring villages in peace, and if you can in any manner, by moderation, small gifts occasionally of brass rods, &c., hasten an amicable intercourse, I should recommend your doing so. Lose no opportunity of obtaining information respecting the natives, the position of the various villages in your neighbourhood, &c.

"I have the honour to be,

"Your obedient servant,

(Signed) "HENRY M. STANLEY.

"P.S.—In the bottom of your ditch put splinters. Keep four or five weak men doing this light job; cut fuel ten days for *Stanley*.

"Give one brass rod per week to each man to buy fish, &c.; in five months these amount to 2580. Give also six cowries per man per week; in five months these will number 15,480.

"Let Mr. Jameson attend to the sick daily.

(Signed) "H. M. S."

"I hereby certify this is a copy of Mr. H. M. Stanley's original letter.

(Signed) "EDMUND M. BARTTELOT, Major.

"CAMP, YAMBUYA VILLAGE, ARUWIMI FALLS."

Since Stanley's departure in June Barttelot tells us he has heard nothing of the advanced column except what he gathered from a Zanzibari who had lagged behind and had returned to Yambuya a few days prior to our arrival. He reported that he had left Stanley at a river flowing north into the Aruwimi, eighteen days' march from here. He said all were well. Neither had Barttelot heard anything at all from Tippu-Tib, or the men promised for our service. Concerning the affairs of the camp, he mentioned there had been difficulty in obtaining food from the natives, who however had not disturbed the white men in possession

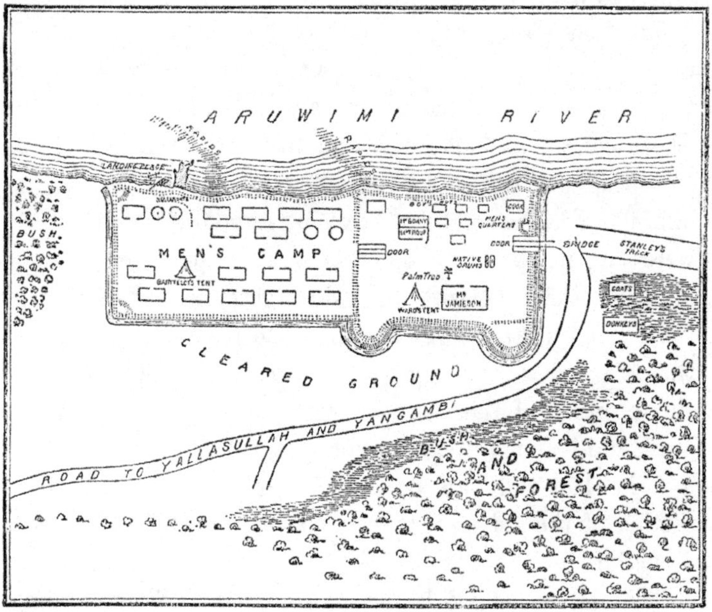

of Yambuya, but had most of them taken refuge on the opposite side of the river.

Such were the chief points of information that we gleaned from Major Barttelot. When we had talked over the news and finished our luncheon, we followed the lead of Barttelot and Jameson up the winding path in the face of the bluff upon which was situated the camp. When we reached the latter it did not strike us as particularly strongly fortified, as the palisading was made of young saplings, which seemed rather imposing, but on examination we could see that it would not afford much resistance to a determined rush of an enemy. I give a plan of it as it then was.

Jameson was occupying a large hut, in which was the remainder of the stores, brought up on the first trip of the steamers, that Stanley had not taken with him. These consisted chiefly of ammunition, the remains of the European provisions, with the surplus personal baggage of himself and officers, a few boxes of mitakos, &c. All these were left in charge of Jameson. Barttelot had his tent pitched nearly opposite the water-gate, in the centre of the men's quarters. A large hut had been constructed on the other side of the camp from Jameson's, which was for the reception of the stores brought up by me, and in this hut Bonny and I took up our quarters, as I had no tent and he had only an apology for one, in the shape of a section of the old mess-tent. Ward pitched his tent near Jameson's hut.

That afternoon we began unloading the *Stanley*, transferring the stores into the hut prepared for them. I had brought up all the stores asked for by Mr. Stanley except the two cases of soap, which I could not obtain, as well as two cases of Madeira which arrived for him after his departure. I copy the list of the stores I had charge of, in order to give a clear conception of the large addition to the supplies already in camp:—

100	cases of	powder.
64	„	Remington cartridges (old).
65	,,	„ „ (new).
10	double cases of	percussion caps.
90	bales of	Zanzibar cloth.
21	,,	handkerchiefs.
3	,,	assorted cloth.
2	cases of	savelist „
7	,,	biscuits.
1	case of	tin ware.
26	bales of	mitakos (brass rods).
27	rolls of	brass and iron wire.
33	sacks of	beads.
13	„	cowries.
8	„	salt.
2	cases of	Madeira.
1	sack of	empty sacks.
Total 493		

I handed all these over to Major Barttelot according to Stanley's orders, receiving a receipt for the same. I had now finished the particular duties assigned to me by Mr. Stanley—that is, having cared for and brought up to the camp all the loads entrusted to me. It now only remained for me to place myself under Major Barttelot's command, and to assume the duties in camp that he assigned to me, and this I did.

I had a long conversation with Barttelot that afternoon. I was sorry to find him far from well; he had had fever, and was suffering from great debility. This was my first talk with him. It did not take me long to discover that he had an intense hatred of anything in the shape of a black man, for he made no disguise of this, but frequently mentioned the fact. His hatred was so marked that I was seized with great misgivings concerning his future dealings with them, more particularly when he would have to handle Tippu-Tib's men. In this same conversation he informed me that the company of Zanzibaris which Jameson had would be handed over to me—in fact, they were my men that Dr. Parke had held for me since their arrival. Ward and Bonny were to keep the 125 men brought up from Bolobo, dividing them between them. Jameson was to look after the sick and be associated with Barttelot, as well as allowed time to go out to add to his Natural History collection. The Soudanese were to remain under the charge of Barttelot.

On the day after our arrival we were startled by hearing shots on the opposite side of the river. Knowing that the natives had no guns, we at once surmised that the shots must have been fired by some of Tippu-Tib's people. Our attention being attracted by this we looked across and saw canoes filled with native men, women, and children, and men swimming down the river, all making a general stampede. It was not difficult to imagine the cause of this hurried flight; no doubt Arabs had attacked their village and driven them out. To ascertain if these were Tippu-Tib's men, Bonny and Ward crossed over in canoes, but learned that the Arabs had already departed. They were told by some of the natives that four of their women had been taken prisoners by the Arabs.

Meantime, the steamer having been unloaded, preparations were made for her speedy departure. A large supply of fuel was all ready for her, to fill her right up now that she was empty. We had all been busily engaged with writing letters home, as this would be our last chance

NATIVE OF ARUWIMI

of communicating with the outside world for some time to come. Mr. Walker, the superintendent engineer of the Expedition, having now completed his duties, was returning in the s.s. *Stanley*, thus completing his two full trips on the Upper River. He had been of inestimable service to us, and we regretted very much parting with this warm-hearted Scotchman. To him was intrusted Major Barttelot's letter to the Emin Pasha Relief Committee, enclosing a copy of Mr. Stanley's instructions (see pages 139—142).

On Aug. 17th all was in readiness for the departure of the *Stanley*, and we went down to the landing-place to see her safely off, and to bid farewell to Walker, Shagerström, and Deman—our old friends. Barttelot had been so seedy that he had consulted Bonny as to whether he ought not to take this opportunity of going home, but Bonny did not recommend his doing so. Therefore he remained at his post. The *Stanley* moved off with the three officers on board, and we watched her till she steamed out of sight.

But to return to the position of affairs in camp. Our garrison, commanded by Major Barttelot, who had the four white officers under him, consisted of 44 Soudanese, 2 Somalis, and 200 Zanzibaris—total, 246. There had been seven deaths since Stanley's departure and before our arrival: 1 Soudanese, 2 Somalis, and 4 Zanzibaris—total, 7.

Each one of the men received, during the first part of our stay, one mitako a piece every Saturday wherewith to buy fish. This was the only ration paid out to them, and, according to Stanley's instructions, was to be continued only five months, at which time he was expected to return. They had the run of large fields of manioc, further up the river near the rapids, which had been planted by the natives before they left the village. From these they obtained their daily food. After digging up the roots they usually ate them raw and made the green leaves into a sort of spinach with palm oil, when they could get this latter commodity. There were two Zanzibari interpreters—John Henry, who had been in the service of the Congo Free State before, and who was known to me, and Bartholomew. The Soudanese interpreter was Assad Farran. We were obliged to appoint some of the best Zanzibaris as headmen, because Mr. Stanley had taken with him all the recognized *niamparas*, as the headmen are called.

The daily routine of camp was as follows: We each took turn as orderly-officer of the day. Reveillé was sounded on the big native drum at 5.30 o'clock. This drum, left by the natives, was a huge log, whole at the ends, with the inside hollowed out so that when struck with two sticks headed with a lump of native rubber, it gave forth a weird, dull sound, audible for some distance. At 6 o'clock the signal for falling in was given, and the men were told off by their various officers to their daily duties. A detachment would have to sweep the camp, while others were sent out to cut and bring in wood, grass, &c. Those who were not on duty would go out to collect their daily ration of manioc. At 12 o'clock the orderly-officer ordered "piga ngoma" [that is, "strike the drum"], and the men would come in for their midday meal and rest. Work, such as it was, was resumed at 2 o'clock, and the retreat was sounded a little before sunset, about 5.30 o'clock. The orderly officer's duty was to see that the camp was kept scrupulously clean, and its sanitary arrangements were all right, and to attend to the camp during the day. Major Barttelot told off the guard from his Soudanese every morning for sentry duty under one of their officers. They furnished the sentry at the water-gate and east-gate, and were responsible for the guarding of the whole camp. At sunset a picquet was furnished from one of the Zanzibari companies in rotation to increase the guard, acting as watchmen, being placed at intervals between the Soudanese sentries. The orderly-officer's duty was to turn out the guard three times during the night, and see that the sentries and picquets were on the alert.

Bonny had assigned to him the special task of buying food from the natives. I gave out to him a certain number of mitakos and handkerchiefs from the Expedition stores for this purpose, and Ward superintended the mess arrangements.

At this time, too, began that dull round of meals that was to last throughout our camp life without improvement. These were eaten in Jameson's large hut. Our staple diet was rice, plantains, and tea. At first we had fish bought from the natives, and sometimes a fowl or two from the same source; these were soon denied us, but afterwards on rare occasions we got a fowl or a meagre goat, and as our other supplies got low we would try a bit of manioc or chiquanga, which we found poor eating and rather indigestible. The natural question arises in the reader's mind, "What became of all the European provisions?" These were not supplied in vast quantities, as will appear.

Prior to his departure Mr. Stanley gave out to *the officers* then *with him* a ration of these stores, that was supposed to be enough to supply them for six months, and a small amount had been given to Ward and Bonny when they were left at Bolobo. I had never received *any* European provisions, except a pound of tea at Manyanga, from the Expedition. Naturally when we, Ward, Bonny and I, reached the camp we asked for our six months' rations, but Major Barttelot refused us, saying as Stanley had left no orders on the subject, he could not undertake to give us anything; Stanley would require of him all the stores left in camp, and he urged that he could not possibly give us these rations. We felt we were right in our claim, and it was urged so strongly that Major Barttelot yielded so far as to promise us *three* months' supply in the same proportion that they had received, and the following is the list of stores that he distributed to each of us:

AN OFFICER'S RATION FOR THREE MONTHS.

$1\frac{1}{4}$ lb. coffee.	1 tin sardines.
1 ,, tea.	1 ,, sausages (same size as sardines).
$1\frac{1}{2}$ tins salt.	1 lb. fancy biscuits.
$1\frac{1}{2}$,, butter.	$\frac{1}{3}$ tin red herring.
2 ,, milk.	$\frac{1}{2}$ lb. flour.
$\frac{1}{4}$ lb. sugar.	1 pot Liebig.
$1\frac{1}{2}$ tins jam.	$\frac{1}{4}$ lb. tapioca.
$\frac{1}{4}$,, chocolate and milk.	$\frac{1}{4}$ lb. sago.
1 tin cocoa and milk.	

CHAPTER IX.

EXTRACTS FROM DIARY.

On August 15th, as I have mentioned, we saw a stampede of natives from the opposite shore, and believed that their village had been attacked by some of Tippu-Tib's men. From the natives we learned that four women had been taken prisoners. It was important for us to know if the attacking party were Arabs, so we sent out some Zanzibaris to find out.

Thursday, August 18th.—The Zanzibaris returned bringing with them four Arab chiefs, the headman, Abdallah, with followers. Barttelot held a palaver with them at which we were all present. They informed us that the 600 men that Tippu-Tib had promised to send to take our loads to Lake Albert had come down some days ago, and had struck Stanley's track with the advance column, and coming to the first *Boma* thought we had all gone, and therefore they all returned to the Falls! These Arabs volunteered to take a letter from Barttelot to Tippu-Tib. We held a consultation among ourselves, and all agreeing to the great importance to the success of the Expedition in getting these 600 men to return here without delay, it was unanimously decided that two white men should go to the Falls, the Arabs promising to escort them there, the chiefs to get a good *dash* [present], and the men to have three pieces of handkerchief each. They were to go back to their camp and return on the third day to pick up the two white men. They promised not to molest the natives on the other side, as we were trying to get on friendly terms with them so as to get food.

Friday, August 19th.—Barttelot informed me that as Jameson wished to keep the company of Zanzibaris he had, they would not be made over to me, but that I would have half of the 125 men from Bolobo. Strange this! Jameson was not to have a company according to Stanley's instructions; the men he had were those Parke held for me, and it was understood they were to be made over to me when I came up. It has been decided that Jameson is to go to the Falls with the Arabs. Ward is to accompany him.

Saturday, August 20th.—Jameson and Ward received instructions relating to their mission to Tippu-Tib. If he consents to send back the 600 men, they (Jameson and Ward) are to return as soon as possible. Ward is to come back ahead with news. Barttelot has written Tippu-Tib a letter explaining the mistake his men made, that Mr. Stanley had gone on, that we were still here with over 600 loads and wished to follow on without delay. He asked him to send the men back, and told him that the powder had arrived for them, and that the agreement made at Zanzibar still held good.

[This letter I had assisted Barttelot to write in English. Assad Farran then translated it into Arabic, and was made to translate his Arabic version into English that we might be sure everything was right. Stanley, in an agreement made at Zanzibar, had promised to pay a certain amount of gunpowder to each man provided by Tippu-Tib, and this was the powder referred to above.]

Sunday, August 21st.—The Arabs arrived, but not all of them: the rest are to come in to-morrow.

Monday, August 22nd.—The remainder of the Arabs arrived with a host of old natives as slaves. They promise to give back to the natives their four women. The slaves with them had been theirs for some time at the Falls, though originally from these parts. They were a peculiar-looking lot of people, all marked with raised cut skin, armed with spears and carrying food and the guns of the Arabs. They seemed quite contented with their lot, were in excellent condition, fine strapping fellows. The Arab chief brought us a fine big fowl, which was very acceptable. It was arranged that Jameson and Ward should start to-morrow with them to the Falls; they (the Arabs) state they will take about four days in the journey to the Falls, mostly through forest.

Tuesday, August 23rd.—Jameson and Ward left for the Falls, taking the Zanzibari Bartholomew as interpreter. I took out all the loads of beads and cowries to overhaul them; they have all suffered from being badly packed.

Wednesday, August 24th.—Had loads out again, trying to improve them.

Thursday, August 25th.—In view of trying to reduce the loads as much as possible, Barttelot and I went through the Fortnum and Mason provision boxes. They had all, or most of them, been opened before to get various things out for rations to white men, so some of them had very little left in them; we re-sorted them and saved two loads. There were twenty-four boxes, and we reduced them to twenty-two, keeping a careful list of what we took out and how the rest was distributed among the other boxes. This took us all day.

[These boxes, it should be remembered, were the ones brought in the first trip of the steamers by Stanley himself; what he had not distributed in giving out the six months' rations to the officers, were left under the charge of Major Barttelot and Jameson. Twenty-four boxes incomplete, or twenty-two full boxes were all the European provisions that remained of those supplied to the Expedition. These were for the use of all the officers, and were only left in camp as it was inconvenient to carry them on. We were to guard and care for them. They were for us all; those who had gone on were to have part, and we, of course, in a future division were to have a share. But indeed these stores were not extensive, and should be looked upon more as medical comforts than anything else: this may be judged by the list of contents given on p. 155.]

In such an Expedition as this every pound weight must be considered, and it would be natural to expect the supplies to be of strongly condensed necessaries of life, such as meat extracts, condensed milk, plain biscuits, &c., which would be unobtainable in Africa, but of what earthly use were *jam*, *red herrings*, *sardines*, and *fancy biscuits* to support life while in the bush, especially when in such minute quantities that the enumeration reads like a list of "samples for stock to be supplied." It must not be forgotten that these would be our only food when our rice failed, for we could place no dependence on the supply of meagre goats, diminutive fowls, and manioc which we obtained with much difficulty even when circumstances favoured us. Taken in the light of the medical comforts even they hardly form a suitable list, and it will be seen that valuable space was occupied by a large supply of articles that could have been dispensed with, such as jam, sugar, fancy biscuits, red herrings, sausages and sardines. Had these been replaced by Liebig's extracts, Brand's essences, or even plain biscuits, tea, and a larger supply of milk and arrowroot, the invalids at least might have had a chance; as we were all ill at one time or another, we should have all benefited by such an improvement. It is well known that Fortnum and Mason are capital caterers for Derby hampers, but they are not quite the firm one would expect to see engaged to supply an African Expedition such as ours, though no doubt they could have obtained the plain necessary supplies had *some one* taken the trouble to give the proper order.

Of course, it cannot be supposed that Mr. Stanley had anything to do with this department, but he doubtless placed these arrangements in the hands of one of those who assisted him before his departure from Europe. It is a matter of considerable surprise that such difficulty was found in supplying the necessaries of life to only ten men. Had it been an officer connected with the commissariat department of Her Majesty's Service who was engaged in attending to supplying the army, we might imagine that this small task would have been more successfully accomplished. This is not complaining of an insignificant matter; so much depends in Africa, health and even life itself, on proper food and medicines for the white men. The importance of this cannot be too often reiterated, for many a time have I known expeditions most inefficiently supplied. If only small quantities of provisions can be sent, let them be of the most essential articles and of the best quality, no such expedition should be burdened with a single *un*-necessary thing. This should be thoroughly impressed upon every one who undertakes to provision African travellers. However, whoever it was that blundered, here is the list of supplies that were in our camp on August 25th:

110 tins condensed milk.	3 tins coffee.
11 ,, cocoa and milk.	5 ,, candles.
34 ,, fancy biscuits.	5 ,, vegetable soup.
5 ,, sugar.	4 ,, red herrings.
31 ,, tea (2 lbs. each).	6 ,, salt.
70 ,, butter.	27 bars soap.
7 ,, chocolate and milk.	20 tins sausages.
42 ,, jam.	3 ,, arrowroot.
18 ,, sardines.	1 tin sago.
23 ,, Liebig's extracts.	

These, as I have said, were intended to supply the wants of all the members of the Expedition, and we were given to understand that our duty was to guard them intact until we handed them over to Stanley.

[I was engaged for several days at this time in examining the stores that had been brought up when I came.]

Tuesday, August 30th.—A holiday to the Zanzibaris; they say it is their Christmas, and want a present of mitakos.

Friday, September 2nd.—Had a walk with Bonny; tried to buy fish of the natives; got some; came back, both feeling seedy. Bonny turned in. I felt the old sensation when I got to bed, shivers, &c. Afterwards fever, as usual.

Saturday, September 3rd.—Fever still on. Bonny also seedy. Some Arabs came in from their camp, stated that the messenger they had sent to Tippu-Tib before Jameson and Ward started had returned, bringing news that the men were ready to come. Barttelot made them a present of two canoes, which the natives had left in their flight when the Expedition first came here. They also received three pieces of handkerchief. During the night the natives tried to take the canoes away, and we were startled by two shots fired, as we discovered, by the Soudanese sentry at the river gate, on the natives attempting to take them away, which they failed to do.

Sunday, September 4th.—Native chief came in saying one of his men had been wounded by shots fired last night; he supposed it was done by the Arabs. Barttelot told him to bring in the man and we would see to his wound. The chief went to fetch him. In the meantime the Arabs had started with the two canoes to take them up stream to their camp, but the canoes were taken from them by the natives.

Monday, September 5th.—Sentry fired at some natives last night, thinking they were coming for canoes again. No harm done.

Wednesday, September 7th.—This evening Ward returned from the Falls, bringing letter from Jameson. They had two days' rough march from here with the Arabs, then reached a village on the Congo where they got canoes. After paddling for two days they reached the Falls. They were well received by Tippu-Tib, who explained to them that Stanley had told him he would find his (Stanley's) camp not far from Basoko on this river; that he (Tippu-Tib) had come up in canoes, passed Basoko with the men promised, that he had sent his men for chop (food), and they had been attacked, he losing four men. His men were anxious to get back as they could not find our camp, added to which their food had given out. Tippu-Tib sent a body of men overland to find the camp and went back himself with the main crowd. He has promised to send at once as many men as he can, but is afraid that he will not be able to get the 600, the number promised, and which he brought down the first time. Jameson returned with Ward in the canoes as far as the village Yallasullah, on the Congo, and is waiting there for some of the men. Tippu-Tib is coming here himself. Ward came on, leaving Jameson, last Monday. He describes the road as very bad, very wet also. He had been marching since morning and was done up. We were glad to see him and get the news, as we are all anxious to move on. If it had not been for the mistake made either by Stanley or Tippu-Tib about the situation of our camp, we should have been a month out on the march by this time. Some of Tippu-Tib's men came back with Ward.

Thursday, September 8th.—Ward seedy to-day. The chief on the other side has not been sending the food promised every day, and is inclined to be cheeky.

Friday, September 9th.—Barttelot and I started this morning to survey the road taken by Stanley from here, reaching his first halting-place. We took the two donkeys for crossing streams two Zanzibaris, and two donkey boys. We found the road fairly good, with a few small streams; had no difficulty in tracing the route as the trees were frequently blazed. Once we got a little at fault owing to his having made a slight *detour* for his donkeys, whilst the column went by the other route, marked also. We passed no less than four rapids, the road being close along the river, and the views were very pretty; went on to a deserted, or rather burnt village, where no doubt Stanley stopped the first night. We returned to camp about 4.30 P.M., having had a very pleasant outing without feeling very tired. Ward is no better.

Saturday, September 10*th.* — Ward suffering evidently from dysentery—a very bad case.

Sunday, September 11*th.*—Ward remains about the same, his illness is getting serious.

Monday, September 12*th.*—About noon Jameson returned from his mission to the Falls. He came back with Salim-bin-Mohammed, and Salim-bin-Massude, an interpreter, with a following of men. [Both Salim-bin-Mohammed and Salim-bin-Massude had come round with the Expedition from Zanzibar.] Tippu-Tib had not turned up, he is waiting for men still. Jameson says he will be here in about ten days, but it is impossible to say with what number of men. Jameson brought with him a lamb and some sweet potatoes—a great luxury. He had a wet journey from the village on the Congo named Yallasullah, where Ward left him, walking through water, heavy going all day. Bartholomew, our Zanzibari interpreter, who went to the Falls with him and deserted, has not been captured yet, nor has the other Zanzibari who deserted from here, but Tippu-Tib has promised to do all he can to get them for us. I am disappointed that Tippu-Tib did not come with Jameson. It is weary work in this camp.

Tuesday, September 13*th.*—This morning I handed back Jameson's company of Zanzibaris which I had taken over during his absence. Barttelot decided to leave the other men I had with Ward and Bonny, as I am to have charge of all Tippu-Tib's men when they arrive. Salim and his men have made an encampment at the back of our camp, outside. The native chief from the other side came over and I held a palaver with him. He brought back a stolen rifle, seven fowls, some fish and honey, and we rewarded him; he went away rejoicing, promising to bring more. I had a good deal of difficulty in getting him to put the things on shore before he was paid. I received a Remington rifle from the Expedition, in fairly good order, but very strong pull.

[This was in place of the promised Winchester, which was never given to me, though mentioned in my contract. This Remington would not have been of very great value in saving my life ; as the pull was strong, anything like an accurate aim was out of the question.]

Wednesday, September 14th.—Ward about the same—still very ill.

Thursday, September 15th.—Native chief brought some fish. Heard shots on the other side, and found that the Arabs who are camped further up, Abdallah's lot, had been over. Beastly shame after their promise to us, and we had given them presents, so they should not molest those natives whilst we are here. Natives are clearing out from the other side after the firing, a general stampede. Found that Salim the interpreter, with Salim-bin-Mohammed, was employing our Zanzibaris to buy them fish, giving them axe-heads for the purpose. Barttelot caught one of our men with a large fish, it was handed to Salim with a remonstrance which led to a palaver between Barttelot and Salim-bin-Mohammed. He (Salim-bin-Mohammed), I think, is all right, but the interpreter Salim is a treacherous fellow and is trying to make mischief. Salim-bin-Mohammed returned the revolver Barttelot had sent to Tippu-Tib, saying that Tippu-Tib thought it would be more useful to him, as he (Tippu-Tib) had no cartridges except those sent! Ward is very ill with dysentery, giving much anxiety.

Friday, September 16th.—Ward still very ill, trying all kinds of medicines. Jameson is attending him.

Saturday, September 17th.—Heavy firing again on the other side of the river, the natives who had returned after the firing of two days ago, left again in their canoes. I suggested to Barttelot, and was supported by Jameson, that it should be stopped; that he should ask Salim to stop Abdallah's men and explain that Abdallah promised not to attack the natives; this he did. Our old friend, the native chief, we saw in a canoe passing in hot haste; we hailed him; he came and had a palaver, and we got him to come into camp. We told him that we had arranged with the Arabs to cease molesting him, and asked him to come over with his people to live on our side. He informed us, poor chap, that the Arabs had killed nine of his people. I am very sorry about the whole business, as the Arabs have behaved most treacherously.

Sunday, September 18th.—Ward getting very low, much anxiety about him.

Monday, September 19th.—I was looking to-day at the small tin medicine-case which I got the other day, the same as supplied to the officers of the Expedition. It contains ten small bottles, most of the contents supposed to be in tablets. The whole case has been ruined by water, most of the tablets dissolved and spoilt, so that they are of no use, and the directions are all obliterated! Jameson tells me that the wooden box which had in it all these cases he has seen swimming in water on board the steamer! The few things I have received from the Expedition are all in similar condition—a saddle without stirrups, with one side of the flaps cut away, a water-bottle that leaks, and a medicine-case almost worthless! Ward is a little better.

Tuesday, September 20th.—Found that some of the men had stolen two tortoises that Jameson had brought from Stanley Falls—a deliberate theft, as they were tied up in the cook's shed. Suspicion pointed to the Soudanese, as the Zanzibaris are not supposed to eat such *unclean* things. Had search made amongst the ashes of their fires and discovered the burnt shells. After some trouble found out the culprits—one Soudanese who cut the string, but three partook in the feast. Barttelot decided to flog the one who actually took the tortoises to-morrow morning, and give minor punishments to the others. I forgot to mention that four Soudanese were flogged last week for having been mixed up in the stealing of an axe and selling it to the natives. Barttelot's boy Muledi flogged for stealing. Ward about the same as yesterday, slowly getting better.

Wednesday, September 21st.—Ward is improving. Salim-bin-Mohammed started this morning for Abdallah's camp and did not return; expect him back to-morrow.

Thursday, September 22nd.—Salim-bin-Mohammed returned from Abdallah's camp. Weary work waiting for Tippu-Tib to turn up with the men. Nothing to do here. Salim the interpreter paid me a visit; he says the chief of Tabora is stopping Tippu-Tib's caravans coming from Zanzibar to the Falls; a consequent shortness of cloth at the Falls. Is very anxious to get white man to drill Tippu-Tib's men. Some of his men say they met Stanley on the road, and that after sixteen days' march we shall get into an open country where we can get milk and sheep, there being plenty of cows. Of course I take this with a big pinch of salt, but it would be a great thing if it were true. Ward is on the mend. Another of the Zanzibaris of Jameson's company died to-day, this making seven men who have died since I have been here. The graveyard has now fourteen men in it. They seem to waste away; there are a number still sick with bad ulcers, but the Expedition has provided nothing for their relief. I am able to do but little with the pot of zinc ointment I got from Dr. Sims at the Pool.

Friday, September 23rd.—Ward very weak, but better. Have seen nothing of Ngungu since Tuesday, and we have had no fish with the exception of a very small piece at lunch to-day.

Saturday, September 24th.—Bonny got some palm wine. Ward mending.

Sunday, September 25th.—Salim, interpreter, brought some palm wine. Abdallah came in from camp further up; he is going to punish the natives further down, who killed Tippu-Tib's men when they first came up to try to find our camp. In the afternoon forty of Tippu-Tib's men arrived from the Falls with messages from Tippu-Tib. They brought back Bartholomew the interpreter, who deserted from Jameson and Ward at the Falls, and also stole cloth, &c., from them there; also 'Mpsa, another Zanzibari, who deserted from the camp after Jameson left and followed him to the Falls; he was made prisoner but got away. Both these men were brought back in chains.

Monday, September 26th.—Salim-bin-Mohammed has been to Barttelot to-day, and told him that Tippu-Tib had sent down to say that he cannot get the carriers; they make excuses, saying that the loads we have are too heavy. Tippu-Tib says that he has sent to Kasongo for some men (it takes nearly thirty days to go to Kasongo and fourteen days to come back to the Falls), and he expects the canoes back in about a month! The gist of all this is that we *might* have a chance of getting men in six weeks' time, but my idea is that we shall get none and will not be able to move from here, and that Stanley will find us here on his return whenever that may be. It is very bad news, and most disheartening to remain here and to be out of the hunt. Barttelot talks of going to the Falls and taking me with him, which would be a change. Bonny is seedy, in bed with pains in limbs; Ward is slowly pulling up.

Tuesday, September 27th.—Had all the men paraded outside of the camp this morning. Bartholomew and 'Mpsa, the two Zanzibaris who deserted and stole from Jameson and Ward, were flogged. The former received 150, and the other 100 lashes. Assad Farran, the Soudanese interpreter, came and informed us that the chief Ngungu had told him that he had heard from other natives that Stanley was ten days from here on his way back. He had been fighting, and had lost twenty men and one donkey. In the afternoon Jameson and I went out to try and meet Ngungu. We were told that he came over to a small hut on the river-side some half an hour's walk from here; we waited at the spot for over two hours, but Ngungu did not turn up. All the men out cutting grass and wood; it is decided to build new houses for them in the Fort. Our trip to the Falls postponed pending confirmation of news about Stanley.

Wednesday, September 28th.—We made two attempts to-day to see Ngungu at the hut, but failed. The village belonging to him is deserted since Abdallah's last raid, and I believe he has gone down stream, only coming up to look after fish. I am afraid our food supply will cease, all owing to the brute Abdallah breaking his word and attacking them. I don't know what to make of the rumour concerning Stanley. I am inclined to think it is not true. Some few men of Tippu-Tib's came in from the Falls to-day, and have gone up to Abdallah's old camp. I don't know what they are up to. I suppose raiding. The whole situation is very unsatisfactory; we cannot evidently get the men Tippu-Tib has promised. We are told he has sent to Kasongo to get men from there, but I don't believe he will obtain them. It has been constantly in my mind lately to doubt the veracity of the statement made by Tippu-Tib, that he came down in the first instance with 600 men in canoes and could not find the camp. I argue that after coming all that long way and getting close here, it is not in keeping with the man to have turned back for the reason he gives, viz.: that the men were tired of paddling, and that four or five of them were killed a little lower down from here! Then again he says he sent Abdallah with some men on shore to try and find our camp; we find these men far up past our camp raiding villages, and to get up to these they must have left the river and passed at the back of our camp, purposely avoiding us, the avowed object of their landing. I may be wrong in my conjecture; anyway there can be no doubt of this fact, that either Stanley did not make it clearly known to Tippu-Tib where our camp was to be, or else Tippu-Tib never sent any men, and does not mean to either.

Thursday, September 29th.—Ward much better, Bonny all right. No news of Ngungu.

Friday, September 30th.—After breakfast a Soudanese came in to say that Ngungu was at the small hut half an hour from here, so Jameson and I started and found him at last. We took one of Tippu-Tib's men as interpreter. Ngungu seemed very dejected. He stuck to his story about the return of Stanley. He says that a small boy belonging to him came in the other day from a neighbouring village, where he learnt that Stanley had been fighting with some natives; that they had killed ten Zanzibaris and one donkey, using bows and arrows; that Stanley had afterwards killed a lot of natives, and was coming back here, and would be here in four days. He was bringing a lot of cattle, sheep, and fowls with him. When Ngungu was asked if he was telling the truth, he replied that *he could not tell a lie;* that Stanley would kill him if he was not speaking the truth. He also told us that he had nearly all his people killed by Tippu-Tib's men down the river, where they had gone from their village to take shelter when Abdallah fired on them. He has no regular home now, and sleeps in different places every night. I felt very sorry for him, poor chap; it is too bad—these Arabs have broken faith with us all round. It is, I consider, a most deplorable position to be in. We have to be friendly with the Arabs, as we expect them to get men for the Expedition; and although we have told them to leave the people alone here, they have not done so, and it is impossible for us to force them to obey us in any way. First, we cannot afford to fall out with them, and secondly, if we did they could give us a good licking in the long run! All the same it goes against one's feelings altogether to see these poor people hunted to death! We asked Ngungu over and over again to bring the remnant of his people to the Fort and live with us, promising him protection in every way; and although he says he will come we see nothing of him, he is mistrustful of every one, and no wonder after all he has gone through lately. We bought a large fish from him. I don't know what to make of the rumour about Stanley; if it is true, I can't make out his returning if he has had only a few men

NGUNGU, CHIEF OF YAMBUYA VILLAGE

killed; he is well-armed, and he has not had time to go to the Lakes and come back. Barttelot and I have decided not to go to the Falls in consequence of this rumour until next week.

Saturday, October 1st.—Ward has had a relapse.

Sunday, October 2nd.—Went down to small hut to try and see about getting fish; saw some men in canoes fishing, but they would not come ashore. Salim the interpreter came in this afternoon, stating that the natives in canoes on the river called to him when bathing, and said that there were white men close here with plenty of men, and would be in camp to-morrow or next day. Later towards evening Ngungu came into camp with one of the Soudanese officers, and stated that he had seen two canoes with twenty Zanzibaris far up the river above the rapids, and had spoken to them; they bought fish of him, giving a knife in payment, which he produced; he further stated that they would be here to-morrow. They told him that Stanley was at a village called Opoy, some seven days from here, and repeated the story about having ten Zanzibaris killed, &c. We are all very anxious for further news, and all kinds of suppositions are advanced among us. Some of Tippu-Tib's men came in from a village called Sogoro, bringing us a sheep and a goat. Salim-bin-Mohammed's native band, three men with drums and three women with bells, played in front of the hut during dinner. Ward getting on well again.

Monday, October 3rd.—Salim's band played again at early break-fast, and each received one piece of handkerchief. In the afternoon Salim wished to know if he should send up to Abdallah's camp to find out if he had heard anything about the twenty Zanzibaris supposed to be coming here from Stanley. Barttelot declined his offer. Later Omaha, a Soudanese officer, came in and stated that Ngungu had told him that Abdallah had stopped the twenty Zanzibaris from coming down! Ngungu had mentioned this yesterday, but Ferhani, Tippu-Tib's man, had not interpreted it to us. Barttelot and Jameson decided to go out to Abdallah's camp, thinking it best that we should find out the truth about these rumours by sending some one.

Tuesday, October 4th.—Orderly officer last night; called Barttelot and Jameson at 1.30 this morning; they left the camp about 2 o'clock, taking twenty Soudanese and two Zanzibaris, the latter to show the road to Abdallah's camp.

[The object of this early start was to reach Abdallah's before he received word of the proposed visit, that it might be possible to see the Zanzibaris about the camp, he not having time to have sent them out of the way. For this reason no previous orders were given to the men who were to go, but when I turned out the guard at 1.30 they were told to make themselves ready for a march, no mention being made even then of their destination. The two Zanzibaris were awakened from their slumbers and ordered to march with the others, as they were needed as guides. Barttelot expected to reach Abdallah's at dawn, before he had warning of his coming. This secrecy was necessary to prevent the Arabs from having a chance to deceive him about the men.]

I am left in charge of camp, with instructions that if they do not return by to-morrow night and send no messenger, I am to ask Salim-bin-Mohammed to send some men to see what is up, but on no account am I to send any of our Soudanese, Zanzibaris, or white men from the camp.

Everything carried on as usual in the camp. This morning I was informed that Salim had made Ngungu prisoner, so I sent Assad Farran to inquire. Found the rumour was not true. Ngungu had been in Salim's camp to see his women, and sold Salim some fish. Another of Jameson's men died this morning. Barttelot and Jameson returned about 5 o'clock; they had been to Abdallah's camp, where they heard nothing about the Zanzibaris reported to have been seen by Ngungu, so we must conclude the whole story is a got-up affair. Barttelot and Jameson report plenty of water along the road. Barttelot and I are to start for the Falls on Thursday.

Wednesday, October 5th.—Making preparations to leave for the Falls to-morrow. Barttelot says we are to take remainder of the bale of handkerchiefs that has been opened for buying goats, viz., forty pieces, but I don't think this will be enough, so I have opened another bale and taken out ten pieces, so we take fifty pieces handkerchiefs and 100 mitakos for buying live stock. I have sewn up all the mitakos in rice sacks, and left them in charge of Bonny.

Thursday, October 6th.—Left our camp at Yambuya this morning about 8.30 with Barttelot for the Falls, taking two Soudanese, one Somali as cook, and one boy each. Salim-bin-Massude accompanied us, and Tippu-Tib's men carried our loads. Jameson was left in charge of camp. We made a rapid, hard day's march to the village of Yambow, reaching there about 4.30. Fair road, through some dense jungle, and plenty of water. Camped the night in the village, most of our things wet, having been out in the rain nearly all day. The village consists of a few huts in a good, clear, flat ground. Tippu-Tib's men in possession of village, the natives quite friendly.

Friday, October 7th.—We had to dry our things, so did not make a start till about 10 o'clock, passed through several small villages, had some palm wine in one of them. Did not make such a long march as yesterday. Camped in bush, having missed watering-place.

Saturday, October 8th.—Started early, about 6 o'clock. Met some of Tippu-Tib's men coming from Yallasullah, who informed us that Tippu-Tib was expected there to-morrow, so we pushed on. Found plenty of water, in fact walked long distances through it. Before reaching Yallasullah, had to cross a river up to one's armpits. Caught in a thunderstorm and heavy rain before getting in to the village, which I did about 4 o'clock, drenched. Found Barttelot had got in an hour before me. Well received by Sheik Seyid, one of Tippu-Tib's men, in charge of village, with large native population, who gave us food and a house, and did all he could to make us comfortable. Tippu-Tib did not arrive, but is expected to-morrow.

Sunday, October 9th.—About noon Tippu-Tib arrived in his canoe from the Falls, accompanied by several of his headmen. We all adjourned, after greetings, to an open shed near the river, used as a palaver house, squatting on mats. Tippu-Tib informed us, through Salim the interpreter, that he had seen five of our Zanzibaris at the village of Yarricombe, a little higher up the river. They had deserted from Stanley's column with four others. They had come down to Abdallah's camp, and he had sent these five on, showing them the road to the Falls. The other four were still at Abdallah's. Tippu-Tib went on to say that the five he had seen at Yarricombe had several tusks of ivory with them, which they offered him, claiming his protection in return, but he said nothing to them, telling his sheik at Yarricombe to look after them until he returned from the Lomami. But now that he knew we were here he would send for them and hand them over to us.

So after all Ngungu's story was true and Abdallah had fooled us. Barttelot told Tippu-Tib about Jameson and he having gone to Abdallah's camp to find out about these men, how he had denied seeing them, &c., &c. Tippu-Tib told us he would punish Abdallah and take him away from the road. We found that the deserters all had had guns, which they had left at Abdallah's. Tippu-Tib despatched one of his men in a canoe to Yarricombe with instructions to bring the men bound in ropes to us here. Barttelot gave Tippu-Tib the gun that Jameson sent as a present; he seemed pleased, but said nothing in the way of thanks. We had some bananas and coffee with them all; one of the sheiks brought out a very good cake! Sheik Nossoro, Ali, &c., &c., were with Tippu-Tib.

Monday, October 10*th.*—I forgot to mention that Yallasullah, where we are at present, is a village on the Congo under Tippu-Tib's control. The natives have a market every day close to the palaver house. This afternoon the canoe returned with the five Zanzibaris, and they were brought to us in the presence of Tippu-Tib. Of course they denied deserting, stating they had been left behind and could not find the road, &c. They were twenty days in a canoe coming down to Abdallah's camp from the time of leaving Mr. Stanley. He had not reached the Lakes when they left, food was scarce, he had had trouble with the natives, at one place they fired on the column with bows and arrows. A white man, who we thought from their description must be Nelson, got an arrow wound, but was all right again by the time they left. Nearly all the Soudanese had died, as well as a number of Zanzibaris, from disease and want. Tippu-Tib asked that they (the deserters) might not be punished this time, and Barttelot said he would do nothing to them till Stanley returned. Tippu-Tib told them this, and added that if they or any other Zanzibaris ran away again and came to him he would shoot them. Barttelot gave Tippu-Tib the ivory they brought with them. Tippu-Tib undertook to keep the men safe at Yallasullah till we returned from the Falls.

Tuesday, October 11*th.*—Arranged through Salim, the interpreter, for the purchase of twelve goats at three pieces of handkerchief each, to be here on our return. We had some good fish here. The people live principally by fishing; the marketing is all done by exchanging. The principal things for barter are small lots of fire-wood, manioc tops, palm-oil, dried fish, pottery, and quanga. There is a market-master who looks after things in general, and sees that so much tribute is paid to Tippu-Tib's sheik from the articles brought in. We left this morning in a canoe for the Falls. Tippu-Tib left in his canoe for the Lomami down stream, promising to be at the Falls in three days, so we postponed all palaver about the promised carriers till then. We paddled up stream, reaching the village of Yarricombe about 6 o'clock, passing two villages, and getting drenched in heavy rain. At Yarricombe, on the south bank of the Congo, we found a mud-house in charge of a sheik of Tippu-Tib's. We saw very little of the native village, as it was dark soon after our arrival. Stopped in the mud-house. There is another village on the opposite bank.

Wednesday, October 12th.—Started early in a canoe paddling on to another village, changed men there and paddled right on to the Falls, which place we reached about 11 o'clock at night. The water is very low. We were comfortably put up in a mud-house.

Thursday, October 13th.—Up rather late. Found the house not bad; it has an enclosure at the back and a verandah in front. Breakfast was brought to us; it consisted of fried potatoes and plantain, with tea, all very well cooked. Sheik Nzige, brother of Tippu-Tib, is in charge during the chief's absence; he paid us a visit. We had a look round the place, viewing the Falls, a very large rush of water, but not much of a fall. Plenty of large rocks, with big sticks set up by the natives for fastening their nets when fishing. It was on the island opposite that the State station was.

Friday, October 14th.—Arab Sunday. To-day we had plenty of goat done up in kabobs, curried fish, rice grown on the place, all beautifully cooked. Our meals, sent to us four times a day, usually consist of fried sweet potatoes and plantains in the morning, curried fish and rice in the middle of the day, then potatoes and plantain again in the afternoon, and curried fish and rice in the evening. Tea or coffee at each meal.

Saturday, October 15th.—Went for a walk over to Sheik Nassoro's house up beyond the Falls, through a native village. He is away with Tippu-Tib. His is a fine place, with a good view looking down on the Falls, as well as up river; it is approached by a pleasant avenue of pau-pau trees. Afterwards we crossed over in a canoe to an island opposite, and visited Tippu-Tib's small father, viz. uncle. He has a good house, and we squatted in the verandah on mats in the Oriental fashion. He entertained us most hospitably, giving us a big luncheon, consisting of Indian corn, plantains, curried fowl, rice, and pau-paus. Everything was excellently cooked, and I enjoyed the visit very much. He was such a good fellow, with a kindly manner, and did all he could to make us welcome. We came back by another island, and then crossed to Nassoro's again.

Sunday, October 16th.—Went over after breakfast to the island opposite, where Deane's station was. After luncheon with an Arab, who treated us well, we had a look at the three Krupp guns that Deane left here after his fight last year. They are of no use, the vent screws destroyed and other pieces missing. There is also a small muzzle-loading cannon. I little thought this time last year that I should see these guns again, especially under such circumstances. I remember well packing them up at Leopoldville, when Deane first went up to the Falls in June 1885. No sign of Tippu-Tib's return yet. Arranged with Salim, the interpreter, to buy some fifty fowls.

Tuesday, October 18th.—Both Salim, interpreter, and Ferhani, who escorted us down from our camp, and who are looking after our comfort here by order of Tippu-Tib, were inclined to be cheeky to-day, and we had a palaver with them.

From this date till Monday, October 24th, we waited at Stanley Falls for Tippu-Tib to return. On Saturday, October 22nd, I wrote a long letter to Colonel de Winton, giving him all the news, hoping this would get through either by steamer *viâ* the Congo, or by Tippu-Tib's caravan to Zanzibar, and while we were at Stanley Falls Barttelot also wrote letters home.

Monday, October 24th.—Tippu-Tib returned from the Lomami. He came and saw us at our house, and made arrangements for our starting Wednesday next, on our return journey to our camp at Yambuya. He tells the same old story about the carriers, that he expects to get them at Kasongo, although we were told through Salim Massude that there had been fighting there. We can't believe Salim, and it is a thousand pities that Barttelot did not overcome his dislike to Assad Farran, the Soudanese interpreter, and bring him. I shall not be sorry to get back to camp.

Tuesday, October 25th.—My impression of Tippu-Tib and the Falls is this. He has very few men here really of his own, though there are several native villages under his control. The great part of his own men are away after ivory, &c., all over the country. A large body are at the Lomami under Raschid, trying to settle down there among the natives for trading purposes. I believe there is some big palaver going on over the subject. Then a lot are scattered over the district Stanley has traversed as well as elsewhere. His son, Sefu, who reigns at Kasongo, has plenty of men, and it is from there they say they will get our carriers, but I don't put much faith in their story. It seems Tippu-Tib's sheiks have many men with guns under them. The place round here has been cleared, and there are several acres of rice-fields under cultivation. On the other side, where Deane's station was, there is excellent flat ground for plantation close to the river, which is laid out, principally in rice; the women do all the labour in the fields, and there are plenty of them, all from the Manyuema district. The sheiks' houses, as well as several others, are of mud.

Wednesday, October 26th.—I gave my letters to Tippu-Tib, who has promised to send them to Zanzibar by special messenger, and says they will reach there in two and a half months. After saying good-bye to Tippu-Tib, who came down to the canoe to see us off, Barttelot and I started on our return journey in Tippu-Tib's private canoe, whilst the two Soudanese and Ali, the cook, were in another with three goats and some fifty fowls. Tippu-Tib made us a present of one large goat. We stopped the night at a village on the north bank of the Congo, sleeping in the verandah of a mud-house. Barttelot received a letter from Jameson written from Yambuya, stating that two prisoners had run away, viz. Bartholomew and another. Bonny had been seedy, which I was sorry to hear.

[In connection with this letter from Jameson, it should be mentioned that, taking all things into consideration, it was deemed advisable that the Zanzibaris in our camp should be disarmed at this time, therefore Barttelot in his reply sent instructions that this should be done. In order not to rouse the suspicions of the men, it was suggested that they should be called upon to give up their arms on the plea that these were not properly cared for, were allowed to get dirty, &c., in their huts, and that they were in future to be kept together under the charge of an officer, and given out when required for the picquets.]

Thursday, October 27th.—Started early and paddled down to Yallasullah, which we reached about 3 P.M.; here we found our five Zanzibar deserters. Sheik Seyid, our old friend, was glad to see us, and treated us well in every way. Here we found twelve goats which Salim Massude had bought for us, but some of them were very small, and I am afraid he has cheated us, charging three pieces of handkerchief for each of them. There should have been three other goats, presents from Tippu-Tib to Jameson (in return for his gun), but they tell us one has died and the others cannot be found.

Friday, October 28th.—Find we cannot get away till to-morrow. Barttelot is going ahead quickly, hoping to reach our camp in two days, and I am to come along with the goats and fowls, as they require looking after. Seyid gave us plenty of fish.

Saturday, October 29th.—Barttelot left early this morning, taking two carriers. I got away about 11.30, having a lot of bother in getting a sufficient number of men to carry my loads. The five Zanzibari deserters carried the fowls, and the Manyuema men my loads. The two Soudanese and the two boys with one Zanzibari and myself looked after the goats. Found a canoe at the river, which was a good thing, as the water had risen very much. Camped at the first long water about 5 o'clock. Tent which Barttelot had lent me did not come in, so slept in a fly. Had great pain with ulcerated tooth, and took opium. It rained hard during the night, and I had to put the fowls inside the fly.

Sunday, October 30th.—Raining hard; feeling very seedy, half dazed, I suppose, with the opium; don't want to try that remedy again. One Zanzibari led the big goat, then came Muledi with the big Billy, and one Soudanese in the rear. I was last of all. I found myself sleeping whilst I walked. Camped in jungle about 5 o'clock, plenty of water in the road.

Monday, October 31st.—Started early, feeling better; had a lot of bother in getting the goats through the water, which had risen very much since I was here last. Reached the village of Yambow about 5 o'clock. We slept in huts; goats all right, also fowls. Found Barttelot had left his two carriers behind here, having gone on by himself.

Tuesday, November 1st.—Up early and ready to start, but rain came on very heavily and kept on till past 1 o'clock, so it was impossible to move on, as the goats would not stand rain. Had to remain all day at Yambow. Found Muledi, Barttelot's boy, had stolen a quantity of sweet potatoes from one of the carriers. They belonged to us. He confessed to the theft, and on my threatening to punish him severely, ran away, and I have seen nothing of him since. I suppose he has bolted to camp. Had all the goats tied up ready for a start early to-morrow.

Wednesday, November 2nd.—Started about 7 o'clock; got the goats away without much difficulty.[1] Stuck hard at work pushing them along. We found a lot more water than we expected. Had to carry the goats over one by one in several places. The two Soudanese worked very well, so did my boy Hamesseke. The goats strayed off the path when we came to the water, and gave us a lot of trouble. We had had to entice them across each stream or rivulet; often they wandered far afield, till I no longer marvelled at the Scripture remarks about dividing the sheep from the goats, and heartily agreed that the latter were consigned to the right place.

Our goat-driving excursion was nearly ended—so nearly, that in my eagerness to reach the camp I determined to march late into the night. Darkness falling with great rapidity, as it always does in tropical countries, soon involved us in fresh difficulties. We found it harder than ever to keep the flock together, and more than once I feared we had lost our way. However we pushed on though I was dead tired, both with the long march and with the remnant of a touch of fever still hanging about me, so that, as I walked, I fell asleep, but was roused occasionally by bumping my head against a projecting branch, as we made our way through the thick jungle. We were marching in single file, Muledi leading the chief goat, and the other men urging the rest of the flock to follow as best they could. I brought up the rear with a Soudanese soldier just in front of me. From one of my "forty winks" I was suddenly roused by a more than usually unearthly yell from the Soudanese. Startled as I was, my first thought was of an attack

[1] The following account of a goat-driving excursion was written out at some length in connection with a Christmas story, so I give it here in full. It was published by permission of the Committee.

from enemies in ambush, but I was quickly reassured by seeing that it was only that we had come to a water place in the cane-brake: the shouts and cries that resounded on all sides were nothing more than encouragers to the goats, who, as was customary, utterly refused to take to the water. In spite of the difficulties of the situation I could not help laughing at the strange contortions of the Soudanese. He was dressed in an old Karki suit with a blue overcoat strapped on his back, and as he emitted from his spacious mouth all sorts of extraordinary shouts he flapped his arms, in a way suggestive of the wings of a bantam cock in a rage. "Yup, yaah! yup, yaaaah!" he cried, yet the more he shouted and yelled the less the goats seemed pleased. The place was little better than a swamp; each step we took plunged us deeper in the mire. When we did get the obstinate brutes started it was only to find ourselves in a veritable slough of despond, in which we floundered hopelessly about, totally unable to find the right way to the opposite shore.

At last, in despair of ever getting out of it without assistance, I fired my revolver; well knowing that we must be within earshot of Yambuya camp, I trusted that my white companions there would hear it and come to my rescue, but no answer came. I called to the Soudanese who carried my borrowed Winchester, and made him discharge that. We quietly waited, straining our ears for a reply, till my patience was exhausted, and I was on the point of giving the command to retrace our steps, if possible, when suddenly I drew a short breath, fancying I heard a distant report. A soldier beside me confirmed my suspicion by a joyous shout. We waited a moment, and then, nearer this time, we heard another shot; the Soudanese guns replied, and a joyous antiphone sounded through the jungle, led by the Winchester. Nearer and nearer came the friendly reports, more and more audible they grew each time, so that I knew they were advancing in the right direction, yet the reverberation in the coverts misled me as to which side they were on. My men now raised shouts of delight, the goats baa-ed vociferously, till the din was perfectly deafening, drowning the voices of our rescuers as well as my cries of "Who's there?" I was endeavouring to call the men to order when suddenly a flash from a lantern gleamed out, and I recognized old Jameson's voice above the tumult. His cheery shouts urged us on, and we struggled as best we could to reach his beacon. We shouted to the goats and tried to keep them together. My anxiety concerning this precious freight was almost inconceivable. What mattered my discomforts, or those of the men, so long as we got the goats safely into camp? After my long march and awful tussle with them; after one man, like the Good Shepherd, tenderly caring for those who are with young, had devoted his entire attention to a certain nanny-goat of the flock, my disappointment would have been unendurable had I lost one of them. On scrambling to *terra firma*, I immediately had them all drawn up in front of me and tried to count them, but they—restless creatures—strongly objected to this proceeding. At last I found my tale complete, and in triumph I entered the camp. I envied no Roman general returning from a victorious campaign when I saw my flock safe within the boma [stockade]. It was with satisfaction that we all gathered around the goats and made mental calculations whether the dozen of them could be made to last us five officers through the next four months.

CHAPTER X.

BACK AT YAMBUYA AGAIN.

THE general course of events will be seen from the foregoing extracts from the diary. But as there are certain facts that are not dwelt upon sufficiently, and which need some enlargement, it is as well here to review the situation. At the time of writing some occurrences did not seem to be of the importance they afterwards assumed in consequence of other affairs.

Very soon after my arrival in camp a band of Arabs made their appearance, and took up their quarters within a stone's throw of our fort, if such it might be called. It was a mistake to allow them to come so close to us, and it would have been wise at the very outset to have urged and insisted upon their removal from this place, which could have been claimed as land that we required for planting or some such thing. But this was not done, and they settled down at our very gates. However, they represented to us that they were of the number who were to accompany us, and so they were allowed to remain. The disadvantages to us were innumerable. The Zanzibaris in our camp were of the same religion and really of the same race as these half-caste Arabs, and were far too ready to be friendly with the new-comers. They were able on account of this close proximity to gather around the Arab fires and discuss the situation from their point of view. They would doubtless describe the hardships they endured, the tiresomeness of being obliged to conform to the white man's rules, the floggings they received for any disobedience. They would complain that their supply of mitakos for buying food was stopped, and of the lack of consideration shown to their grievances. The men who were to carry our loads would be Manyuemas. Many of this race had come with the Arabs, and they would see a terrible hardship in giving up their comparative freedom under Arab rule to be placed under the strict surveillance of the white men. These men would be able to spread reports—exaggerated, no doubt—among their comrades, and this would deter the latter from undertaking to act as the carriers Tippu-Tib was to supply us with. Here alone was a disadvantage in allowing the Arabs and their followers to learn too much of the inner working of our camp; and there is no doubt that grievances were magnified by being discussed.

Meanwhile the Arabs might point out to our men how much better off their followers were, and greater discontent would arise. Then, too, they saw that the Arabs were obtaining food without difficulty, and we could get almost none. The new-comers interfered with the supply of fish and palm-oil which we had, until their appearance, been able to obtain from the natives; but now there was competition in the market, and not only that, but the Arabs were able to force the natives to supply them instead of us, and would then sell at a higher price to us such things as they wished to part with. Moreover, they increased our already great difficulties by their treatment of the natives. They raided these poor creatures in order to obtain their supplies or to punish them for some fancied offence. They would not hold to their promise not to molest Ngungu, for instance, and treated the natives most cruelly and shamefully. It should not be forgotten that we had been told by Stanley in his instructions that we were to endeavour to keep on friendly terms with the natives, for the purpose partly of obtaining food through them. But it was also important that we should keep on good terms with the Arabs, who were to supply us with porters, and we were consequently placed in a very difficult position, where tact, diplomacy, and firmness alone could enable us to steer safely between these two dangers —a difficult position for any one except Stanley, who knew well how to manage under such circumstances. There were complications also that at times placed us in extreme danger, for it must be remembered that had the Arabs in anger taken matters into their own hands, they could have quickly made an end of our portion of the Expedition, either by starving us out or by an attack on our not-too-strongly fortified and none-too-well garrisoned camp. Such were some of the facts of the situation that in my opinion had something to do with influencing Tippu-Tib in not supplying the men necessary for our advance, as it may be surmised that he was consequently half-hearted in trying to overcome the obstacles in the way of obtaining men for our service.

On our return from our visit to Stanley Falls we were forced to look the stern facts of the situation in the face.

There was, we now knew, no immediate prospect of the arrival of the promised carriers. Tippu-Tib was going to Kasongo, an excursion that must occupy all of six weeks, and he would stay there to *try* to get men. Hence we found that we must sit down, mustering as much patience as we could, to wait for six weeks at the very least, and who could guess how much longer? We were forced to acknowledge that the outlook was black enough and most discouraging, especially as our twelve goats would not last for ever, and we had no supply of food for ourselves that could be depended upon, still less for our men; these were already decreasing gradually in number, and there was no apparent prospect of our condition improving, unless Stanley should put in an early appearance and happily terminate the difficulties of our situation.

I had now seen Tippu-Tib several times and had become well acquainted with his personal appearance, while at the Falls I had an opportunity of seeing his quarters and his method of managing his sheiks. Stanley, in his *Through the Dark Continent* (p. 387), writes of him as he appeared when he first saw him near Nyangwé. "Last came the famous Hamed-bin-Mohammed, *alias* Tippoo Tib, or, as it is variously pronounced by the natives, Tipo-Tib or Tibbu-Tib. He was a tall, black-bearded man, of negroid complexion, in the prime of life, straight, quick in his movements, a picture of energy and strength. He had a fine, intelligent face, with a nervous twitching of the eyes, and gleaming white and perfectly-formed teeth. He was attended by a large retinue of young Arabs, who looked up to him as chief, and a score of Wangwana and Wanyawezi followers, whom he had led over thousands of miles through Africa. With the air of a well-bred Arab, and almost courtier-like in his manner, he welcomed me to Mwana Mamba's village, and his slaves being ready at hand with mat and bolster, he reclined *vis-à-vis,* while a buzz of admiration of his style was perceptible from the onlookers. . . . He was neat in his person, his clothes were of a spotless white, his fez-cap brand-new, his waist was encircled by a rich *dowlé,* his dagger was splendid with silver filigree, and his *tout ensemble* was that of an Arab gentleman in very comfortable circumstances."

Since 1876 he had aged somewhat, but Stanley's description suited him to a T, if one only added that his black beard had whitened. He was a man of fine presence, and we could easily see that he was respected and admired by his followers, and in some ways his conduct commanded the admiration of the white men as well. He was a shrewd, intelligent man, and could easily read the characters of those around him, and knew how best to deal with them. His influence was strong enough to make him a powerful chief. He spoke Swahili, the Zanzibar language, but little or no Arabic; but he could read the latter understandingly, for the Arabic and Swahili characters are identical.

He had established himself now firmly at Stanley Falls, and made himself a comfortable house fitted in true Oriental style, and in his presence one was wise to bear himself as best he could with Oriental courtesy. His private residence was a large mud-house, surrounded by a high palisading, with a remarkably fine garden attached. We were not allowed to enter his house, but were received usually in a large building, which I always called "the

TIPPU-TIB AND SHEIKS.

receipt of customs." Here, in a large room, he would occupy the seat of honour, surrounded by his sheiks, each in his place upon the divan that ran round the room; they all sat in Oriental fashion with their slippers on the floor beside them. In this place he managed his business with his dependent chiefs, receiving their reports, giving them orders, and consulting them upon certain matters. Here, too, he held palavers with us, treating us with Oriental politeness.

It was during the absence of Tippu-Tib that Major Barttelot and I occupied ourselves in writing letters to send home. We asked Nzige for writing materials, and those he supplied were very Arabic in design, especially the pens, which were sticks of wood split up such as are familiar to Eastern travellers, but with these we were able to write our letters. Major Barttelot was writing, among others, to Mr. (now Sir W.) Mackinnon, and suggested sending him an outline of the action he proposed to take in February, provided Stanley did not by that time turn up; but as it seemed such a remote contingency, he decided, I believe, not to mention it. He had already told me his scheme, which was, that if he could not get the men according to Stanley's instructions to carry up our loads, he would try to obtain men to go up with him to look for Stanley, for he would be convinced that Stanley was dead if no news of him arrived before February. The loads which we were not to carry would be left at the Falls with Tippu-Tib. To this plan I paid little attention at the time, as I did not imagine that it would ever have to be seriously discussed, because I believed that long before the date he mentioned could arrive, Stanley would be with us or would have sent us some further definite instructions.

CHAPTER XI.

LIFE IN YAMBUYA CAMP.

Thursday, November 3rd.—Found great improvements had been made in the camp whilst I had been away—new houses made for all the men, and our quarters have a boma to separate us from theirs. Bonny has built a house for himself. The place looks quite smart. [Bonny deserves great credit for the buildings and improvements in camp; the new quarters for the men were better constructed and more comfortable, the arrangements of the inner camp were superior to what they had been; and his efforts to improve the sanitary condition of the camp and the comfort of both men and officers were most praiseworthy.]

Friday, November 4th.—Bartholomew and the other prisoner who ran away have been recaptured since Jameson wrote, and 'Mpsa was brought in to-day by Tippu-Tib's men. Sheik Ali Mahommed, one of Tippu-Tib's most influential men, paid us a visit yesterday. Salim-bin-Mohammed is going away, and there is some talk of Ali Mahommed coming in his place. Abdallah has sent back six guns belonging to the deserters; we have the five men I brought back, and two others out of the nine. No further news of Stanley. Salim Massude, I am thankful to say, has returned to the Falls; there has been a row between him and Salim-bin-Mohammed. Found Barttelot's boy who stole the potatoes had come on here; he was punished yesterday.

Saturday, November 5th.—Settling down to the monotonous life of camp again; nothing to be done. I propose taking out all the stores in my house when Bonny shifts into his, just to see that

VIEW OF YAMBUYA CAMP.

they are all right. Just before dinner Bonny caught my boy and his in the act of stealing manioc flour from the house; had them both put in the guard-room, and will have them punished to-morrow. There is a talk of all Tippu-Tib's men going from our camp. I am very glad of it; they do no good here, and stop our supply of fish.

Sunday, November 6th.—Salim-bin-Mohammed came to say good-bye, and started for the Falls with his band gaily playing. I am feeling seedy again.

Monday, November 7th.—Feeling altogether out of sorts. Bartholomew and 'Mpsa, who ran away—the former twice, the latter three times—received 75 and 150 lashes each, according to orders. Had out loads and altered arrangement of my house so as to be able more easily to see any one entering whilst I am away at meals.

Tuesday, November 8th.—Hear that Tippu-Tib's men, the few still here without any responsible man in charge, made some remarks to the effect that they would not allow our men to go across to the other side to buy food. This is rather good. Barttelot sent for one of the men, and has insisted on his taking one of the Soudanese over to-morrow to buy fish, &c. Feeling better, if anything, still very depressed. Barttelot seedy.

Wednesday, November 9th.—Barttelot has evidently got a touch of fever. Omaha, a Soudanese, went over to the other side with one of Tippu-Tib's men to try and buy fish, &c.; returned only bringing some palm-oil. Tippu-Tib's man says the natives don't like Omaha, as he threw some stones at them some time ago, and that if we sent another man we would have better success. They brought the news from the natives that Stanley would be here in three days. I only hope it is true, and I shall be rejoiced if this is a fact. I am sick of this camp existence; nothing to do from morning till night, under not very pleasant circumstances. Had out the 100 cases of powder this afternoon; found them in fair order. Some white ants about, but nothing to hurt.

Thursday, November 10th.—A Soudanese went over to the other side with one of Tippu-Tib's men to try and buy fish, &c.; got some palm-oil. No further news regarding the rumour of Stanley's return, except that there are donkeys with him. Barttelot had a pretty strong attack of fever again last night; cooler this morning, and has taken two good doses of quinine; fever returned, however, towards night, but not so intense. I had an overhaul of my private stores; gave a bottle of pickles and one of sauce to the mess.

Friday, November 11th.—Rumour about Stanley's coming is now contradicted; they are Arabs who are coming down from some far outlying village. Barttelot better, though still seedy. Nassibu from Yambow sent us in a small basket of rice. Ngungu, native chief, came over to Arab camp and also to gate of our camp; first time I have seen him for a long time.

Saturday, November 12th.—Ward went over to the village on the other side with some of Tippu-Tib's men to sketch, returning to dinner. Barttelot much better. Bonny has started building a mess-house and house for Barttelot.

Sunday, November 13th.—Some Arabs came in from a village about an hour from here on the road to Yambow, asking for medicines for Nassibu.

Monday, November 14th.—This morning an Arab came in from Yambow with dried fish and a spear as a present for me. I had seen him on my way from the Falls. Bought fish for mess, and gave him fifteen mitakos from my own stores for the present of a spear. (*Arabs never give presents for nothing.*) Feeling seedy again.

Tuesday, November 15th.—Jameson laid up with rheumatism. I am feeling quite ill. Nassibu came into camp to pay us a visit.

Wednesday, November 16th.—Jameson about the same; suffering a good deal of pain, with symptoms of bilious hemoturic fever. Not well myself. Opened provision boxes, under Barttelot's supervision, as our three months' rations were up on the 14th. Ward, Bonny, and I each received forty days' rations, that is up to the time that the others are rationed, viz., 28th December. We received each one tin milk, one tin butter, one tin Liebig, one tin jam, half-a-pound tea, and one pound of rough salt, which I bought from Liebrechts. The following were divided between the three of us:—Two tins cocoa and milk, two tins sardines, two tins sausages, two tins biscuits, one pound tin flour.

Thursday, November 17th.—My birthday. We were going to kill a goat, and have a *dinner of sorts*, but Jameson is still sick, and I am far from feeling well, so the feast is postponed.

Friday, November 18th.—Feeling very seedy, in great pain, chill accompanied by slight fever, so turned in. Jameson is very seedy too; properly knocked up. Some palaver on, I hear, with Tippu-Tib's men, who are interfering with our *special commissioner*, who is buying fish from the natives. This kind of thing I knew would arise. Tippu-Tib's men buy from the natives and then offer to us at double the figure. Barttelot has taken the matter up, and is sending word to the Falls to complain to Tippu-Tib about it.

Saturday, November 19th.—Not much better. Jameson has got jaundice. I gave him some gallic acid that Dr. Sims gave me. Ward left this morning for the Falls, taking six Zanzibaris and one Somali, Chama. I gave him out fifteen pieces of handkerchief and 179 mitakos, for buying coffee and onions.

Sunday, November 20th.—Felling better, but still weak. Jameson about the same; has taken one dose of gallic acid, but it seems to have done him no good.

Monday, November 21st.—Still seedy. Jameson the same.

[Similar entries of fever, weakness, &c. until Sunday, November 27th, when I began to improve slightly.]

Monday, November 28th.—No fever; feeling better; Bonny gave me some medicine that relieved me. What with recurring fevers and no appetite I feel very exhausted. Got from Barttelot six teaspoons of brandy, two days' supply. Bonny had asked for a larger quantity of brandy for me from the two cases, but Barttelot gives me this from his own bottle.

Tuesday, November 29th.—Feeling much better; able to eat some Brand's essence and some arrowroot from my private stores. Jameson much better.

Wednesday, November 30th.—Ward returned to-day from the Falls, bringing with him some onions, potatoes, and forty eggs! The latter have been given to me, a great treat in my present wretched condition. They were for Jameson and myself, but the former would not have any of them; in his condition he could not well eat them, so all were given to me. Ward found that Tippu-Tib had gone to Kasongo some few days before he reached the Falls to try to get the men for us. Won't be back for three months! Ward got a present of a goat also, but the Zanzibaris lost it on the road. Have been eating a little to-day, living on milk diet, arrowroot, milk and rice, and rice-water from my private stores, feeling better. Bonny's birthday. Barttelot puts in orders to-day that Bonny is to examine the sick every morning in future.

He takes over the medicine-chest from Jameson. We were all under the impression that according to Stanley's letter of instructions to Barttelot, Jameson was to see to them, but the cry has always been that there are no suitable medicines. I have given from my own store of medicine over and over again to some of the men—but I have so little.

Thursday, December 1st.—I received the eggs to day; they were delicious. Received from Barttelot three spoonsful of brandy, one day's supply. Still on milk diet, feeling better, but *very, very* weak. Forgot to mention that during my illness Ngungu, native chief, came in and had a palaver with Bonny. He has promised, now that all of Tippu-Tib's people have gone, to sell us fish, &c.; he stated to Bonny that Tippu-Tib's people had told him that they would shoot him if he sold to us whilst they were here!

Friday, December 2nd.—Slowly pulling up, I think; need feeding up. Last night half a goat was stolen from Ward's house where it was hung up; suspicion points to Soudanese sentry.

Saturday, December 3rd.—Feeling about the same. Jameson is all right again, but his colour is still bad. The thief discovered, the Soudanese sentry who was on duty at the fort-gate; the meat found hidden in the roof of his hut. He says Muledi, Barttelot's boy, helped him in the theft, telling him where the meat was.

Sunday, December 4th.—Feeling better, took two or three turns on the road; very shaky. Soudanese sentry received 150 lashes. Muledi has been expelled from camp, and has to carry a load for several days [this punishment is to carry a load up and down daily near the camp].

Monday, December 5th.—Feeling stronger every day. Barttelot discovers that the Soudanese told a lie when he implicated his boy Muledi in the theft of meat. The Soudanese says this himself now! I hear that three more men died last week; there are now thirty-one graves in the yard.

Tuesday, December 6th.—Stronger every day. Employed all day in putting mitakos into two boxes, which I have made long enough by scooping some of the wood off the ends.

Wednesday, December 7th.—Getting on well. Another Soudanese died to-day. A lot of Soudanese and Zanzibaris seem to be in a bad condition; the wet weather, along with their weakness, is telling on them. There is a miserable crowd at Bonny's every morning;

he is doing all he can for them with what medicines he has at his disposal. Sores are plentiful, and a number of the men are wasting away to skeletons.

Thursday, December 8th.—I went in to meals with the others for the first time since my illness.

Friday, December 9th.—Pulling up. Ngungu came into camp to-day. Another Soudanese died to-day.

Saturday, December 10th.—Another Zanzibari of Bonny's company died to-day. Two of Tippu-Tib's men with some women came in from the Falls, bringing some fowls.

Sunday, December 11th.—Feeling much better.

Monday, December 12th.—Orderly officer for the day. First time I have been able to be on duty since my illness. Birth in camp last night—one of the goats had a young one; mother and child doing well. Another Zanzibari of Bonny's company died to-day.

[The goat whose birth is mentioned above became a great pet among us; would sometimes come shyly into our tents, and was a source of amusement on account of its frisky gambols. He was not eaten during my stay in camp, I believe.]

Tuesday, December 13th.—To-day the feast postponed from my birthday took place. Killed a goat, and I gave a bottle of prunes from my private stores for luncheon. At dinner we had a real *roly-poly* pudding. I supplied the jam, Jameson and Barttelot gave the flour. Ward made it, and it did him a deal of credit. It was very good, and reminded one of home.

Wednesday, December 14th.—New order about the Zanzibaris falling in in the morning commenced to-day. They are to fall in after 8 o'clock, this enabling them to get a meal before going to work, and also avoiding the wet grass. They work till 1 o'clock, fall in again at 2 P.M., and work till 5 o'clock. The suggestion was Bonny's. These men never can stand the wet or cold, and as there is really nothing of importance for them to do, it seemed absurd, bearing in mind the amount of sickness there is amongst them, to send them out at 6 o'clock to sit shivering in the cold and wet.

Thursday, December 15th.—The Zanzibaris don't like the arrangement about hours that commenced yesterday. They complain of having to work through the hottest part of the day—12 to 1 o'clock—and would sooner have the old hours, so they have gone back to them. I still maintain that it would be better not to send them out till 8 o'clock in the morning, as there is nothing particular to do.

Saturday, December 17th.—Another Zanzibari of Bonny's company died to-day. In the afternoon twenty-eight men of Tippu-Tib's arrived in camp to await the arrival of Salim-bin-Mohammed, who, they say, is returning here at the end of the month. I am sorry they are coming back; our men are much better without them. They bring no news of Tippu-Tib, who ought to be at Kasongo by this time. Neither have they heard anything of Stanley at the Falls.

Monday, December 19th.—Another Zanzibari of Bonny's company died this morning, and another of Jameson's company; the total of deaths now among the Zanzibaris and Soudanese since this camp has been formed is thirty-eight. The Zanzibaris commenced building houses outside of the fort for Stanley's men. Barttelot decided on this owing to the rumours concerning Stanley's return.

Wednesday, December 21st.—Hear that Tippu-Tib's men brought in Ngungu as a prisoner here because he will not come and live on this side! In the afternoon some natives tried to catch two of our Zanzibaris; they succeeded in getting one. As he wore a dress something like Tippu-Tib's men, we suppose they took him for one of them, and wanted to retaliate for the capture of Ngungu. Tippu-Tib's men brought Ngungu out and made him shout across to his men on the other side to send the Zanzibari back; they had taken him over to the other side. There was a lot of unnecessary excitement; the man was sent back shortly afterwards, all right, though scared.

Thursday, December 22nd.—Tippu-Tib's men exchanged Ngungu for another prisoner, so as to enable Ngungu to go and fetch his people to build on this side. I don't know what will turn out over this business, but it seems cheek on the part of Tippu-Tib's people to cause this palaver, as they say on our behalf, without our asking them to do it. I should have fancied that if we wished him to come over on this side we could have set about it ourselves.

Saturday, December 24th.—The big goat which I brought from the Falls was killed early this morning for Christmas festivities. It is a real fine one. To-day the men have a half-holiday, and to-morrow is a general holiday. My own private stores are low enough, but among them I found a tin containing an untouched ham, a gift from an old comrade I had seen on my way up the river six months before. This I laid aside, and choosing the largest and best of the few jars of pickles I still possessed, and adding to these a bottle of sauce, I presented them as my contribution to the mess. Barttelot and Jameson gave two tins of flour and a pot of jam, while Bonny and Ward had added what they could to the general fund. After our usual smoke, flavoured to-night with cheerful stories, we perambulated the bounds to see that all was in order; then we lay down to dream of the good old times, when Christmas Eve had meant so much to us as youngsters, but we well knew that here in the heart of Africa we should wake to no such jovial experiences as still lived in our memories.[1]

Sunday, December 25th.—Christmas Day dawned clear and bright, and its cheerfulness was reflected in our faces as we turned out to greet each other with the same familiar words that our friends so far away were repeating under such different circumstances. Our mess-table wore its gayest air. Jameson and Ward had sat up till the small hours, engaged in a joint conspiracy, the result of their "plotting and planning together" being a laughable series of Christmas cards, which we found prominently displayed on our plates. The two that fell to my share I have treasured carefully. Jameson's card for me was a water-colour sketch, where, in the upper portion, I am depicted in the Slough of Despond, amid a thicket of trees, with struggling, baa-ing goats in most absurd positions; my own attitude expresses utter despair. Beneath is a different scene, where I, clad in the famous cashmere coat, Stanley's parting gift to me, am reclining in the only easy-chair the camp could boast, with a kettle and a cup of tea beside me, the very picture of comfort, in spite of my exhausted air. The background is formed of the precious Emin Relief stores. He even carefully lettered the boxes, and printed a Christmas greeting at the top.

Ward had made for me, with great skill, a most excellent likeness of the head of the famous goat, off which we were to dine that day. The expressive eye, the sardonic grin, the twisted horns, were all carefully delineated; Ruskin himself could not but admire

[1] The following account of our festivities was written out for a Christmas story published by permission of the Committee.

the delicate lines of the faithful reproduction, which did credit to Ward's artistic skill. The whole picture brought to mind the familiar old carol—

"A boar's head in hand bring I,
Sing merrily, oh, sing merrily,"

though we must needs substitute "goat's head."

I, too, produced some gifts for my comrades. For some time past I had noticed poor Jameson's struggles with a burnt-out pipe. He had tried to block up the hole at the base of the bowl with chiquanga or manioc, but before he had time to take a second good pull at the pipe this vegetable plug would dry and fall out, a most annoying condition of affairs. It was, therefore, with much delight that I found in my private stores an unused pipe, which I gave him as a Christmas present. It is needless to add that he was well pleased to be able to have his "calumet of peace." He soon had it "lit up," and his broad smile of contentment was a pleasure to behold. Ward and Bonny were very grateful for the pound of coffee apiece that I was able to give them from the stores I had bought for my own use on my way up the Congo. They seemed to feel that I had fully repaid them, the former for his beautiful sketch, and the latter for his thoughtful care of me during my first serious illness. It may be thought by outsiders that my gifts were small, but coffee was so much of a luxury that it was fully appreciated.

Our breakfast was made of cold goat and the ham I had given. I was much amused to hear the remarks made as Ward carved the thin slices of the latter and laid them on each plate.

"My dear fellow," said Barttelot, politely, "that is quite enough for me."

"Don't give me too much, Ward," exclaimed Jameson, with equal politeness. Presently they all sang the praises of the ham in chorus—"We've had nothing so good for many a day." "Perhaps—yes, really, they did think another slice, a thin slice, would be acceptable." A little urging brought them all up for another "help," and I saw with a certain satisfaction that my contribution had met with high favour. I know I enjoyed my share immensely. When we rose from table there was little left except the bone, which would supply us with a tempting soup, for with true economy we never wasted a scrap of food: we had already learned its value too well.

After breakfast the Soudanese and Zanzibaris were paraded in front of Barttelot's hut, and each man received two mitakos (brass rods used as money) and a handful of cowries as a Christmas dash (word for present used by Zanzibaris and others). We each were favoured by state visits from the Soudanese officer and the Zanzibari Niamparas (headmen), who expected individual gifts; we had so little of our own to spare that we could only give them three mitakos apiece, while a few important men received a small bit of tobacco. My own boy, Hamesseke, and the mess-cook, Juma, came in for their share of presents.

Dress parade, such as it was, ended, I took a turn about the boma, and paid a visit to my favourite donkey, a fine little white fellow, on whom I looked with special affection. On my way up the river I found, at Kinchassa, three donkeys had been left for our use, and I immediately chose this one for my own. They were put on board the little steamer, but it was very close quarters for all of us, so these animals were stowed in the bow, very near the boilers. It was hot enough for them, poor things, but I tried to make them as comfortable as possible, and admit I showed great partiality for "Monsieur Blanc." I never shall forget their glee when they were safely landed, they caracoled about like veritable Arab steeds, kicked up their heels, and stretched their legs after being cramped up so long. My little fellow was a thoroughbred, cleanly made, and in the end I discovered, in spite of spare fare, he was sound in wind and limb, and would have passed the examination of a "vet." I grew much attached to this pet of mine, and often visited him with some delicacy in hand to tempt his appetite, so that he was soon devoted to me. I found it very hard to part with him nine months later, even though he went into the hands of an old friend—a far better fate than Jameson's donkey met, for he, alas! died. On that sad occasion the Soudanese sent a deputation to Barttelot to beg the remains of the late departed because they had a penchant for donkey steaks. Their petition was granted, on condition that the feast should take place at a distance from the camp, and they retired to enjoy their banquet.

After our Christmas breakfast all was bustle and excitement: the men were more cheerful, and an air of conviviality pervaded the camp, as if some great event was about to take place. Strolling across the inner boma, where stood the officers' quarters and our precious relief stores, I entered Ward's hut, and found its owner busily engaged in the culinary art. He had produced from somewhere a small drawing-board, and with an empty bottle as a rolling-pin he had set to work, with sleeves rolled up and befloured hands, to make the Christmas puddings. I roared with laughter, and began chaffing him so hilariously that he shouted laconically, "Get!" but as I was by no means ready to accede to his polite request he added, "I'll never finish this if you stay bothering with your jokes, so you just get!" It was an amusing sight to watch his vigorous pounding and rolling of the pastry, and I had no intention of leaving him to pursue this new vocation in peace.

As soon as I could see after my hearty laugh I set to work to investigate what he was preparing, and found all his attempts took the form known as "roly-poly." For lunch he was making a meat pudding—the inevitable goat forming its interior. This was to be followed by banana pudding; a most delicious dish it proved a few hours later, when we sat down to luncheon, and demolished it with marvellous rapidity. Not only had we the well-known sauce of a good appetite, but the jokes, flying round "promiscuous-like," added much to its flavour. Well do I remember, as the last pieces vanished, hearing Jameson exclaim, by way of au impromptu grace after meat, "Thank God for my good lunch," and we all echoed his words with a hearty Amen. In the afternoon we met as usual by the big native wooden drum, on which we sat kicking our heels, smoking, and yarning—Barttelot full of hunting reminiscences of days spent down near his home in Sussex; Jameson telling of his sporting adventures in the Matabele country and in the Transvaal; Bonny giving us accounts of his various campaigns in Zululand, the Transvaal, and two phases of the Soudan; Ward adding descriptions of his strange adventures in New Zealand, Australia, Borneo &c. While so pleasantly engaged the time slipped away right cheerily, till, in spite of our hearty luncheon, we felt that the savoury dinner promised us would be most acceptable. Some hint of this brought Ward to his legs, and, with a rush, he departed to see that Juma, our Zanzibar *chef*, was cooking properly everything that Ward had so carefully prepared. The latter might well have felt proud of the result of his volunteer

MR. HERBERT WARD.

labours, for he had most kindly undertaken to add to-day to his usual duties, by attending to the kitchen department. It was no wonder that a broad smile of satisfaction pervaded his features as we marched in to our Christmas dinner. Barttelot took his usual position at the head of the table, and we all fell into line, quite ready to attack the well-arranged menu, which was as follows:—

MENU.

GOAT SOUP, À L'AFRICAINE.

GRILLED GOAT STEAK, À LA YAMBUYA.

ROLY-POLY PUDDING, *à la Ward.*

WINE LIST.

Tea *noir.*

Cognac, *une liqueur à chacun.*

This last was a most unusual treat, and we sipped it slowly, as between the pulls at our pipes we drank healths after dinner. Stanley had given Barttelot and Jameson a bottle of brandy each before the rest of the white officers, forming the last contingent of the rear-guard, had reached the camp; but he had left in addition two cases of brandy that were never opened during our stay at Yambuya. Barttelot on this occasion produced the remnant of his bottle, which amounted to a small wineglassful apiece. The chief toast at our banquet was "Absent Friends"; this was drunk in silence, as we thought not only of those at home but of our comrades who had advanced into the jungle, and whom under Stanley's leadership we had expected back with us before this date. We could only wish that they had been as fortunate in their Christmas celebration as we felt we had been that day; we hoped for their speedy return, or that we might soon be able to join them. We pondered often on the probable causes of their silence. We hardly dared believe that our great leader might never return. We thought constantly of them, and our minds were full of unanswerable queries. Had it been necessary for Stairs to use his Maxim gun? Had Dr. Parke found the task of caring for the invalids of the wayworn camp more than he could manage? Had Jephson had insurmountable difficulties in getting his men to carry the sections of the boat through the thick jungle? Was Nelson wearied out by his allotted tasks? Had William, Stanley's white body servant, had the romance of adventure all knocked out of him? Who had fallen by the way? Where were they at this joyous season? Such questions as these suggested themselves and remained unsolved. We fancied they had already joined Emin, and were even now homeward bound, yet their long-continued silence made us daily more anxious for their welfare. We knew well that they would be thinking of what had become of us, for we were to have followed them in the August previous, yet here we were forced to remain inactive, waiting for the promised carriers to transport the bulk of the relief stores. Little wonder, then, that an undercurrent of sadness pervaded our Christmas cheer, and though the air of festivity to-day seemed to brighten the prospect, yet by contrast it served to deepen the gloomy forebodings that were so soon to follow.

However, on this occasion our songs were cheerful; we even found voice for the comic as well as the sentimental, Bonny's fine tenor being in great demand in our impromptu concert. Our stories had a brighter tone than usual, and we strove to make the best of the situation. The men, too, seemed to catch the infection from us, and around their camp-fires were singing right merrily. Poor fellows, they had no romance of chivalry, no expectation of fame, no dream of doing good to poor beleaguered fellow-creatures, to keep up their drooping spirits during those weary days of waiting!

We sat late smoking and talking together. Those silences that would fall as we thought of home and friends were only broken by the voices of the men singing, or by the rushing of the river below us. We could see the rapids glistening in the clear moonlight, that touched up with its splendour the singular scene before us. The quietness was not even disturbed by the strange cries of the parti-coloured parrots that in great flocks had flown high over our heads hours ago on their way to roost. We found the pauses in the conversation getting more and more numerous, and the stories less and less absorbing, and as Christmas night, even in Central Africa, must needs end some time, we at length knocked the ashes out of our pipes and started on our final rounds for the night. We saw that the Soudanese sentries were at the gates, and replied to their challenge with the curt "Ameen" (Friend), and heard their response, "Ameen allairi" (Pass, friend). We found that each intervening post had its appointed Zanzibari watchmen; so, with cheery "good-nights," we parted, each to dream of the happy days of long ago.

Little did we think that before another Christmas dawned two of our comrades would have bravely met their fate, nor that all traces of our camp in the wilderness would have faded. Neither did we imagine that the rest of our party, who had so cheerfully volunteered to attempt the relief of Emin Pasha, would now be sadly returning home, forced to admit that their own efforts had been in vain, yet hoping, in spite of some misgivings, that Stanley and his advanced detachment had been more successful.

Wednesday, December 28th.—Early this morning, about 3 o'clock, we heard shouting, and then a rifle-shot from the direction of the first rapids, where Tippu-Tib's people have built a hut; also heard natives shouting and singing in canoes. On inquiry we find the Arabs had captured a canoe yesterday from some of the natives on the other side, and brought it close to their hut. They put some of their men in it to guard it during the night, but these must have gone to sleep, as the natives came down in another canoe, stabbed one of Tippu-Tib's men, and retook the canoe. Barttelot and Bonny went to the hut after breakfast and saw the wounded man; he had a nasty stab, and Bonny does not think he will live. One of the Zanzibari boys, who had belonged to Jephson, Kahamese, died to-day.

Thursday, December 29th.—The man belonging to Tippu-Tib's camp, who was stabbed, died early this morning. Another Zanzibari boy, Obedi, died to-day; this brings the total up to forty-one. There were only seven or eight in the graveyard when we came up on the second trip of the *Stanley* on August 14th. Tippu-Tib's headman here came in to inform us that four of his men had been taken by the natives at a small village on this side. He wants our aid to get them back, but Barttelot has declined. It is their palaver which they have brought on themselves. We have perfect quiet when they are not here; just as surely as they come they make trouble with the natives. I wish to goodness they would clear out, or be cleared out!

Friday, December 30th.—A dead body is stuck on some wood or stones in the river opposite the camp. Men were sent over to investigate; they say it is one of Tippu-Tib's men, but not belonging to the lot here. Probably one of those from higher up river. Tippu-Tib's headman here informs us that the natives have *made soup* of the four men they captured yesterday! After breakfast we heard firing up river, and saw some canoes rushing from this side to the opposite side. Later in the day Tippu-Tib's headman informed us that after Ngungu had been set free, he went to a village lower down on this side, and instigated the chief there to retake the canoe and capture the men; but it looks as if he were trying to justify his firing of this morning at two canoes of Ngungu's that came up alongside as usual. He wounded one man. I wish Tippu-Tib's lot were out of it; they ought never to have been allowed to interfere with Ngungu at all.

Saturday, December 31st.—Here we are at the last day of the year again! I have been eleven months from England, ten months in this country. I wish we were on the move, but all the same have a lot to be thankful for.

Sunday, January 1st, 1888.—Yambuya Camp, Central Africa. Happy New Year to all! We had an excellent breakfast of goat chop. Luncheon, goat and stewed prunes; the latter I supplied. Barttelot and Jameson went for a walk higher up and saw natives who came over in canoes to this side, in reference to releasing the men they have of Tippu-Tib's. Nothing definite has been arrived at. For dinner, goat and roly-poly pudding; Ward and I finding flour, and I the jam.

Monday, January 2nd.—Barttelot and Jameson went over to native village on the other side to arrange about getting back the men they have as prisoners belonging to Tippu-Tib. They managed to get two of the men back; the other, the natives say, escaped. Some men were sent this morning to Abdallah's camp to try and buy fowls.

Tuesday, January 3rd.—In a conversation with Barttelot this morning, he referred again to his idea of going up to look for

Stanley, taking with him two officers and about 100 Zanzibaris and Soudanese, with only ammunition in the way of loads. I would say here that last October when I was at Stanley Falls with Barttelot, he mentioned to me that if Stanley did not return by February, that he would go up, taking Jameson and myself to look for him, as he (Barttelot) would come to the conclusion that something had gone wrong with the column, should they not be back by the above-mentioned time. He seems convinced in his own mind that the column has met with some disaster, says he does not expect ever to see Stanley back, and such-like cheery ideas. In his conversation with me this morning he seemed to wish to have my opinion on the matter, which I gave somewhat in these terms:—I told him I did not share in his ideas about disaster having happened to Stanley; true, he had been away a much longer period than he stated when he left here on June 28th. Barttelot informed me, he had told him he expected to be back early in November, viz., about four months from the date of his departure, and it is now over six months, and will be seven at the date of Barttelot's proposed move. I pointed out to Barttelot that if we were to believe what the deserters had told us, it would appear that Stanley had met with some difficulties—bad country to go through, trouble with the natives, and last but not least, trouble with his own men in getting them to carry on the loads, and also desertions among them, and that instead of taking two months to go up to the Lakes, he would have taken a good three. When he got there (Kavalli) he proposed making a camp, leaving all his white men and bulk of Zanzibaris, he proceeding, with a few men, in his boat across the lake to Kibiro, and thence by land, perhaps on as far as Wadelai. I don't know the distance it will be from his camp at Kavalli to Wadelai. Say he takes two weeks to do this. He would stay some time with Emin, and would not be back at his camp, I should say, much before five or six weeks have elapsed from his departure from it. This would account for four and a half months he has been away; if he was coming down at once it would leave him one and a half months to get here, making up the six months he has been away up to date. I also stated that even if Mr. Stanley had been stopped by numbers of natives, I could not see what such a small force as Barttelot proposes to take could do to help him out of his difficulties. I do not see the necessity why at this present time any caravan should be sent up, and certainly not one that can be of very little service to Stanley should they meet him; in fact none unless he is short of ammunition. It is not at all unlikely that Barttelot may miss Stanley, or some of his party, on their way down, with definite instructions as to what we are to do. In this case all the men would be away, with the exception of the sick, and a few others left at the camp here, and the consequence would be that the Expedition would be split up in three sections, scattered over Central Africa, and those left at the base would be unable to carry out the orders of the chief. Barttelot's written instructions left him by Stanley are clear enough. He was to follow on with all the officers, men, and stores *when* he had the carriers promised by Tippu-Tib, and in the event of Tippu-Tib sending only a part of the requisite number, a list was given of loads which he might not dispense with.

["*It may happen also that, though Tippu-Tib has sent some men, he has not sent enough to carry the goods, with your own force. In that case, of course, you will use your discretion as to what goods you can dispense with to enable you to march. For this purpose you should study your list attentively, viz.:*—1. *Ammunition, especially fixed, is important.* 2. *Beads, brass wire, and cowries rank next.* 3. *Private baggage.* 4. *Powder, caps.* 5. *European provisions.* 6. *Brass rods, as used on the Congo.* 7. *Provisions (rice, beans, peas, matamas, biscuits). Therefore, you must consider after these, sacking tools, such as shovels (never discard an axe or a billhook), how many sacks of provisions you can distribute among your men to enable you to march, whether half the brass rods in your boxes could not go also, and there stop. If you still cannot march, then it would be better to make marches of six miles twice over (if you prefer marching to* staying for our arrival), *than throw too many things away.*"—Stanley's Written Instructions.]

I stated to Barttelot that I believed Stanley had been delayed, owing to unforeseen circumstances, as well as to the causes already mentioned, adding that he might have waited at the Lakes some time expecting us. That as we had waited so long, we might wait a little longer, so as to give time, should all the delays, &c. have occurred as I surmised, to get news or see him here in the next few weeks. Barttelot lays great stress on our not having received messengers from Stanley, telling us not to come up if Emin is coming down with him. But if he has been delayed in getting right up to Wadelai, he would think we were well on the road by the time he had seen him. On the other hand, he may have sent, and the men were not able to come through. I have every confidence that after great difficulties Stanley has reached his destination, having taken a longer time than he expected; that he has looked for our coming up, and that if he has Emin with him, as I am inclined to think, he is on his way down here again. If Emin is not coming away, I could understand Stanley sending down to urge us to come on with the Relief Stores. Stanley has four good white men with him, as well as some picked Zanzibaris, and some four hundred rank and file, all well armed; and he should be quite able to hold his own against any natives. He has done so before with less men, and I have great confidence in him for such work, and hope to see him, or a part of his column under a white officer, back before many weeks are over our heads. Holding these views, which I have written down as stated to Barttelot, I am entirely against a small body of men going to *look* for Stanley, as quite unnecessary and useless, against the interests of the Expedition, taking the men away on a wild-goose chase, when they will be likely to be required for work which would benefit the Expedition. Barttelot says he is going to call a council on Saturday or Sunday, to hear the opinion of the other officers on his project. I hope sincerely that the majority will hold the views I have taken, as I am convinced, having only the interests of the Expedition at heart, that Barttelot's project can do no good to Stanley, even if all my surmises are wrong, and will only end in crippling the Expedition. I don't think he (Barttelot) would be doing the right thing in leaving the camp, of which he has the command.

Wednesday, January 4th.—Zanzibaris sent to Abdallah's camp returned, having been able to purchase four fowls. Heard that Ngungu had gone up river, taking with him seven guns captured from Tippu-Tib's men.

Thursday, January 5th.—Turned out very hot. Jameson says it was 92° in the shade. Messenger of Tippu-Tib's arrived from the Falls, brought two fowls and some pine-apples. Says that

Salim-bin-Mohammed will be here in three days. Barttelot seedy.

Saturday, January 7th.—Nassibu, Tippu-Tib's headman of village of Yambow, arrived with small following; brought three goats: made a present of one, and we bought the two others.

Sunday, January 8th.—Nassibu came in from a neighbouring village to pay us another visit. Yesterday he came to see me in my house, and had a chat through the interpreter. He remembered me on my way to Stanley Falls. He says Tippu-Tib has not come back from Kasongo. I received a fowl from him as a present, which I returned in the shape of a piece of handkerchief. Another Zanzibari of Bonny's men died to-day, making a total of forty-two.

Monday, January 9th.—Salim-bin-Mohammed arrived in the afternoon. The only news he brings is that Tippu-Tib has not yet come back from Kasongo; he does not expect him till about the 15th of February. Expects some men down about the same time, but does not know how many; the water being low, they cannot come in big canoes, and would have to use small ones. They might come down sixty or seventy at a time, or perhaps 100, if they have a sufficient number of canoes. By what Salim says they could not be here at the earliest before the middle of next month. He states that Tippu-Tib is trying to get the men, telling them they are coming down here to fight the natives; he will not tell them the truth, that they are wanted to carry loads for us, till they reach here. This is not very hopeful news, as the men may refuse to carry loads when they get here. This might cause trouble. When asked if the Lomami palaver with the natives had been settled, Salim said that it had, and all their men had returned to their homes, and these men had also been promised us when the Lomami business was finished. Bonny bought a goat that they had brought, and Salim presented Jameson with a big goat (as a present from Tippu-Tib), as well as with some "ghee" (kind of fat) and some coffee-beans.

Tuesday, January 10th.—Heard through Salim-bin-Mohammed that two more Zanzibari deserters were coming down and would be in camp to-morrow. Barttelot talks of going up to a village past Opoy, where Abdallah is, on Stanley's road, some twenty days from here, to see if he can get any news. No council has been held, and by his going as now proposed to this village, one is led to suppose that the wild-goose chase to go and look for Stanley has been abandoned. I am very glad for the sake of the Expedition. As far as I can make out there is very little chance of our leaving here before Stanley comes down, which I most sincerely trust will be soon.

Wednesday, January 11th.—Ward's birthday; had extra provisions from private stores in honour of the event. In the afternoon Barttelot, in the course of conversation, as we walked up and down our promenade, mentioned to me something about a piece of paper Stanley had given him, not signed, in which he stated that the loads might be taken up by the Zanzibaris in the event of Tippu-Tib's men not turning up, the Zanzibaris to make twelve journeys, or rather taking twelve days to get all the loads to one point, from the camp. This is what I understood Barttelot to say the paper contained; but I have not seen it, and this is the first time I have heard that Stanley made such a suggestion. It was not in accord with the general written instructions.

The idea of our moving out with the bulk of the loads with the Zanzibaris only, has never been entertained, never mentioned before, and would have been in my opinion perfectly insane. Bearing in mind that Stanley expected to be back in November, suppose we had started at the end of August or beginning of September, with about a hundred Zanzibaris—all that would have been fit to carry—with five hundred loads (supposing we had dispensed with two hundred), it would have taken twelve days to get all these loads one day's march of four miles. At this rate we could not have gone far before we should have met Stanley coming back had he kept his time; and as he has not, we would have been on the way up still if we had any men left.[1] Barttelot told me in this conversation that he felt sure that Stanley would blame him for not having come up with the loads, and try and damage his reputation at home. Bonny seedy. Nassibu sent Jameson a live antelope, but it died from fright shortly after its arrival in camp.

Thursday, January 12th.—Nassibu came in with an Arab, who, he says, has just come down from Abdallah's camp past Opoy.

[1] This was a calculation made at the time, but it does not work out quite right. See fuller remarks on this subject in Chapter I.

This Arab says there were eight deserters in all; six had died, and two he left at Abdallah's, they being too sick to come down. He says they have heard nothing more up at Abdallah's about Stanley.

Friday, January 13th.—Two men came in from Nassibu's. The Arab who came down from Abdallah's camp yesterday is going to the Falls. Another Zanzibari of Bonny's company died to-day, making a total of forty-three.

Sunday, January 15th.—Another Zanzibari of Jameson's company died to-day, making a total of forty-four. Discussion about supply of food. Our fish supply has stopped altogether since the Arabs returned, as usual. Bonny has been able to get a few goats lately. We kill one a week, which lasts two days, the remaining five days we live on rice and plantains.

Thursday, January 19th.—A Soudanese died this morning, making a total of forty-five.

Friday, January 20th.—Another Zanzibari of Bonny's company died, making forty-six. Nassibu came and paid us a visit.

Tuesday, January 24th.—Another Zanzibari of Jameson's company died this morning, making a total of forty-seven. Salim-bin-Mohammed paid a visit to camp, gave some news about an Arab who had been to the Lakes. Some natives came into camp with one of Tippu-Tib's men, from village lower down river.

Wednesday, January 25th.—Another Zanzibari of Bonny's company died during the night, making a total of forty-eight. Had all the cases of powder out of store, and examining them found them all right.

Friday, January 27th.—Tippu-Tib's men attacked a village lower down river, captured four men including chief, twenty-six women and children, and three canoes. This is the old palaver about the natives having captured guns and not returned them. The palaver, Salim-bin-Mohammed tells us, will now be settled, as Ngungu's people opposite are afraid, and sent men over on hearing the fate of the village below, and wanted to be friends. Salim adds, the guns are going to be sent over, but they will have to pay ivory for the redemption of the women. Salim's band played at dinner.

Saturday, January 28th.—Another Soudanese died this morning, making total forty-nine.

Sunday, January 29th.—Received 6 oz. of tea from Expedition stores, thirty-three days' supply!

Tuesday, January 31st.—Jameson's donkey died during the night. The Soudanese begged the carcass for food, and it was allowed them on condition they ate it away from camp.

Wednesday, February 1st.—The Zanzibaris who were sent to Yallasullah to buy oil and fowls returned to-day, bringing one goat, ten fowls, and some palm-oil. They brought two letters for Salim-bin-Mohammed from Stanley Falls. Salim tell us that ten canoe-loads of men have come down from Kasongo, but have gone on to the Lomami to settle some palaver there, and will come on here afterwards! Tippu-Tib has not yet returned from Kasongo. Our men heard rumour at Yallasullah that a white man had been seen on an island far up, fighting the natives. The natives here have succeeded, while we were at dinner, in cutting away two canoes that Salim had captured from them.

Friday, February 3rd.—Some Arabs arrived from Stanley Falls with a small body of Manyuemas, who had just come down from Kasongo, and were *en route* for Abdallah's camp (on Stanley's track, far up). They state that Tippu-Tib was still at Kasongo when they left. He was getting men for us, the difficulty being to get a sufficient number of canoes to bring them down. Some of these Manyuemas added that the men did not wish to come, and were taking their canoes out of the way. We met these men on the *esplanade*, and got this out of them through John Henry, our interpreter. Later in the day Salim-bin-Mohammed came in to see Barttelot. By what I can gather from Barttelot's conversation with me afterwards, Salim has not received any letter himself from Tippu-Tib, but expects one in a few days! But he has been told that Tippu-Tib will be here about the fifth day of the next moon, which should be the 16th of this month, with about 250 men; and that he (Salim) is going to build a house at the back of our camp for him. Salim stated to Barttelot that there is a road from Stanley Falls right up to Lake Albert, that is to the Unyoro end, and that they have Arabs right along it. He says it is a better road than the one Stanley took, and that it would take about three months to reach the lake by this route. This is like everything else connected with Tippu-Tib's people—rumours, contradictions; statements, contradictions; keeping back news, then springing something quite new, at different periods. There have often been rumours that they had a road to the Lake, direct from the Falls, but no definite statement has been made before by Salim to this effect. Barttelot talks about going to the Falls to meet Tippu-Tib, if he can get Salim-bin-Mohammed to accompany him. Salim informed Barttelot he was going to clear out the natives opposite in a systematic manner, early to-morrow morning, as they would not give back all the guns, six in number, which they had captured. He was going to send his canoes up at night, so as to be in position to attack at dawn to-morrow; having also arranged with Abdallah's men to turn the natives, should they try to escape in their canoes up river, for down river his men were in possession. I feel sorry for the natives, and only wish they would give Tippu-Tib's men a licking, which is far from likely, as they are scared at the sound of a gun. Another Zanzibari died—total, fifty.

Saturday, February 4th.—Heard some firing, so supposed Salim's people had attacked the natives; impossible to see much owing to the fog. Later in the day, when it lifted, we could see canoes far up near the third rapid trying to get up, being fired on by the Arabs. Very few, if any, escaped down river. Tippu-Tib's men set fire to the main village, leaving a few straggling huts standing. In the afternoon Tippu-Tib's men returned, as usual, to our landing-place, having captured one fine large canoe, along with some smaller ones. They got very little loot out of the village; only a few wooden native beds, pots, some fishing-nets, and a bar of rough iron. The native chief, Ngungu, has not been seen. I should be sorry if they killed him, as I look upon him as an old friend. Third anniversary of Jameson's wedding-day. Big goat killed; had a roly-poly pudding—this was the only way we could utilize the one-pound tin of flour given to each of us. Soudanese prisoner escaped from guard-room while we were at dinner, taking with him the rifle of another man and twelve rounds. Men were sent to inform Salim-bin-Mohammed, and a reward was offered if any of his men could capture him.

Sunday, February 5th.—News from Tippu-Tib's men further up puts quite a different complexion on the fight of Friday. It appears that when the natives fled from their village opposite, men were posted in canoes near the upper rapid to turn them. The natives got the best of it; they killed ten of Tippu-Tib's men, a canoe capsized, and the guns were lost in the river. The natives, we hear, are now camped higher up the rapids. Another Soudanese died this morning, making a total of fifty-one.

Wednesday, February 8th.—Another Zanzibari of Bonny's company died to-day, making fifty-two. I have been feeling seedy with fever for the last few days.

Thursday, February 9th.—The Soudanese prisoner who escaped from the guard-room on Saturday last, was brought back to camp to-day by some of Salim's men. He had wandered as far as Abdallah's first camp, where they captured him.

The case of the Soudanese deserter came up for discussion after luncheon to-day, and we had to decide what punishment should be meted out to him. It was argued that the Zanzibari deserters had been flogged, and that this man ought to receive the same punishment. But it was held by some of the officers that the Soudanese were engaged as soldiers, and were under military discipline, therefore this case should be dealt with as desertion in an enemy's country. The end of the discussion was that two of us (I being one) held that he should not be shot, but the others, being in the majority, carried their point, and he was therefore condemned to suffer the extreme penalty.

Friday, February 10th.—The Soudanese who ran away was shot this morning, all the men in camp being drawn up. Another Zanzibari of Bonny's company died this morning, making total of fifty-three.

Sunday, February 12th.—Barttelot and Jameson decided yesterday about going to the Falls on Tuesday.

Monday, February 13th.—A Zanzibari died to-day, making a total of fifty-four.

Barttelot, at luncheon, read out his plans for trying to get Tippu-Tib to give us men. He and Jameson are going to-morrow to the Falls, accompanied by Salim-bin-Mohammed to await Tippu-Tib's arrival from Kasongo.

He proposes to try and get the men as laid down by Stanley to carry the loads. Failing this, he will propose that Tippu-Tib should give us two hundred men to carry loads, and, say, another two hundred men to come up armed, without loads. If he cannot get Tippu-Tib to give us men under either of these conditions, he proposes, as a last resource, to ask for, say, twelve hundred men, or as many as he can get, to carry no loads, but to go up armed, under the immediate charge of Salim-bin-Mohammed. He proposes to guarantee Tippu-Tib five thousand pounds over and above what Stanley promised to pay the men for their services. In the event of Tippu-Tib furnishing the men under the third proposal, Barttelot proposes to send all our loads here over to Stanley Falls, leaving a white officer there with all the sick, to look after the interests of the Expedition. Our opinions were asked on the third, or last, proposition, as the other two were agreed to. The general idea seems to be entertained by the officers and others, owing to his having been away so long, that Stanley is in difficulties. He expected to be back in the early part of November. It is remarked generally the absence of any news either. And we are all surprised that no messengers have been sent to inform us of his whereabouts or difficulties, if he is in any. Barttelot goes further, and has stated more than once that he thinks that Stanley has been killed, and never expects to see him again. Jameson thinks that Stanley must be in a hole somewhere, unable to come or send down, and that we ought to lose no time in starting up to hear some news of his party. He, like Ward and Bonny, thinks that, owing to Stanley's long absence, we should be quite justified in taking some steps to find out where he is.

There is no doubt that if we heard that Stanley was in a corner, and wanted assistance, we should not hesitate to get men to go up to his relief, not bothering about loads, except powder and ammunition, and other things necessary to buy food, &c. But we are quite in the dark as to Stanley's position. We are justified in surmising he must have met with difficulties, and may be delayed, owing to most of his men having deserted. I myself am not disposed to think that any serious disaster has taken place. The delay in return could be accounted for in many ways, and, whilst admitting that we cannot sit down here in camp for an indefinite period, waiting tidings, yet I am doubtful as to the advisability of going without any relief stores; but if Tippu-Tib's people will not carry them, there is an end to the matter. Neither do I like the idea of making over the loads, which it really is, to Tippu-Tib. Stanley had said, over and over again, that his loads are of the greatest importance to the Expedition, and without them he would be in the position of Emin, wanting relief, if anything happened to them. It would be impossible, nevertheless, to keep up the camp here, unless a body of men were left, and that could not be done, as all will be required. The idea of leaving a white officer at the Falls with the loads would be the right thing, if he could do anything to protect them; but that is quite out of the question, for when once the loads are placed at Stanley Falls, if Tippu-Tib plays false, they are his, and the white man is of little use, and might be at the Lakes or anywhere else. Whilst I do not like the idea either of leaving loads at Stanley Falls, or of our going up without them, I cannot well vote against getting men as a last resource to go up and obtain some intelligence as to the whereabouts of Stanley's column (that is, only when Tippu-Tib states he cannot get men as carriers). I should most reluctantly go with such a column. I hope still, most fervently, that before any such column has been formed we may have news of Stanley; or, better still, see him back with his officers and Emin, all well. I do not see why we should not. It will be a good six weeks before we could get away, and any day I hope to hear of Stanley. I have every confidence in him, knowing what he has done and can do, and that, although he may have had a lot of trouble with his men, &c., he will turn up all right, and I hope that no other column *without loads* starts from here.[1]

Tuesday, February 14th.—Barttelot and Jameson left to-day for the Falls with Salim-bin-Mohammed. They took eighteen Zanzibari carriers and the Soudanese interpreter, Assad Farran. The Zanzibaris are going only as far as Yallasullah on the Congo, and will return here. (They take with them a lot of cloth and mitakos.) Barttelot has left me in charge of the camp, giving me the following written instructions:—

"*To Mr. J. R. Troup, Aruwimi Falls,* 14*th Feb.*, 1888.

"SIR,

"During my absence and that of Mr. Jameson at the Falls, the command will devolve on you, and in especial the command of the Soudanese, over whom I request you will keep the strictest

[1] It should be remembered that the above arguments were written at the time, and were even in print before Mr. Stanley wrote his letters from the East Coast.

supervision, and be down on them for any misdemeanour. Mr. Jameson's company of Zanzibaris will be merged into Mr. Bonny's company, and, acting under your orders, you will take command of both companies. I have projected a plan for the drawing in of the boma, which I shall be glad if you will get carried out, viz., from the guard-room west corner to west corner of water-gate; if possible dig a trench on the western side of the new boma. I am taking three niamparas (headmen) and fourteen carriers, of whom all will return except one niampara, whom we take on to the Falls, but they must return to Yallasullah at a later date, on receipt of a letter from me to that effect. Should you hear of any news of Stanley during my absence, please forward a letter at once to the Falls. I will arrange with Seyid at Yallasullah for transmission. Should a steamer arrive at the Falls during my stay I will acquaint you, and it will be best to send a white officer to Yallasullah to meet it. There should always be two white officers in the camp, and no officer should leave the camp without your permission and on urgent business. Should I not return and you can hear no news of me or of us, after three months you may deem us prisoners, and must act in the manner you deem best, and as circumstances may dictate for your own, the loads (powder and ammunition), and the men's safety. The boy Muledi will bring you one load of wood and one load of manioc every day for the use of No. 13, Private Alma del Shagi, Soudanese. I am, Sir,

"Your obedient servant,

(Signed) "EDMUND M. BARTTELOT, Major."

A Zanzibari of Jameson's company has been absent from camp. Bonny sent three men after him; they had not returned at night.

CHAPTER XII.

SPECULATIONS AND REFLECTIONS.

In connection with the foregoing extracts from the diary, the following comments seem necessary :—

Stanley had now been away from the camp seven and a half months, and with the exception of the information gleaned from the deserters, we had heard nothing of his column and its whereabouts. He had calculated that his march to and from Wadelai would occupy five months, so he was now two and a half months overdue, and we were naturally most anxious about him and his companions in consequence. The commander of our camp entertained the gravest fears, even saying that he believed Stanley was dead, and therefore considered himself justified in taking what steps he thought best, and in such action he would have been supported by some of the other officers. One party, headed by Barttelot, held that something must be done; leaving out of the question the importance of carrying up the loads, they said a move of some kind would be expected from this Rear Column; they ought to go up to look for Stanley; that it was impossible for us to remain for ever in this camp, especially under the present circumstances. I held, and was not wholly unsupported, that no such move was called for—certainly not at present. We had never received news that Stanley was dead; neither had we any authentic rumour that he was in difficulties in which he needed our assistance. Had we any such information we ought *then* to consider whether we should not push on at once, regardless of all else, to his assistance. As to the fact that no messengers had come down, it was possible that they might have been stopped by the Arabs, just as we knew the Zanzibari deserters had been. It was not even certain that these very Zanzibaris had not been the bearers of messages which had fallen into Tippu-Tib's hands, and possibly he knew much more about Stanley than we did. I urged also that if, as they surmised, Stanley was dead and his column destroyed, what was the use of our going up after him? Where he had failed with the force at his command, *we* had no chance of success. But supposing he was alive, we were not justified in ignoring his instructions, and starting out on an expedition of our own. He had laid the greatest stress on the importance of the loads left with us; our camp simply existed for the preservation of these very stores; if these had not been of importance all the men could have gone on with his column. Placing such great weight on the care of these loads, he had even gone so far as to indicate that rather than sacrifice too many of them we were to remain where we were, guarding them *until he returned* to our camp. As it was, we had received no positive news of serious disaster—certainly no news of his death; under these circumstances we were bound to follow his instructions, and to remain faithfully in charge of the stores confided to us. I firmly believed he had sent messengers down to us; that they had not been able to come through; that he was carrying out his plans; and as to the delay, I protested that time was of no account in African exploration. He undoubtedly had been delayed by one cause or another; but I had faith in him as the man who would eventually overcome difficulties, and that if we waited we should hear from him, or he would come back as he said he would. Any expedition sent up under the present circumstances would fail, I believed, and the stores in such event would be sacrificed. More than once in conversation with Major Barttelot I have argued on the presumption that he followed out his scheme. "Suppose," I said, "you do go up without the loads and should happen to meet Stanley, his first question would be, 'Well, Major, where are my loads?' What could you reply? When you informed him that you had left them behind and had come up in search of him, I can well imagine the warm conversation that would ensue."

It always struck me as rather absurd that we should set out to *look* for Stanley, especially with such a ragged lot of men as ours, who had by suffering, &c., become entirely demoralized, with no heart left in them. It would have taxed a man even of Stanley's resources and influence to re-establish confidence or enthusiasm among them. In imagination I saw how ridiculous we should all appear when we did find Stanley, and having explained why we were without the loads, I could hear him ask, "Well, now you have found me what do you propose to do?"

During the first part of the time we were in camp matters were in a very unsettled condition. We were daily expecting to be able to follow Stanley, and yet could not do so as the arrival of the carriers was daily postponed, therefore it was impossible to anticipate that we should remain stationary here for any length of time. But as the days went on, we found that Tippu-Tib held out no immediate prospect of supplying us with porters. By the middle of November he started on his excursion to Kasongo with the avowed purpose of getting us the men, and we knew that this trip would occupy nothing short of six weeks. Under these circumstances, we now knew that we should be obliged to remain at Yambuya for two months more, unless we had news of Stanley. The number of able-bodied men were already reduced to a mere handful, and they suffered more or less terribly from ulcers and all kinds of sores that are prevalent in this climate, but we had nothing provided to alleviate their pain. In fact our portion of the Expedition was entirely destitute of many most necessary medicines.

I, fortunately, had with me a two-ounce bottle of carbolic acid, which was of inestimable service as far as it went. There were no counter-irritants of any sort, which caused great inconvenience. As the white officers suffered greatly, it may well be imagined how much worse off were the men. The large body of Zanzibaris and Soudanese were totally without assistance for their ailments, and through this very cause were reduced in strength and numbers to a most pitiable condition. It will be seen how much all were suffering in these respects.

Tippu-Tib kept us dilly-dallying for the men, and as matters stood there seemed no immediate prospect of their being supplied to us. It will be seen that it was with the object of making renewed efforts in this direction, and even proposing new schemes, that Barttelot with Jameson set out in the hope of meeting the Arab at the Falls. We had been tantalized by rumours of Stanley's approach, but as usual these were all false. We, however, did gain some information from Salim-bin-Mohammed about a road of which we had heard vaguely before. In my diary I recorded his mention of a road well known to the Arabs, who were stationed all along it; this route led from near Stanley Falls across country to Lake Albert, and it might have proved a satisfactory means of communication with that district. The Arabs strongly urged that if they went with us we ought to take this road, but to this suggestion we could not yield, because it was Mr. Stanley's intention to return by the way he went, and we were instructed to follow him.

There is one thing in connection with Tippu-Tib that should not be omitted. In the agreement made by Stanley at Zanzibar, the State promised to send out two white officers to act as Tippu-Tib's secretary, &c., at Stanley Falls, and it was naturally supposed that they would arrive at the station a month or so after the Arab was placed in possession. A certain personal interest was given to this expectation of their arrival, because while going up the Congo Tippu-Tib complained that some of his loads had been left behind, but he was assured that the steamer which brought his assistants would also bring his loads. There was yet another reason why that steamer would be awaited anxiously by the Arabs. It seems that it is the custom among the Arabs and natives, when the latter swear to be friends, that a daughter of, or some woman belonging to, the native chief, is handed over to the Arabs, and is looked upon as a token of friendship; when the two parties meet in future, the woman is shown in this friendly capacity. Such a woman had been given by a native chief to Salim-bin-Mohammed, but while the Expedition was at Matadi she had run away, or had been inveigled away by some one, and in reply to his complaints concerning this he was told that every effort would be made to get her back, and she would be sent up on the steamer if she was found. But the arrival of the steamer was long delayed. We learned afterwards that the two officers sent from Brussels got as far as Leopoldville, where one of them died, and the other was so ill that he had to go home—a considerable time elapsed before their successors arrived, and it was a year before the promised steamer and men reached Stanley Falls. It is not surprising that, in consequence of this long delay, the Arabs entertained grave suspicions that the promises made by the white men would not be fulfilled.

CHAPTER XIII.

EXTRACTS FROM DIARY (*continued*).

Wednesday, February 15th.—Bonny with some men commenced alterations of camp, according to Barttelot's instructions. Zanzibaris returned with the runaway from a distant village. He can give no account of himself; we think he is a little off his head. Put him in guard-room till Barttelot returns.

Thursday, February 16th.—Another Zanzibari died, total fifty-five. Bonny had out some of the Zanzibaris trying to teach them how to use a gun—it was ludicrous to see them, some were even afraid to fire; they are only porters, engaged at Zanzibar, taken from plantations, and are not soldiers.

Friday, February 17th.—Bonny had the men out again. Heard that natives had taken two Arab prisoners.

Saturday, February 18th.—Birth in camp. Goat gave birth to twins, mother and children doing well. Bonny continued course of firing with the Zanzibaris. Natives came across in canoes; they know Salim-bin-Mohammed is away, and that we will do nothing to them. They were fishing, close to our bank, we tell them not to be afraid of the firing, as we would not hurt them They seemed to understand. Natives have taken two more Arab prisoners.

Monday, February 20th.—Hear Tippu-Tib's men captured some native women yesterday. Some of Tippu-Tib's men came in from down river to-day; they report that in their wanderings inland they came across some natives, who told them that some white men had been up their river in a steamer, and showed beads and brass wire that they had received from them. They say that the place where they saw the natives is about sixteen days inland from here, on the opposite side of the Aruwimi. This must be the Itembezi river or Leika, running into the Congo between the Aruwimi and Upoto. Another Zanzibari died to-day, making a total of fifty-six.

Tuesday, February 21st.—The Zanzibaris who accompanied Barttelot and Jameson to Yallasullah returned to-day, with the exception of three; two they took on to the Falls, and one remained behind sick at Yambow. Letter from Barttelot, says rumour at Yallasullah, that Tippu-Tib has got 1000 men for us! Believe it when I see them! Another Zanzibari died, making a total of fifty-seven.

Wednesday, February 22nd.—Bonny had already been obliged to put some of the sick in huts outside the camp for sanitary reasons; two Soudanese were to-day put out also, these were all cared for by men who were told off for this service. Burnt two huts outside camp, which had been run up without permission, as we had found a Remington in one, and discovered it belonged to John Henry, the interpreter. On investigation, found that he was in the habit of scaling the boma at night and sleeping in the hut. He ran away to a neighbouring village, but was brought back and confined in the guard-room for the night.

Thursday, February 23rd.—John Henry received fifty lashes this morning. Nassibu paid us a visit; says there are men on the way down from Yallasullah, gives same rumour about natives seeing white men on steamer.

Friday, February 24th.—A Soudanese died during the night, total fifty-eight. In the afternoon about seventy men from Kasongo arrived; they say eighty more are coming to-morrow; they passed Barttelot and Jameson on the river as the latter were going to the Falls. After eight months' waiting some men have come at last! This, they say, is the first contingent of a lot more coming from Kasongo. If this is the case, there will be no necessity for Barttelot to propose the extra money scheme for men *not* to carry loads.

Saturday, February 25th.—Nassibu paid us a visit with some other Arabs. Some more men from Kasongo arrived this afternoon. Decided to send messengers with a letter to Barttelot to tell him some men have turned up here; I have no proof that he knows it.

Sunday, February 26th.—Ward went over to Nassibu's village. Wrote Barttelot acknowledging the receipt of his of the 18th from Yallasullah, informed him that seventy men arrived on Friday last from Kasongo; others arrived yesterday. The headmen stated that there were a large number at the Falls, who would follow on down here. I told him that we were rejoiced at this good news, which would avoid the necessity of the loads going to the Falls, or of mentioning anything to Tippu-Tib or Salim-bin-Mohammed, regarding the proposition of extra payment for fighting men. I added that the men informed me they had seen him on the river, so I suppose he will be returning. I sent two Zanzibaris with the letters to the Falls this morning, instructing them to go with all speed. Found Barttelot had given instructions to Bonny about buying canoes; but though I was in command of camp during his absence, and always had charge of disbursing mitakos and handkerchiefs for buying articles needed, I had been told nothing of this, and was annoyed at this counter-order, which was not written down. Two more Zanzibaris died to-day, making total sixty.

Tuesday, February 28th.—Had fourteen Soudanese out at target practice—five shots each at two hundred yards, at a board about the size of a man; none of them hit it, but this is not to be wondered at, considering the hard pull of the Remingtons. Nassibu paid us a visit.

Wednesday, February 29th.—Had the remainder of the Soudanese out at target practice. Found the marker yesterday did not signal properly—the target was hit twice. Salim-bin-Mohammed returned from the Falls this afternoon, bringing letter from Barttelot to me.

From Barttelot to Troup.

"*Singatini.*[1]

(Despatched by Salim-bin-Mohammed, *February 24th.*)

"Dear Troup,

"We got here last Monday, and were received almost enthusiastically by Nzige. Two hundred men are coming with Salim-bin-Mohammed, but they are not to move till Tippu-Tib

[1] Singatini, or Zinkatini, Arab name for Stanley Falls.

comes, which event may happen any time within the next fifteen or thirty days. They do not look to me much like carrying loads. I suppose our men turned up all right? When they were told they were not coming on here they grew very sulky. Jameson and I, unless you send for us, intend to await Tippu-Tib's arrival here. Salim-bin-Mohammed says there will be no difficulty in getting 500 or more fighting men in excess of the 600 carriers. No news here of Stanley. Would you kindly have all the brass coil wire made into half loads, as that and beads are, I believe, the only useful stuff up at the Lakes. The little father (Tippu-Tib's uncle before-mentioned) is away, gone to Kasongo with a young bride. We are in the same old house. Assad Farrau is useless, and disgusting to look at. I hope the Soudanese are giving you no trouble. It has rained here every day since our arrival. Nzige does us royally this time—is getting us a milch goat. I can obtain no news of any sort. Remember Jameson and self to you all.

"Ever yours truly,
(Signed) "EDMUND M. BARTTELOT."

(It will be seen he had not then received my letter.)

By his wishing to have the brass and iron wire as well as beads made into half loads, I presume that he does not propose taking any of the other loads up, except ammunition.

Thursday, March 1st.—Had Soudanese out volley firing, general drill. Nassibu paid us a visit. Salim-bin-Mohammed paid me a visit on his return; brings no fresh news; says about 200 men came with him, plenty more coming down with Tippu-Tib from Kasongo, who Salim thinks is on his way from there to Stanley Falls. He says he has heard no news of Stanley. A hundred of the men who came down before Salim have been sent to the Lomami, but will return soon. On my asking him if the men would carry loads, he said "Yes." I observe that he is more reticent than usual. I asked if he were going up with us if we started; he replied that he could say nothing till the return of Tippu-Tib, who would settle everything. The sick Zanzibari left by Barttelot at Yallasullah returned to-day.

Friday, March 2nd.—Heard of outbreak of small-pox among some of Salim's men; arranged with Bonny that he should talk to

Salim about having the sick man removed from his camp, and warned all our men not to go into the Arab's camp.

Saturday, March 3rd.—Salim has removed the cases of small-pox some distance in the bush from us. There are no disinfectants of any kind among our medical stores, except a few drops of carbolic acid.

Monday, March 5th.—Salim-bin-Mohammed paid us a visit. Nassibu also came in, and brought a small goat for the fifty mitakos he received from Barttelot a long time ago. After luncheon Nassibu-bin-Seif and another Arab arrived from Abdallah's advanced camp at Opoy; they paid me a visit along with Salim. Nassibu-bin-Seif brought one of our Remington guns, which his men found in a deserted village near Abdallah's camp. He says that the two Zanzibari deserters at Abdallah's are still very sick, having bad ulcers on their feet, and unable to come down here. They (the two Zanzibaris) have stated to him that they left Stanley with five others in a canoe, at a point where there were big rocks in the river standing high up. The river was very broad there, with many large islands. Stanley could not get his boat through, having to take it overland. These seven Zanzibaris got a canoe and came down river, but were upset on the journey; five of them were captured by the natives and eaten, only these two reaching Abdallah's camp. Nassibu-bin-Seif went on to say that these two Zanzibaris state, that it took them one moon to come from where they deserted Stanley at the large rocks to Abdallah's camp, and that it had taken Stanley five moons to reach the large rocks from here. They say the villages up there are as large as Basoko at the mouth of the Aruwimi; that the natives have shown fight sometimes, but often clear out to the other side of the river on the approach of the white man. Several other Zanzibaris have deserted, but they, as well as plenty of sick men left on the road, have fallen into the hands of the natives. Food was plentiful at the rocks—sugar-cane, plantains, Indian corn, and manioc. Stanley had met some Arabs near the rocks who had come from Kibonge, south of Stanley Falls, to get ivory, which is plentiful also. Nassibu-bin-Seif adds, that the natives round Abdallah's camp have five more Remingtons, and that he will try and get them for us. He states also that the Zanzibaris say that Stanley was wounded by an arrow, and that some of the Arabs found twenty pieces of handkerchief in a native hut near

Abdallah's. As I considered all these statements most important, I took great pains to cross-examine him through the interpreter, but he in no way altered his story. Knowing the approximate date of the arrival of the Zanzibaris at Abdallah's, we calculated the time, and found that their six moons tallied almost exactly with the amount of time that elapsed between Stanley's departure and their return to Abdallah's. This fact made us give their story more credence. Ward and Bonny agreed with me that a messenger should be despatched at once to Barttelot with a letter containing all these details.

Tuesday, March 6th.—Wrote Barttelot, sending him all the news obtained yesterday; as Nassibu-bin-Seif is going himself, this will give Barttelot an opportunity of examining the Arab for himself and comparing the statements.

Wednesday, March 7th.—Another Soudanese died this morning, making total sixty-one. Salim-bin-Mohammed and Nassibu-bin-Seif paid us a visit; the latter does not go to the Falls till to-morrow. My letter to Barttelot, Salim has sent by one of his own men. I sent an additional letter to Barttelot by Nassibu-bin-Seif.

Thursday, March 8th.—Another Soudanese died, total sixty-two. Had all the loads of beads out, and sampled them. Salim telling me in what districts the different kinds would pass as currency, and mentioning their value as compared with mitakos and handkerchiefs.

Friday, March 9th.—Another Zanzibari reported dead this morning, making total sixty-three.

Saturday, March 10th.—Another Zanzibari died, total sixty-four. Two Zanzibaris that I sent to the Falls thirteen days ago with a letter to Barttelot, returned this afternoon, bringing an answer.

From Major Barttelot to Mr. J. R. Troup, at Yambuya Camp.

"*Singatini.*

(Despatched by Feragi.)

"*March 5th*, 1888.

"Dear Troup,

"We got your letter on the 2nd March, at midnight, and we both thought it contained news of Stanley at last, or a mutiny in the camp, or some other very serious matter; but we were agreeably disappointed at finding nothing of importance in it. I am afraid the arrival of seventy men, and their word that more are coming, is not of sufficient importance to prevent the necessity of carrying out the original plan of seeing Tippu-Tib, for which purpose Jameson is on the point of proceeding to Kasongo. I am sorry to hear the death-rate continues high. Whatever men have arrived here are, I am given to understand, placed entirely under the command of Salim-bin-Mohammed, and are not to be utilized by us till we hear to that effect from Tippu-Tib himself. It will be best therefore for you to have nothing to do with them till you receive definite instructions from me on the point. I have had some trouble and been put to some expense concerning the departure of our messengers, so please do not send any letters except of real importance, such as news of Stanley, or any trouble in the camp. I may be back any day after Jameson's departure, or on the other hand I may require Ward's presence here. That man Suedi Wadi Barosi was sick when he left Yallasullah. Our treatment here since our arrival has been that of favoured guests, and Nzige has done everything in his power for us, as regards the Expedition and our own personal comfort and amusement. I hope you are all well. Jameson and I unite in messages to all. Should I require Ward I will send for him.

"Yours ever,
(Signed) "EDMUND M. BARTTELOT.

"P.S.—There are no men from Kasongo here, nor have we seen any since our arrival here, though we are told some are coming, but numbers and purpose uncertain. Should a fresh issue of tea be required, please issue it from that open case of tea for two months, dating from February 29th.—E. M. B."

[A comparison of the two letters, dated respectively February 24th and March 5th, will show that his statements were slightly confusing. In the former he speaks of two hundred men he has seen—in this he refers to seventy, who had gone on to Yambuya, and that he had seen no others. He dwells on the unimportance of the information I had sent him about the carriers who had come in, but I felt bound to tell him of their arrival, as I had no proof that he knew they were on their way.]

Monday, March 12th.—Some of Salim's men came back from down river; they say they captured Ngungu—our Ngungu—and will bring him in to-morrow. They say also they have killed four chiefs down below.

Tuesday, March 13th.—Salim's men brought Ngungu in a canoe from down river, where they had captured him. He was warmly received by our men, and shook hands with us all round. I was very glad to see him again, and wished it were under more happy circumstances. I have taken a fancy to him, and hope he won't be badly treated. He went up to Salim's camp, and afterwards went down river in charge of some of Salim's men, returning again. I believe Salim insists on his coming with his people to live on this side.

Wednesday, March 14th.—Saw four native canoes fishing near the rapids, so I suppose peace has been concluded between Ngungu and Salim-bin-Mohammed. Ngungu came into our camp during the day, and I made him a present of two mitakos. I forgot to mention that Ward, who was a very good linguist, and already had a fair knowledge of Swahili (Zanzibari language), had now improved every opportunity to increase his acquaintance with the language. Salim-bin-Mohammed had often paid him visits, giving him instruction, so that now Ward was able to assist us on occasion as interpreter. It will be seen by the following how his knowledge proved of great service to us. Salim this afternoon, on one of his usual friendly visits to Ward, complained about our Zanzibaris constantly interfering with the women of his Manyuemas. Ward thinking this a matter of importance sent over to my hut for me, and repeated in English the gist of Salim's remarks—that the Manyuemas had complained to him about this, and threatened to knife or shoot our men if they caught them enticing their women away. Salim thought it right to tell us, so as to avoid any ill-feeling between his men and ours. I promised Salim that all the Zanzibaris would be cautioned at afternoon parade not to interfere in any way with the Manyuema women. I told him that if after this warning they did so, the best thing for the Manyuemas to do was to tie them up and bring them to us, and we should punish them severely. Salim seemed quite satisfied with this arrangement, adding that one of his men had gone out with a gun to catch a Zanzibari, whom he suspected of taking his woman away from him. Whilst at luncheon we were informed that the Manyuema in question had caught Muni Chandi, one of our Zanzibari niamparas, in company with his woman, and had tied him up. After luncheon he was brought in by the Manyuemas, and we held a palaver over the business. The Manyuema gave a detailed description of his finding the Zanzibari and woman together; he seized Muni Chandi, and took his knife, shirt, and cap. The Zanzibari begged him not to take him before us, and offered to pay some cloth if he would not do so. Muni Chandi did not deny the accusations, so we ordered him to pay the man the cloth he promised so as to settle the palaver. He went and brought a doti of white cloth (a doti is about a fathom of cloth, and is measured by the distance of the extremes of outstretched arms). I should mention here that most of the men had sold all their garments to buy food with, but this man, who was a head-man, being a great dandy, had managed to have a *wardrobe*, though not very extensive. The cloth was handed to the Manyuema, but he was not satisfied, wanting more. Muni Chandi declaring he had no more to give, we told the Manyuema that he must be satisfied, as he had also the knife and shirt. A long palaver took place between the Manyuema and his companions, and they eventually decided to take what had been tendered, and to all appearance the palaver seemed amicably settled. But shortly afterwards the Manyuema man returned with another, each with a gun, and after an altercation with a Soudanese sentry at the east gate got into the camp. Bonny called me from my house, and on my going out I found the Manyuema man, who had seized Muni Chandi, sitting on the wooden drum. On asking him what he wanted, he got up in an excited manner, and gave me to understand that he had come down to shoot Muni Chandi. I replied that the palaver had been settled, and ordered him out of the camp, along with the other, and told the Soudanese sentry not to allow him in again. They went away, only to try and come in again, but were not allowed, and after a noisy harangue outside they were persuaded to go away by Omaha, the Soudanese officer. Ward went up to Salim-bin-Mohammed to tell him that after settling the palaver, evidently to the satisfaction of all parties, these two Manyuema men had come into our camp with guns, threatening to shoot one of our men, which was a gross outrage, and that if they tried this kind of thing again the Soudanese sentry would enforce the standing order, which had always existed since Salim came here, that none of his men should be allowed to come into our camp until permission had been granted by the officer in charge. If his men interfered with our sentry, they must put up with the consequence, as the sentry must obey his orders. Salim, Ward says, told the Manyuemas all this, and though they were very excited, seemed to let the matter drop. At the afternoon parade all the Zanzibaris were cautioned about interfering with the Manyuema women, and were distinctly told that if they did, the Manyuemas would tie them up and bring them in to us, when they would be severely punished. Muni Chandi was put in the guard-room, not so much as a punishment, but to ensure his safety until this matter blew over. We decided, under the circumstances of his having paid cloth, and being deprived of a portion of his wearing apparel, that he had received sufficient punishment so we only severely reprimanded him. It might have been a nasty business. The Manyuemas are as much to blame as our men. They must have known about Muni Chandi and others going with their women for some time, but now they get up this palaver for the purpose of obtaining cloth. But, of course, it would never do to allow this kind of thing to go on, after it had been brought to our notice. I think after this the Zanzibaris will take this lesson to heart; this event has scared them. Salim behaved well in telling us about it, as we never could allow any row between his men and ours. It is to our interest to have a good feeling existing between us and these men who were living at our gates.

Thursday, March 15th.—Just before luncheon some Manyuemas came in with two goats for sale. Bonny gave two pieces of handkerchief for each; the men returned during luncheon to exchange the pieces of handkerchief as they stated they were soiled. Bonny changed them. They went off, but shortly after returned again saying two pieces were not enough for the goats, so the animals were handed back to them when they gave up the handkerchiefs. Then they came back again and said they would take the two pieces after all. Shortly after this there was a disturbance with some men wishing to come through the gate without leave. One of them tried to push roughly past the sentry, knocking the gun from his hand. Fearing there might be a repetition of yesterday's proceedings, sent Ward up to Salim to inform him of the occurrence, and to state that he must put a stop at once to this troublesome conduct of his men by punishing them, or there might be serious consequences. He sent word down to say he would inqu re into the matter. Later on he sent some Manyuemas down with one of our headmen with a message that if we could identify the culprit we were to punish him as we liked. The sentry picked out the man, and I sent him to Salim with a message that it was his duty to inflict the punishment; but he was again sent back, and we were told to do what we liked with him. I told the man that he had been handed over to us for punishment by his chief, and I asked him for some explanation of his conduct. He said he was coming about the goat palaver again. It was explained to him that we had no objection to his, or any of the others coming into our camp, but that the sentry was there to challenge all strangers, and must get the permission of the officer on duty before he could allow any one to pass. The sentry must obey his command, and if he interfered serious consequences might happen. We told him we wished to keep up the good feeling between his countrymen and our men, and hoped there would be no more misunderstandings, that though Salim wished us to punish him, we would not do so, as perhaps he did not understand the Soudanese sentry, and therefore we would send him back to Salim with this admonition and explanation for his and all the Manyuemas' future guidance. The man was much scared, and seemed much surprised at our clemency, as he no doubt expected severe punishment. We have decided to place a Zanzibari beside the Soudanese sentry at the gate, as the latter does not understand the language of the Manyuemas, and this circumstance may cause unnecessary friction. The Zanzibari will act as a sort of interpreter, and will also come down to inform us of the application for admission. This seems the more necessary because there are so many new men in the Arab camp who do not understand the reason why the Soudanese stops them. I cannot help remarking that it is absurd for Salim to send us his men and to expect us to punish them. I need hardly say I did not fall into the error or trap of taking upon myself the responsibility of correcting his men. In the first place, it is not the right method of procedure; and in the second, it would only be heaping coals on the fire, intensifying any ill-feeling, and end in open rupture. It is these small sparks which are so easily fanned into a big flame, and might lead to a general conflagration. After these outbursts of the last few days we have deemed it advisable to keep a sharp look-out over our men, as well as to take a few additional precautions for our safety.

Friday, March 16th.—Another Zanzibari died this morning, making total sixty-five.

Tuesday, March 20th.—Found roof of powder-house letting in water, and put men on to-day to re-thatch it. Asked Salim to send one man to the Lomami for goats, as we are getting short. An Arab came in from down river, having palaver with natives.

Wednesday, March 21st.—Salim informs us to-day that one of his men returned from the Falls last night, bringing news that a large body of men were on the eve of arriving at the Falls from Kasongo, but that Tippu-Tib was not coming with them. Jameson had not left for Kasongo when this man started, namely, the 27th of last moon, or 8th March. Salim says that the Arabs whom Stanley was supposed to have met at the *Rocks*, came from Kibonge, only a little higher up the Congo than Stanley Falls, and would only take nine days to go from there in canoes to the *Rocks* [?]; while Stanley, by the Zanzibari account, took five moons to reach the *Rocks!* [I give this for what it is worth. There is a river going north-east from near Kibonge.] Another Zanzibari died to-day, making sixty-six.

Saturday March 24th.—Another Zanzibari died to-day, making sixty-seven. About 4 o'clock Barttelot returned, looking very ill; he seemed to have lost flesh. He tells me that both he and Jameson have been very sick off and on at the Falls, and says he had very bad food, and even went so far as to mention suspicion that the Arabs had tried to poison them! Jameson went off to Kasongo, some little time before Barttelot left the Falls, to hurry up Tippu-Tib with the carriers. Barttelot is awfully disappointed, like all of us, that Tippu-Tib has not returned. He has come back with a new project. Jameson and he have evidently talked the matter over, and he has come to the conclusion, without consulting us, to send a white man down to the coast for the purpose of cabling to Mr. Mackinnon, Chairman of the Emin Pasha Relief Committee, to inform him of the state of affairs and what he has done respecting the mission of Jameson to Kasongo, to offer Tippu-Tib more money for carriers and fighting men. Barttelot has got one canoe from the Arabs, and made arrangements for getting another to take the white man, with some men, to Bangala. He has selected Ward for this mission. It is a long way to send for very little benefit to the Expedition. On the contrary, it means the temporary, and perhaps permanent, loss of thirty of our men, who are to accompany Ward. However, Barttelot has made up his mind about it. I made my report to him concerning the affairs of the camp during his absence.

Sunday, March 25th.—Barttelot is very ill; after luncheon he had to turn in, with strong symptoms of fever. I went in to see him; he seemed much upset. He is very anxious to go with Ward as far as the village on the Congo, from whence he will start; but I have persuaded him that he is too ill to undertake the march, and suggested that I should go instead, and he agreed to this. Ward is busy making preparations for his journey.

Monday, March 26th.—Barttelot slightly better, insisted on getting up; he was very excited, and it did him no good. Later in the day he spoke in the course of conversation about the necessity of leaving a white officer at Stanley Falls, in the event of there not being sufficient carriers provided to take up all the loads, and he did not wish to leave Bonny. Bearing in mind that I had been informed by Bonny that Barttelot wished to leave me, but hardly liked to ask it, I had thought the matter over from the point of view of the benefit of the Expedition, and I came to the conclusion that it would be best if I should offer to remain. Of course I knew by Barttelot's remarks that, counting Bonny out, I was the only one he had to leave, so I told him I would stay in charge of the loads. Barttelot informed me that I had taken a great load off his mind; that he had wished me to remain, but disliked to disappoint me. I helped Barttelot during the day, making out drafts of letters to M. Vankerckhoven, of the State station at Bangala; Captain Vangèle, State Commissioner for the Upper Congo; Lieutenant Liebrechts, Chief of Leopoldville; and M. Camille Jannsen, Administrator of the Congo Free State at Boma. In these he informed the gentlemen named the object of Ward's mission, and asked their assistance to expedite him in every way. At Barttelot's request I wrote also personal letters to all those

State officials who were well known to me, urging on them the importance of helping us in this matter. Barttelot read me over the telegram he was sending to Mr. Mackinnon. It was a *précis* of events; he told him of Jameson's mission to Kasongo, when he was expected back, and that he (Barttelot) should start with some loads should the carriers turn up in June. He also asked them to arrange with the State to have a steamer at Leopoldville to bring Ward back with the reply. He also read me his letter to Mr. Mackinnon with details supplementing the telegram. He gave Ward all his instructions. He was to take thirty Zanzibaris and five Soudanese in the canoes as far as Bangala—a distance of 500 miles down the Congo; these men were to be left at this point, and the letter he had to M. Vankerckhoven, chief of that station, contained an urgent request that if he had his small steamer, to send them back to us without delay. Ward from Bangala would get a relay of men from the State station to paddle him down the other 500 miles to Leopoldville; he would then proceed overland to Matadi, and thence by steamers on the Lower Congo to the mouth of the river at Banana. He would then proceed either to San Thomé or St. Paul de Loanda, according as the mail steamer for one or other arrived earliest—as these were the nearest stations from which a cable message could be sent to Europe. He was to wait for a reply, and on receiving the same was to return with all despatch to Yambuya camp. In the event of the arrival of the carriers, and consequent departure of the Rear Column, before Ward's return, Barttelot promised to make arrangements with Tippu-Tib to provide Ward with men to follow on by quick marches and overtake the column.

Tuesday, March 27th.—Gave Ward by Barttelot's orders handkerchiefs, mitakos, and cowries for purchasing food for himself and men on the way down, also one sack of rice and eight cases of Remington ammunition. Ward was allowed to pick out twenty-five Zanzibaris, and Barttelot handed over to him five of the Soudanese. Most of us took the opportunity to send letters home by Ward. Barttelot asked Ward to bring up on his return some cases of tinned meats, champagne for medical comforts, and other like necessary articles. I was with Barttelot when he gave these final instructions. Ward asked for reassurance that men would be provided him in case of the departure of the column

NATIVE OF YANGAMBI

before he returned, and Barttelot told him that this other matter would be satisfactorily arranged. Ward and I start to-morrow; I am going with Ward as far as the Lomami, for the double purpose of helping him in fixing his canoes, and to endeavour to purchase goats there for the camp. Barttelot wishes me to return as soon as possible, as he has arranged with Ngize, Tippu-Tib's brother at Stanley Falls, to send two hundred loads to that station on the first of next month, and I have to go on there with them. All our arrangements were now complete for an early start on the morrow.

Wednesday, March 28th.—Wet morning, heavy rain; could not get away till late. It was very difficult to get the poor weak men started carrying the loads that we were obliged to take; and owing to the slipperiness of the path through the jungle they were constantly falling and dropping their loads. They were too weak to replace these themselves on their heads, and we were continually obliged to give them help in this matter. Ward and I walked behind them for this purpose. On account of our slow progress we were unable to reach the usual camping-ground at the village of Yambow. The Soudanese with some of the men with the lighter loads got on to the village, but we could not leave the stragglers, or we should never have got them on at all, so we camped in the jungle.

Thursday, March 29th.—Started early, camped at the far village. We were here obliged to lighten the ammunition loads by taking out half and tying them up in handkerchiefs, &c., and distributing these among the other men. Ward, in order to brace them up, gave them some rice out of the sack he had, as well as some Indian corn we were able to buy, so they had a good feed—the first for many a day.

Friday, March 30th.—Started early, reached the village of Yangambi on the Congo about sunset. From the hill behind it we had a glorious view of the river, with all its different islands, while almost at our feet rose the smoke of the village. Yangambi is a little further down the Congo than Yallasullah, where we formerly struck the river, and is under the control of the Arabs, like all the villages in these parts. All our men did not come in.

Saturday, March 31st.—Found Seyid had not sent the second canoe from Yallasullah, so sent messenger to say we had arrived, and to tell him to come down without delay. He sent word back, saying he would be down to-morrow morning early with canoes and paddles. The canoe Barttelot got is here. Ward has bought a thatched roof for it from the natives.

Easter Day. Sunday, April 1st.—Seyid did not turn up till noon, bringing canoe and paddles, and one goat—the latter I bought—and we started shortly afterwards. We had purposed to lash the two canoes together before starting, but as Seyid was so late we decided to postpone doing this till we reached the Lomami. Ward put all his loads into the canoe with the thatched roof; we both went down in this with some of the men, while the other followed with the remaining men and my loads. Never saw such a start. The Zanzibaris cannot paddle a bit, not able even to keep the canoe straight, so we went round and round, sidling down with the current. To make matters worse, shortly after we started we were caught in a thunderstorm. The rain came down in torrents, and the Congo was soon lashed into a tempestuous sea, making progress dangerous. I wonder we did not capsize. As it was, we were obliged to pull up until the storm somewhat abated. The other canoe, having no sheltering awning, worse luck! was filled with water, damaging my few things. The men had to get out and bale it to keep from getting swamped. We reached the Lomami at its confluence with the Congo about sunset. There is an Arab settlement under Raschid, Tippu-Tib's nephew, with a large native population, who mostly live on the river in their canoes. We were well received by Raschid, who provided me with a small hut, Ward preferring to test the sleeping accommodation of his covered canoe. Raschid promised to have the two canoes properly lashed together for Ward, and also gave him a covering for the other canoe. He entertained us in his own house to a most sumptuous repast that evening, with curried fish, rice, fowl, plantain, maize, and coffee. One was struck with the perfect manners of this well-bred Arab. Ward, being able to speak Swahili pretty fluently, had a long palaver with him, and Raschid promised to do everything in his power to assist him on his journey, and assured me he would get me some goats to take back to the camp. In this way the evening passed very pleasantly. He also enabled us to procure food for our men. Ward was naturally desirous of improving their condition, and generally cheering them before starting on their arduous journey.

Monday, April 2nd.—Raschid said it would take nearly all day to get the canoes properly lashed and fixed, and asked Ward to stop till to-morrow, so there was nothing to be done but to wait for the accomplishment of this; it was perhaps for the best, as it gave the men a chance to rest and get another good feed. In the afternoon we went to see the canoes, and found that Raschid's carpenter had made a very good job of them, having lashed them together and patched up the one which leaked, and made a neat flooring for it. While we were looking at them a canoe arrived bringing a letter from Barttelot to Ward. We were still Raschid's guests, and he entertained us well; anything we asked for was placed at our service. Ward and I had a long talk together that night regarding his arduous journey and the causes for it. We were both of the opinion that it would not be much benefit to the Expedition, only causing a deal of trouble and expense, because the Committee could only reply, "Carry out Stanley's instructions." We discussed the subject of the journey down nearly 1000 miles to Leopoldville by canoes, unless he was lucky enough to catch a steamer at Bangala; but the most dangerous part would be in the first portion of the voyage, as between here and Bangala there were a number of hostile tribes, especially in the Monogiri Channel, where Deane had been attacked before. There was an added danger in this canoe trip, because the men did not know anything about paddling, and panic would surely ensue if they were surrounded by hostile natives. However, Ward would have thought less of this had he been going on a mission that was likely to be more beneficial to the Expedition.

Tuesday, April 3rd.—Everything being in readiness, Ward started about 8 o'clock this morning. I saw him off, wishing him good luck and a safe journey down as well as a speedy return. The canoes I was glad to see went much steadier lashed together, which cheered me a good deal, as I was much concerned because the Zanzibaris were such bad paddlers. There is an Arab post further down before he reaches the mouth of the Aruwimi, and Raschid has given him words of commendation to the headman there. Raschid asked me to remain as his guest for a few days, as he is obliged to go out to fight some natives up the Lomami who have eaten two of his men, and he cannot consequently get me the goats till his return. This I consented to do, sending a letter on the subject back to Barttelot by the messengers who brought the letter to Ward, and who had not yet returned. I forgot to mention that I had seven Zanzibaris and one interpreter, Bartholomew, with me. Raschid put me in the way of getting plenty of food for them. Poor devils, they were not averse to stopping here, under such agreeable conditions, for an indefinite period, so we settled down as comfortably as we could to await the arrival of goats.

CHAPTER XIV.

RASCHID AND HIS HOSPITALITY.

Though the accounts given by the two sets of deserters from the advanced column did not seem exactly worthy of credence, it is rather remarkable that their story after all tallied in so many details with the course of events, since described by Mr. Stanley. There were two sets, as I have said, of these men; the first, those we met at Yallasullah, five out of nine who had deserted Stanley, told us of the scarcity of food, of desertions, of fights with the natives, the wounding of a white man, and that nearly all the Soudanese had died, all of which points Stanley mentions.

The other party originally consisted of seven deserters, five of whom were captured by natives, and two only got as far as Abdallah's camp. Their story seemed to us of more importance. Nassibu bin Seif, an Arab sheik, who had seen them, repeated to me what they had said. They had left Stanley at a place where there were big rocks in the river, and that Stanley was here obliged to carry his boat overland. They had been with Stanley for five moons, and it had taken them one moon to come down the river in a canoe as far as Abdallah's camp, where they arrived in December. This time agreed as nearly as we could calculate with the number of months Stanley had been absent up to that date. They said, according to the Arab's account, that Stanley had been wounded; but it is probable that the deserters had only mentioned the wounding of a white man, the same one doubtless that, after the description given by the Yallasullah deserters, we had guessed was Nelson, but from Stanley's letters we now know that Stairs was the victim. These deserters spoke of there being plenty of food when they left Stanley, which, according to their five moons, would have been in November. Stanley in his letter says that after passing through a famine-stricken country, he reached Ibwiri on November 12th, and found abundance there.

According to the Zanzibaris, it was before they reached "the Rocks" that Stanley met the Arabs. From Stanley's letters we learn he met Ugarrowwa's Manyuema men on August 31st, and reached a point opposite Ugarrowwa's camp on September 16th, and did not emerge from the Arab district until November. The deserters said the Arabs were from Kibonge. This place is eight days' canoe voyage above Stanley Falls on the Congo, and is a large village occupied by one of Tippu-Tib's sheiks with Manyuema followers under him. It was from near Kibonge, we were told by Salim-bin-Mohammed, that the Arabs started in canoes to go to "the Rocks" on the Aruwimi. It should be noticed, that by Stanley's map the Aruwimi reaches its most southern point not far from the river Lendar. Near Kibonge there is a river coming down in a south-westerly direction, emptying into the Congo. It seems possible that by going up this river, called on Justus Perthe's map *Munduku Lilu*, and by making some short overland journey, the Arabs might reach the Lendar and descend it to the Aruwimi, into which river it empties in the very district said by Stanley to be under Arab control.

It must be remembered that none of us white men had ever seen this second party of deserters, who had been stopped at Abdallah's camp. This Arab Abdallah had followed in Stanley's track as far as a village called Opoy, which we were told was about a month's journey from Yambuya. I have mentioned in my diary that a visit to this place for the purpose of verifying these rumours was proposed by Barttelot, but for certain reasons was never carried out. It would have been well if we had seen these two deserters, as they might have given us some clues that would have proved serviceable, and we might have gained further information while at Opoy. At all events, by going to that place we would have at least been able to have familiarized ourselves with that part of the country first traversed by Stanley, and through which we should have to pass when we set out to follow him.

I have mentioned in the diary that after Barttelot told me his intention was to leave a white officer at the Falls with the loads, I volunteered to undertake this task. One reason given was, that as Ward would be away, and as Barttelot refused to put Bonny in charge, of course Jameson would not be asked to stay, so I was the only one to do it. I heard that Barttelot did not relish the prospect of asking me to accept this post, especially as he knew that I had felt it rather hard that on so many occasions I had been put in charge of difficult tasks which prevented me from going on with the main body of the Expedition, and he believed I should be much disappointed if I was again left behind. But as I knew that Barttelot would leave me anyway, I decided to make the best of a bad matter and gracefully accept the inevitable; moreover, there were other reasons besides this that influenced me in this decision. To begin with, I did not approve of Barttelot going without the bulk of the loads, neither did I think it right of him to leave them at Stanley Falls, which was not in my opinion a politic move; but as there was no moving him from his decision, I saw it was wise that, as the steamer had not arrived with the white officers for that station, a white man should be left in charge, and I must candidly confess that I believed that the loads left with me at Stanley Falls would have a far greater chance of reaching their destination at Wadelai than would any of those taken up by Barttelot. I fully expected that Stanley would turn up, and in that event he could easily get the loads at Stanley Falls, while I also firmly believed that the men going up with Barttelot would meet with some disaster, and that their loads would be lost. Added to all this there was another reason why I should be selected for this task, for I must admit that I had never quite recovered from my first serious illness in November; the great difficulty experienced was, that through the lack of any nourishing food and proper tonics, my system had been greatly reduced, and I had been unable to get back my strength under these circumstances. Yet I should not have hesitated at any moment to have tested my powers of endurance if I had been called upon to advance to assist my leader, or to follow his commands about the taking up of the loads. Hence it was for these combined reasons that I accepted the situation.

I should mention here that Barttelot, after his return from Stanley Falls, where he had been to see Jameson off to Kasongo, seemed much upset, and distressed in mind.

It will be noted that the decision to send a white officer to the coast was made by Barttelot and Jameson while they were away at the Falls, and that no other member of the Expedition was taken into their counsel in this matter, although this was a step of importance, and according to Stanley's instructions none such should be taken without consulting Ward and myself. That our opinions were not asked is shown by the fact that Barttelot engaged canoes and made all necessary arrangements before his return to Yambuya. When he did mention it to us it was as a fact that he intended should be accomplished, and he did not consider our approbation necessary. Some of us felt convinced that the project of cabling to the Committee asking what was to be done was unnecessary. There could be but one answer to his telegram, and that was, as I have said, "Follow out Stanley's instructions." It cannot be denied that Ward's journey would be, to a considerable extent, a perilous one, and it was asking a good deal to tell him to make it. It was my opinion that if a white man must be sent to the coast, Ward was the best one who could be selected for the undertaking. He knew Bangala well, and if he once arrived there he would be able to obtain men easily to take him on. He was well known on the lower river, and could thus make the best arrangements for a rapid journey down. He had been very ill, but by this time he had picked up wonderfully, and looked in better health than any of the rest of us. The men we had taken with us we had selected ourselves, and it will be seen from the diary how we both attempted to improve their condition in order to brace them up for their further journey. In regard to feeding them, we found in Raschid a ready helper, and we were able to manage fairly well for their comfort. Raschid indeed was most obliging in every way; his arrangement of the canoes took a weight of anxiety off my mind, for I could see Ward started under fair auspices.

For myself, Raschid made most satisfactory arrangements. He apologized that the hut placed at my disposal was so miserable; it was a small mud erection containing only two rooms. But while he was in residence I always dined with him, and after his departure my meals were served the same in his house, and my comfort was attended to by young Seyid, one of his Arab followers, and by Raschid's chief wife, who, it must be admitted, was by no means charming in my eyes. She was evidently of Zanzibari extraction, and I doubt if she could ever have been considered a beauty. As for Raschid, however, he was the handsomest Arab I ever saw; a tall, well-built young man, a thoroughbred Arab, of light complexion scarcely more than olive, with sparkling eyes, a well-formed nose, and small mouth, altogether very delicate features. He carried himself well, with a self-reliant pose and an air of command. I was at once struck with the awe which he inspired among his subordinates, who looked upon him as a veritable chief and commander. If all the stories we heard were to be

CANOES ON THE BEACH AT THE LOMANI RIVER.

believed, he must be very strict in demanding obedience, and was ready to inflict most arbitrary punishments. Yet with it all he was very quiet, and had a self-possessed manner. He was always robed in spotless white, wearing the usual long, flowing Arab garment with a pure white turban that set off his fine face well. The settlement he had made at this point, the junction of the Lomami with the Congo, had a permanent air about it; he was evidently here for the purpose of thoroughly establishing the Arab authority over the natives in the surrounding country. His present palaver was with the people up the Lomami, who had captured and eaten some of his men, and it was his intention to set about subjugating them. Raschid had with him 200 Manyuema followers and some 600 natives, whom he had previously conquered, and who were going with him to fight their fellow-natives. No doubt some tribal feud had been judiciously fanned into a blaze, and he had had little difficulty in persuading these naturally bloodthirsty natives to join in the fray. It was a fine sight to see the "gathering of the clans," and it took some time to get them all on hand. They came down in huge canoes capable of holding fifty men, the sterns of some of them ornamented with green boughs. The river re-echoed with their war-songs as they came in. They passed like a naval review in front of Raschid's palaver house that faced the river, two or three canoes abreast, and then turning came up to the landing-place. All the time the natives in them shouted and sang, executing marvellous war-dances as they paddled along, till in their wild antics one wondered why the canoes did not upset. They were all in their war-paint, their faces ghastly with chalk and their limbs streaked with the same material, making them look like a company of hideous clowns. They all wore native knives slung round their necks. These weapons are beautifully made, and many of the sheaths are elaborately cut and ornamented. Some of the knives are sickle-shaped, with projecting points of most curious design; these are thrown boomerang-fashion, and inflict dangerous wounds. The chiefs were distinguished by their red feathers, and the whole scene had a decidedly warlike aspect; everybody was intensely excited, having worked themselves up to the highest pitch. All night long recruiting parties gathered at the scene of action, and by early morning the attacking party made ready for the fight. The natives in some cases were too thickly crowded in their canoes, so Raschid's own followers sorted them out into others, while the sheik stood quietly watching the embarkation of his forces. The natives and the Manyuemas were started off by degrees, and when all was in readiness he stepped into his own canoe and set off with them. The natives kept up the excitement with their war-songs, while in almost every canoe was a small wooden drum, with some skilled performer who squatted beside it playing away for dear life, while the paddlers of the canoes executed their most difficult can-can steps and shouted at the tops of their voices, making a tremendous din. Before Raschid left he very politely came up to bid me good-bye, where I stood on the shore watching the scene. I saw them all get away, and waited until the last canoe-load rounded the point on its way up the Lomami, then turning I went back to the settlement. Raschid's house, where I spent most of my time, was a large mud building, well built, and well arranged, far superior to any of those I had seen at Stanley Falls. It was surrounded by the usual high bamboo palisade. On entering it one found oneself in a large antechamber, with a bedroom leading out on the left. The place was well furnished. The most interesting ornament of the outer room was a mirror about two feet square, which exercised a strange fascination upon the native imagination; it was usually covered with a cloth of some kind; the reason of this I was not long in discovering. The day before the natives had swarmed up to the house, looking in at the door and window of the bedroom where I was seated, trying to get a glimpse of me, for no doubt very few, if any of them, had ever seen a white man before. It was quite amusing to see them pushing and fighting to get near the window, bringing up their women in some instances. I think I know now how the hairy man and the fattest woman on earth, who act as side-shows at the circus, must feel when they are the cynosure of neighbouring eyes. These spectators, however, were most dramatic in their admiration, gazing hard, lifting their hands in surprise, then starting back in a sort of awe-struck manner. Some of them were audacious enough to push into the room. So it went on, till they became so troublesome and numerous that the shutters had to be closed in order to get rid of them. Meanwhile those who were in the outer room were revelling in the sight of the magic mirror; they believed it to be a species of white man's fetish, and regarded it with a superstitious fear. They went through strange motions before it, and were deeply puzzled by the result; but at last as there seemed no prospect of their departure, the curtain was drawn over the mirror. One man, perhaps a chief or favourite, was treated to a private view, and I amused myself for some time in watching his antics. He approached it, the image advanced; he lifted his arm, so did the man in the mirror; he moved his hand, and was surprised to see the reflection do the same; then he touched different parts of his body, his eyes dilated, and he put up his hand to his mouth with the usual native motion of surprise. He even went so far as to talk to see if the image would do the same. His face all the time was a perfect study. He was completely nonplussed, and could not convince himself that he was all right.

In the bedroom was an enormous four-post bedstead, evidently the handiwork of Raschid's excellent carpenter; it was hung round with gorgeous thin red curtains, a sort of mosquito-netting. Here, too, was Deane's camp-table, and the Stanley Falls station bell, with a few other relics of that place. My meals were always brought in to me here by Raschid's wife, who announced that all was ready, and then stood by to see me eat. The *menu* consisted of the usual curried rice, fish, fowl, and coffee, and the cooking deteriorated during Raschid's absence. All the women were most gorgeously arrayed, to show their high rank as wives of a chief.

The market-day at this place came round on April 7th, and I found some amusement in watching the canoes that began to come in the night before, in order to be on hand early in the morning; there was a brisk business done, with the usual roar of voices. I tried in vain on this occasion to buy goats. Seyid told me he expected

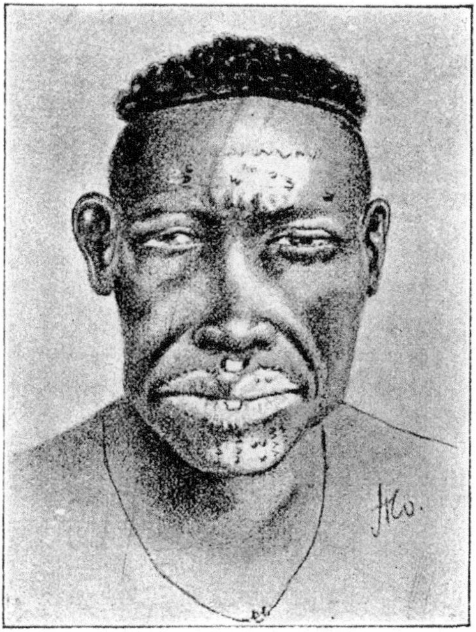

NATIVE OF LOMAMI

some at the market, but none came in; so he instructed the natives who attended to send in some as soon as possible. These market people displayed the usual interest in the "only living specimen," but I was getting used to being stared at.

During the day one of Raschid's followers was sent back badly wounded; in fact he was terribly cut up, and I was called in to doctor him. Of course I could do little for him, as I had nothing; but in searching for some remedy, one of Raschid's people produced from among the articles taken from Stanley Falls a good-sized bottle, which on investigation I found contained carbolic acid, and I was able to persuade them to give me some of the precious fluid.

Raschid had sent a letter to Seyid, asking him to tell me, with his salaams, that he could not be back before three days had elapsed, and I was requested to wait until he returned. This delay was annoying, as I felt that I ought to be back in camp, for Barttelot was expecting me there to help him about the loads. However, it could not be avoided, so I had to wait for Raschid, or at least until I could get the promised goats; therefore I stayed on until Raschid came back on April 9th, having had apparently a "glorious victory!" He was as usual most courteous to me, describing his recent experiences, and even drawing out on a bit of paper his plan of attack upon the native village up the Lomami, and entered into a long explanation of it. The interpreter was never allowed to be present at these interviews in the house, but I managed to understand Raschid's Swahili. He at one time also gave me a full description of his attack on Stanley Falls station, and assured me that if Deane had only had more Houssas, they would have continued fighting *till this day*, because the Manyuemas were very much afraid of these State soldiers. He told me a good deal about the Manyuemas and Zanzibaris. Of the two races he said, "The Manyuemas are good soldiers but never do as carriers, while the Zanzibaris are good carriers but can not fight." This latter part of his remark was entirely confirmed by my experience of the Zanzibaris, for I was well aware that they were not soldiers. Few of them knew how to hold a gun, and for this reason the Congo Free State employ Houssas as soldiers, and we never relied, to any extent, on the Zanzibaris in case we were attacked. I found, however, that Raschid, with all his shrewdness, could not distinguish between the Soudanese and Houssas; he expressed his surprise that we had so few soldiers.

It may not be very clear why Ward found it necessary to go to the Lomami, for it naturally would strike one on looking at the map that our overland journey to Yangambi might have been avoided had we gone down the Aruwimi, but this was out of the question, because at the junction of the Congo and the Aruwimi there was a village called Basoko, which was inhabited by very warlike natives, who had never been subdued by the Arabs, so that the latter did not consider it safe to expose themselves to attack by passing this point. Had our two canoes, rowed unsteadily by the Zanzibaris, gone down the river, we should have fallen an easy prey to the natives, who are skilful paddlers; hence, to avoid any risk, we were obliged to make the tedious overland journey. In passing down the Congo, which near the confluence of the Aruwimi was full of channels, Ward would hug closely the opposite shore, and thus escape the danger of running the gauntlet of the savage tribe at Basoko.

During my stay with Raschid I found his conversation very interesting, and both I and my men enjoyed the comfort of being his guests. This was all very well in its way, but I was daily getting more anxious about the goats, which were not forthcoming. At last, three or four days after Raschid's return, I decided I could wait no longer, but must go back to the camp without them; so I told him I would leave on the 13th, and he promised to send the goats on when they arrived. He very courteously refused to listen when I suggested paying him in advance with cloth for them, but he asked me as a special favour to give him a bit of brass wire, for which he evidently had some particular use, so I promised to try and get it for him when I returned to camp. I had already given him a cartridge-pouch and belt, with some needles, thread, and coffee, from my private stores, with all of which he was well pleased; the coffee he promised to return to me when the canoes came down from Kasongo.

It is worth mentioning here that all the geographical facts about the Lomami and this district, which I gathered from the Arabs, have been fully confirmed by the recent explorations of M. Delcommune, of the Congo Free State.

When the time for my departure arrived, I found Raschid had placed at my disposal his own private canoe, with an awning and good floor, rather more luxurious than the ordinary canoes. For all this I was particularly thankful, because I was feeling rather shaky, after having been down with an attack of fever the day before. Raschid came down to see me safely started. I had on board with me, beside my other men, Dahoma, a boy I had known long ago as wash-jack on the Congo, and my own servant, Hamesseke. The latter had, while at the Lomami, met an old friend from Zanzibar, who presented him with a small kid, which we took along with us; but, sad to say, it was lost in the bush on the way to camp. Dahoma and all my other Zanzibaris had been able to pick up some small delicacies, such as salt, smoked fish, &c. &c., and they were all looking much better for the improvement in their food and the change from the camp.

My state barge was manned by natives under one of Raschid's headmen, who was detailed to escort me safely as far as Yangambi. It was evident that the paddlers were determined to make known to their fellow-countrymen that they had on board a *great* man! They sang and danced their fancy steps as they paddled along, and the drum in the stern was beaten at a great rate, adding to the din. All this noise and rocking about was exceedingly trying to me in my present state of health, but I had to look upon it as an honour. The train of canoes that followed to escort me increased as we advanced. No pause was made to change the paddlers of my canoe, but relays of men came out from the villages in their canoes and jumped aboard, grasped a paddle, and the late holder of it returned in his place to the shore. So the day wore on under such extraordinary circumstances, until at last, late in the afternoon, we reached Yangambi.

CHAPTER XV.

EXTRACTS FROM DIARY (*continued*).

Friday, April 13*th.*—On arriving at Yangambi I was amazed to meet Barttelot, looking terribly ill and very disturbed. I asked anxiously what was the matter, and he told me that there had been serious disturbances at the camp created by Salim-bin-Mohammed, arising out of the purchase of a knife by one of our men from a Manyuema. Salim came down to the camp and had words with Barttelot about this, and accused our men of stealing from the natives. Barttelot says he told Salim he had done everything he could for him, even trying to get Tippu-Tib's loads up, and the woman left behind at Matadi, but Salim would not believe it. The Manyuemas grew very threatening, and so did the natives, till things had come to such a pass that Barttelot determined to go to the Falls for the purpose of asking Nzige to take Salim away, for if he did not there might be trouble. He was now returning to the camp with a letter supposed to contain the recall of Salim. He was evidently much excited over the recent events in camp, and could not give me a clear statement of what had taken place. He told me, however, he anticipated that during his absence Salim would have made some attack in the camp, which had been left in charge of Bonny alone. His one idea now was to push on to Yambuya, as he had promised to be back on Sunday. We spent the night at the Sheik's house at Yangambi, where I was thankful to turn in, as I was suffering from a bad attack of fever.

Saturday, April 14*th.*—Left Yangambi early with Barttelot *en route* for our camp. He had with him a small body of Zanzibaris, among them John Henry as interpreter. I had seven men, with

Bartholomew as interpreter; hence as Assad Farran was away at Kasongo with Jameson, Bonny was left at the camp without an interpreter. It rained heavily all day, making our march most wearisome, but we went at a great pace. When we arrived at the usual halting-place a great number of our men had not come in. We discovered afterwards that a good many had lost their way.

Sunday, April 15th.—Barttelot left early, wishing to get into camp to-day; he pushed on without waiting for his men who had not turned up. I asked him if I was to look after them for him, but was told not to bother about his and only see to my own followers. I did not start till much later. In the first place, I was completely knocked up by a violent turn of fever, the result of yesterday's soaking; and secondly, I waited for some of my men. However, I got started at last, but could not go far, and was obliged to camp in the bush under cover of my waterproof sheet. I was unable to get anything to eat that night, as the men with the provisions were still missing. I had a miserable night, for in addition to a return of fever, I had strained myself badly by a fall, when making my way over some slippery logs.[1]

Monday, April 16th.—Started early, but made slow progress. Did not get into camp till about noon; there I found Barttelot and Bonny. I was told that there had been more palavers with Salim-bin-Mohammed during Barttelot's absence; the natives and Arabs had been even more threatening than before, so that there was very nearly an open rupture. Bonny even thought it might become necessary to shoot Salim and Ngungu, for the former, he understood, had threatened to cut down our men whenever he met them in the bush. It seems possible that owing to the absence of the interpreters there was some misunderstanding between Bonny and Salim. After hearing all this I was somewhat surprised in the course of the afternoon to have Salim come in to ask me if the men with me had brought in his letter[2] of recall, which Barttelot said had been put in a box that one of his delayed men was carrying; but this man with some of the others, including John Henry,

[1] The letter recalling Salim never turned up to my knowledge; certainly he never was recalled.

[2] Mr. Stanley (*In Darkest Africa*, vol. i. p. 246) describes such a forest as the one I here passed through. The fall I received at this time gave me reason to remember such a scene, as it caused me an injury the effects of which I shall feel all my life.

had not yet turned up. Barttelot sent some Soudanese and Zanzibaris to look for them, with special instructions to find John Henry, for some of the men said he had bolted. Bonny has told me pretty much the same as Barttelot did about the palaver during my absence. He says that Barttelot and Salim had high words, arising out of a controversy between Barttelot and a Manyuema. Affairs reached such a point that Barttelot and Bonny came to the conclusion that the best thing to be done was to get Nzige to recall Salim, and Bonny suggested that Barttelot should go to the Falls for this purpose, leaving him in charge of the camp. Bonny urged that this excitement not only placed their lives and the camp in danger, but the Arabs might attack me on the way home, and might make a prisoner of Jameson, who was at Kasongo; thus we should all be made to suffer because a palaver over so small a matter had been allowed to reach such a pitch. Therefore Barttelot went to Stanley Falls on this mission. During my absence Barttelot had been occupying my big hut with the stores, and now on my return told me he would remain there, and that I was to occupy Ward's hut, so I spent the night in those quarters.

Tuesday, April 17th.—I removed the remainder of my things out of my late house into Ward's. Feeling very seedy, and because of the strain I had given myself, I had great difficulty in getting about.

Wednesday, April 18th.—Orderly officer for the day. Still difficult for me to move about. Salim-bin-Mohammed left to go down the Aruwimi in canoes, with a lot of men. Busy making out fair copies of particulars of cloth and mitakos given out for buying food; also a list of rice, biscuits, and ammunition issued.

Thursday, April 19th.—Handed Barttelot the accounts I had made out.

Friday, April 20th.—John Henry was brought in; he was found at Yambow; his load was brought in yesterday by another man. Had out all the stores from my late quarters, checked them off with Barttelot.

Sunday, April 22nd.—Still much troubled with the severe strain, resulting from my accident. Consulted Bonny about it; he ordered me complete rest, and said I had been very foolish to walk about while suffering so. Barttelot this morning addressed the Zanzibaris in reference to John Henry. There had been some discussion about the punishment he was to receive, but I was too ill to take part in it. I heard that it was decided that he should have three hundred lashes.

Monday, April 23rd.—John Henry received three hundred lashes this morning. From my hut I could hear what was going on, and I was much disturbed by this, as I did not approve of such a severe punishment.

Tuesday, April 24th.—Barttelot went out to a neighbouring village.

Wednesday, April 25th.—Orderly officer for the day, but I could not perform all the duties, as I was totally unable to walk, so I was obliged to give in to Bonny's suggestion that I should stay still while he would do my duty for me. I was informed that John Henry died to-day.

From this date onward I suffered terribly, and on May 5th noted in my diary, "I am utterly helpless." Bonny attended me, and did all in his power to alleviate my sufferings; but as there were no counter-irritants obtainable, he had great difficulty in doing so. I improved a little after the application of one remedy he applied; but I had been weakened by repeated severe attacks of fever, and this pain, with the mental anxiety about the affairs in camp, reduced me to a miserable condition. During all this time I was obliged to lie on my back in my dark small hut, the only breaks in the wearisome monotony of the day being when Bonny came in to sit with me, and to tell me about what was going on in camp. He kept me pretty well informed of the course of events. Thus matters went on, with nothing to be chronicled but Bonny's reports, and the details of my suffering.

Tuesday, May 8th.—Feeling a little better to-day. In the afternoon heard the joyful sound of a steamer's whistle, so they have heard at last of our situation here. Bonny came in to tell me that a little steamer was in sight. I could hear sounds of great excitement outside among the men, but I could not move to find out anything. It was some time before I could get any further intelligence, as every one was so excited; but after a while Bonny returned, and informed me that Lieut. Vankerckhoven had come up in the *A. I. A.*, and with him Mr. Werner, his engineer. They soon after came into my hut, and I was right glad to see them; and miserably ill as I was, it was a source of consolation to know that Ward had reached Bangala safely, and I was delighted to think that M. Vankerckhoven had responded so promptly to the request made of him to bring back the thirty-five men who had been sent with Ward. He brought me a letter from Ward, saying he had got down to Bangala in five days—a marvellous journey this! The steamer *Stanley*, which had been at that station, had just left for down river, and he was on the point of starting in canoes, hoping to overtake it at Equator station. All the men came back with the exception of one Soudanese, who died at Bangala. Vankerckhoven had heard no news of Stanley *viâ* Zanzibar and Europe, and was ignorant until Ward's arrival that we were still at Yambuya, having supposed that we had started towards Wadelai long ago. On asking him why the State had not sent up a steamer to Stanley Falls with the white officers for that station, he informed me that Captain Vande Velde, who had been sent out from Brussels along with M. Stillmann, had reached Leopoldville, where he had died of fever. M. Stillman had only got as far as Lutete, and had been obliged, owing to illness, to return to the coast. Vankerckhoven had heard that new officers had arrived at Leopoldville, and were expected up on the return trip of the *Stanley*. I was sorry to hear of so many deaths of old friends down river. Vankerckhoven was most kind to me, and immediately on seeing my condition, had sent down to the steamer for a bottle of brandy, a teaspoonful of which I was able to take diluted with water. I shall never forget as long as I live how much good this tonic did me. This was one of the special things Bonny had ordered for me. And it was from the hands of my old comrade Vankerckhoven that I received the medicine I so much needed. He added afterwards to his gift some other stores. He did not forget Bonny either, and I was rejoiced that my kind nurse was now receiving something for his own personal comfort. Werner came in and looked after me, as I was quite unable to do anything for myself; his kindness at this time was fully appreciated. It was a pleasure to see these new faces after having for nearly a year seen only the same people over and over again. It was a relief to have a change after the monotony, and to hear something about what was going on in the outer world. All this brightened me up a little, though I felt completely out of it, unable as I was to do anything. They all dined together, and I could hear them, as I lay in my hut, talking merrily over a good dinner, as Vankerckhoven had supplied some acceptable additions to the usual meagre bill of fare. Vankerckhoven, Werner, and Bonny came in again after dinner and chatted to cheer me up.

Wednesday, May 9th.—Cheered by visits from our guests; feeling a little better. I should have mentioned before that Vankerckhoven deserves great praise for having come up so promptly with the men left by Ward at Bangala; he certainly should receive some recognition from the Committee for his services.

Thursday, May 10th.—I hear that the steamer is going up to Stanley Falls to-morrow, Vankerckhoven having arranged with Salim-bin-Mohammed to take the ivory he has collected here up to that place for him.

Friday, May 11th.—Vankerckhoven and Werner left in the *A. I. A.* for the Falls to-day; they are coming back here again. Bonny tells me that Barttelot is very anxious that I should go home. Bonny informs me that Barttelot wishes him to give me a certificate that I am unfit for further service, adding that if he does so I shall be forced to go whether I like it or not. Barttelot said that in the service when a medical officer in charge gave such a certificate it was compulsory for the patient to go. Bonny tells me that if I don't get better before the steamer goes down he will send me back in it. At present he finds there is not much improvement in my condition. I have no desire to go home if I can be of any further service to the Expedition, even by staying at the Falls to look after the loads.

Saturday, May 12th.—Bonny informs me that Barttelot is going to the Falls on Monday next.

Monday, May 14th.—Barttelot left this morning for the Falls with the idea of waiting there for Jameson's return from Kasongo. Bonny is left in charge of the camp, for I am about the same in health—still stretched out in bed, entirely unable to do anything.

For the next few days nothing of importance occurred, except the return of the carriers who went with Barttelot, and who brought back a letter saying he had heard from Jameson, who was at Kasongo, and mentioning that he had caught the *A. I. A.* at Yallasullah, so he was going in her to the Falls.

Sunday, May 20th.—Bonny tells me that whilst some of the men were out as usual getting manioc in the distant plantations, they

MR. W. BONNY

were set on by some Manyuemas, who took a knife and cloth away from one of the Zanzibaris, and threatened the others if they took any manioc. Bonny, on hearing this, sent up to Salim-bin-Mohammed to ask him to inquire into the affair. He sent word back to say that he knew nothing of it, and would send out to get the Manyuemas in and ask them about the matter. Later he sent word down that he had got the Manyuema who had taken the knife and cloth and had given him a severe flogging, and now sent him down to Bonny for him to do what he liked with him. Bonny gave him a lecture and sent him back to Salim-bin-Mohammed, thanking him for having so promptly punished the man, and thus saved trouble, as it might have led to a serious palaver. Bonny and I naturally thought that Salim might have secretly told the Manyuemas to stop our men from getting manioc, as it is becoming very scarce after a year's run on the plantation by so many people —the natives, our men, and the Manyuemas. If this had been the case matters would have assumed a grave character, as a conflict could not have been prevented over the food supply. Happily our fears ware dispelled by Salim doing the right thing.

Monday, May 21st.—Bonny has been talking over with me my present critical state of health, and my inability to move; we also discussed whether I could be of any further service to the Expedition by staying here. We weighed the matter calmly over, and he came to the conclusion that it would be best for me to return home, and much as I regret to leave the Expedition, it is clear to my mind that I would only be an incumbrance instead of an assistance in my present condition, so it was agreed that it was best for me to go. Bonny wrote to Barttelot, saying he had advised me to take this step, and enclosed a certificate. I wrote also, saying I agreed under the circumstances.

The certificate Bonny sent was in the following words—

EMIN BEY RELIEF EXPEDITION.

"*Yambuya Entrenched Camp,*
"*Aruwimi River, Central Africa,*
"*16th May*, 1888.

"I have to-day advised Mr. J. Rose Troup to leave this country the first opportunity for England. He is suffering from general debility, consequent of his long residence on the Congo, and the absence of tonic medicines, &c., while with this Expedition. He is unable to move about, and our camp not being entirely free from danger, he would be better away where he could receive proper treatment. He is now recovering from a second serious illness, in the first of which I despaired of his life.

(Signed) "WILLIAM BONNY,

"In medical charge and commanding *pro tem* Yambuya Camp."

Wednesday, May 23rd.—Sent over our letters to Barttelot at the Falls.

During the rest of this week very little of importance happened. We heard that Jameson and Tippu-Tib had returned from Kasongo with 400 carriers. I have omitted to mention that the death-rate among our men continued very high, and by this time more than 100 had died.

Wednesday, May 30th.—Barttelot returned to-day from Stanley Falls; Jameson had started with him, but had stayed behind; will not be in till to-morrow. Bonny tells me that Barttelot has not received the letters we sent to him on the 23rd; so we made copies, and Bonny gave them to him. Later on Barttelot sent me a note regarding my going down.

Thursday, May 31st.—Barttelot and Bonny are busy reducing the loads, because Jameson has agreed with Tippu-Tib that they are to weigh only forty pounds. Bonny has hurt his hand, so can do but little. Jameson returned about midday, and I was right glad to see him. He paid me a visit almost immediately on his arrival, and I was pleased to see him looking so well after his long and tiresome journey. He was sorry to see me reduced to such a pitiful condition, and did his best to cheer me.

Friday, June 1st.—Most of the 400 Manyuemas have arrived, I hear. Muni Somai, an Arab sheik, is in command of the force. He came in and paid me a visit in company with Salim-bin-Mohammed. He is an enormously big fellow. He and Salim began talking away together, evidently about me. Shortly after they called, Assad Farran, the Soudanese interpreter, came in, having returned with Jameson from Kasongo; he was also looking very much better for his trip. I can hear a great commotion going on outside; Barttelot and Jameson are very busy with the loads. Bonny is helping all he can, but owing to his hand being still bad, he is unable to do much. The latter came in and paid me a visit, bringing me a small bottle of Madeira, for Barttelot has opened one of the two cases I brought up for Stanley, because he does not intend to carry them up. I had a return of fever, due to all this excitement.

Saturday, June 2nd.—I find that my boy Hamesseke is not to go down with me. This is a great inconvenience to me, as I am unable to do anything for myself.

Sunday, June 3rd.—The other officers are working hard, making final arrangements. I feel so much out of it; it is distressing to be so helpless.

Monday, June 4th.—Bonny came in, and told me that the s.s. *Stanley* and the *A. I. A.* were in sight, but could not come up to our landing-place, because there was too little water, so pulled up some little distance below our camp. Our letters were sent ashore. Bonny tells me he asked Barttelot whether he had received any letters from the Committee or any news of Stanley, but he replied in the negative.

Tuesday, June 5th.—The steamer people came up this morning to see us. I was delighted to find old friends among them, Capt. Vangèle, Lieut. Baert, Capt. Shagerström, and Deman, and they introduced to me Lieut. Bodson and M. Hencke, all from off the *Stanley*: Vankerckhoven and Werner of the *A. I. A.* came up also. They tell me that the *A. I. A.* was on its way back from the Falls with Tippu-Tib on board, with a good many of his headmen, when it fell in with the s.s. *Stanley*, and together they came into the camp. Capt. Vangèle, Commissaire du Haut Congo, came up in the *Stanley* to install the officers, MM. Bodson, Baert, and Hencke, at the Falls station, but turned aside to visit Yambuya. Later on I had a call from Tippu-Tib, who came in with his interpreter, Salim Masude, and through him expressed his regret at seeing me so prostrated. Capt. Shagerström after a while appeared on the scene again, bringing two small boxes, with a letter sent to me by an old friend, Olsen, and for their contents I was very grateful.

Captain Vangèle was anxious to push on to Stanley Falls with the officers for that station, but Barttelot pressed him to remain a few days, so that he might be present at the signing of the agreement with Tippu-Tib; and also that he might take the surplus loads down to Bangala, which Captain Vangèle told me Barttelot intended to send to that place.

Wednesday, June 6th.—I received eight bottles of Madeira from Barttelot, for my journey down, out of the case of Stanley's, which he has opened. He also sent me my directions for going down river, and gave me papers necessary to secure my passage home. From Captain Vangèle I received every assurance that my comfort on board the *Stanley* would be carefully attended to.

Thursday, June 7th.—Tippu-Tib, I hear, to-day signed the agreement about the four hundred men he was to supply; it was witnessed by Capt. Vangèle on behalf of the State. I believe this took place at Salim's house, where Tippu-Tib has his quarters.

Friday, June 8th.—A big palaver between Tippu-Tib and Barttelot over the weight of loads. They were weighed in front of Tippu-Tib, who found some of them exceeded forty pounds, so he remonstrated with Barttelot, saying his men would not carry them. I was told this by Bonny and the State officers who were present. Shortly after I heard a most tremendous commotion outside, the well-known shrieks and yells of the Manyuemas, with the rushing sound of a stampede of men. The white men who were with me hurried out, and I sent my boy Hamesseke to see what was the matter. He returned after a time, and from what I could gather there had been a big row, which he described in broken English, helped out with Swahili, and most graphic gestures. He said, "Major, he make big palaver with Tippu-Tib. Manyuemas say loads too big, not fit to carry." He added that Tippu-Tib was very angry, and he heard Tippu-Tib had said, "If Major *piga* [strike] Manyuema, Manyuema he *piga bunduki* [strike gun—that is, shoot him]. The loads had to be reduced in accordance with Tippu-Tib's demands. Later in the afternoon I was carried down in a hammock; Bonny kindly superintended all the arrangements for me. I was lifted off my bed into the hammock, and saw daylight again for the first time for over six weeks. Bonny had some picked men to carry me, and he accompanied them, instructing them how to take me down the steep path.

They managed well, but I was thankful to reach the end, and have the hammock, with me in it, laid in the bottom of the whale-boat. At the landing-place I parted with Bonny, who had down to the very last moment attended me like a brother. I cannot dwell further on this scene, my wretched state and the forebodings of disaster completely unmanned me. Some of my old Zanzibaris came down and kissed my hands, and so did some of the Soudanese. I was soon afterwards rowed to the *Stanley*, and when we got alongside there was considerable difficulty in getting me hoisted on board in my hammock, and lifted to the upper deck, where quarters had been thoughtfully provided for me by Captain Vangèle in the pilot-house. My boy Hamesseke is to accompany me, for which I am thankful. We are to start for Stanley Falls early to-morrow morning, with two of the State officials for that place. The *A. I. A.* with M. Vankerckhoven leaves at the same time. M. Baert, who has been appointed secretary to Tippu-Tib, is to remain behind at Yambuya Camp with Tippu-Tib, until Major Barttelot's caravan starts, and will return with Tippu-Tib overland to Yangambi, purposing to go on from thence by canoes to the Falls.

[I should here mention that my diary during this last week, on account of my illness, is very sketchy, with rough notes added from time to time while on my voyage down river, and I have been obliged to compile the above entries from these in order to give the sequence of events.]

It was from Captain Vangèle that I first heard of the loads being sent to Bangala, I was not made aware even how many loads there were, or of what they were composed. Whatever they were, they were given to Captain Shagerström, who received from Major Barttelot full instructions concerning them, as he afterwards told me. I was far too ill to have the care of these business matters.

We left Yambuya in the *Stanley*, on June 9th, and reached the Falls on the 15th. I had bad attacks of fever during nearly the whole voyage. We spent two days at Stanley Falls; and as Tippu-Tib with Lieutenant Baert had not then arrived, Captain Vangèle decided he could wait no longer, so he left MM. Bodson and Hencke at Stanley Falls, arranging with M. Vankerckhoven, who was staying behind in the *A. I. A.*, to settle matters for them with Tippu-Tib as regarded the site of the station, &c. &c.

We left Stanley Falls on the morning of the 18th, and about one o'clock, as we were going down the Congo, we met the Dutch House trading-steamer, the *Holland*, coming up. When we got within hail we found that she had Tippu-Tib and his secretary on board, having picked them up on their way in canoes from Yangambi. M. Greshoff of the Dutch House, Tippu-Tib, and Lieutenant Baert came over to the *Stanley*. I was glad to see my old friend Greshoff again, and the first question I asked him was whether he had any news of Stanley, thinking it might have come round by the East Coast, but he had heard nothing of him. Tippu-Tib came in to see me as I lay in the little saloon, for I had been transferred from the upper deck, as it was too cold for me there. Through his interpreter he intimated that he wished to speak with me. He said that the contract made at Stanley Falls was to the effect that the loads should not exceed forty pounds, but this agreement had not been kept, he asserted. When he objected, Major Barttelot accused him of breaking faith with the Expedition. He was surprised at this, as he (Tippu-Tib) was only asking that the contract should be kept. He added excitedly, "You are going home to the big white men, ask them if I have ever broken faith; ask Boula Matadi (Stanley); ask Cameron, ask Wissmann; and hear what they will say." He went on to inform me that had he known at the Falls when he arrived with the men from Kasongo, the facts that he had since heard from Salim-bin-Mohammed, when he met him at Yambuya, he would not have given us the men, and he now regretted very much that he had done so. He says he had great difficulty in getting the men to go at all, as the Manyuemas who had been round our camp for months had spread evil reports all over the country.

[This is the essence of what I took down of his conversation immediately after, as I felt that this was a most important statement. It is hard to decipher all that I wrote, as I was so very weak at the time.]

The man fairly shook with rage, and I was very glad when this most painful interview was ended. Lieutenant Baert came in to see me just afterwards, and told me that Barttelot, Jameson, and Bonny got away with the men on Tuesday, June 12th. Soon after I bid good-bye to Baert, Greshoff, and Tippu-Tib, and we went on our way down river. In my diary I wrote, "I am feeling a little better, but am *so* weak." But Vangèle very kindly had obtained a case of champagne, which was coming up on the *Holland*, and gave me from it all that I wished, bringing it in as a medicine to me whenever he thought I needed it. No doubt this was in a great measure the means of restoring me to better health, for this is one of the best remedies that can be given to those suffering from intense prostration after fever. Whilst talking of Captain Vangèle, I wish to say that I look back with extreme pleasure and thankfulness to all the careful attention he paid me throughout the time I was in his company.

CHAPTER XVI.

TIPPU-TIB'S CONDUCT.

During my absence at the Lomami, it will be seen, a serious dispute had arisen between Major Barttelot and Salim-bin-Mohammed. It was caused by a very small matter, but the patience of the Arabs had been severely tried for a long time, and when the palaver began, the small question soon dropped out of sight, and the matter became very serious. A little later, while he was away at the Falls, still another palaver took place with Salim-bin-Mohammed over our men being attacked by the Manyuemas in the manioc fields. This was a more serious ground for a quarrel, yet Bonny was able to smooth matters over satisfactorily without much difficulty, though we felt that after the recent occurrences there was much danger in any petty dispute.

Two of the officers were absent. I was far too ill to be of the least assistance in case of trouble. The only two white men remaining to keep matters straight in camp were Barttelot and Bonny. Our garrison was now reduced to not much more than a hundred men (for thirty-five were away with Ward), and few of these remaining were able-bodied men. So it cannot be denied that we did not

TIPPU-TIB.

hold an unassailable position in what might be considered an enemy's country.

Jameson, by his visit to Kasongo, no doubt did much to influence Tippu-Tib to make more haste to get the promised men. Jameson and Tippu-Tib were on friendly terms, and the former never seemed to fully believe that Tippu-Tib was treacherous. We had missed the companionship of Jameson much. He was always to the front when anything was going on, ready for work, and most cheerful.

When Barttelot went to the Falls for the purpose of waiting there for Jameson, he left Bonny in sole charge of the camp, for I was not able to lift my hand.

Captain Vangèle gave me a detailed description of the events that took place shortly after Tippu-Tib's arrival. It should be borne in mind that an agreement had been made with the Arab chief that the loads should not weigh more than forty pounds. This was a slight modification of the suggestion that Barttelot had months previous made in my presence, that in order to induce the men to serve us, he proposed to reduce the loads by one-half. I had strongly urged that this should not be done, for though it might persuade the men to come, I knew from former experience that when any one once began to reduce a load for a carrier, the man would never be satisfied however much was taken off, and the question invariably caused a dispute. But Barttelot having now agreed that the loads should weigh no more than forty pounds, would have been wise to have been extremely careful to reduce them uniformly to that weight even at the greatest trouble. It was upon this subject that the dispute arose. Tippu-Tib, surrounded by all his sheiks, and in the presence of Jameson, Bonny, and the State officers, made his objections known. According to Vangèle's description, Barttelot was exceedingly angry with Tippu-Tib. When the Arabs understood it there was a tremendous row. The sheiks rose in wrath; their great chief had been insulted in their presence, before the white men; they shouted, yelled, and danced about, and their cries were echoed by the 400 Manyuemas outside the camp. They demanded that instant justice should be done. Captain Vangèle understood Swahili, and in the midst of all this excitement he overheard Tippu-Tib call out to the interpreter, "Don't translate what they say,"—for the sheiks were thoroughly roused, and were using terribly threatening language. Tippu-Tib did his best to pacify his followers; but at one time, in Vangèle's opinion, matters had arrived at such a crisis that it only required the accidental discharge of a gun to have brought on a general fight. The Manyuemas would have rushed in, and no doubt not a single white man would have escaped; for in their terrible wrath nothing, not even Tippu-Tib and all his sheiks combined, could have stopped them. It was a miraculous escape, and there is little doubt that had not the State officers and the steamers been at hand, the consequences might have been serious.

Barttelot had long before this been warned that his life was in danger. I had myself pointed this out to him, and more than once had told him that he ought never to venture out alone in the bush, as there was great risk of his being attacked, and it is my belief that he knew the truth of this. I told him also that I had heard rumours of threats upon his life. When my boy repeated to me at this time the report that Tippu-Tib had told the Manyuemas that if they were struck by the Major they were to shoot him, I could not place much dependence on his statement, but I thought it right to repeat it to Jameson and Bonny, as well as to the Belgian officers, and probably they told Barttelot of it. I have not the least doubt that he was well aware when he started that he was taking his life in his hands; and it is exceedingly doubtful whether any further warning would have been of use.

As to Tippu-Tib being accused of treachery, taken in the light of all the events that are now known, it is totally absurd. Had he been treacherous, what would have prevented him from making an end to our branch of the Expedition? Why had he not instructed his men to attack us? Why had he not told them long before, that whenever the next palaver arose they should make an end of it by discharging a gun? Or why did he not tell them to pick us off one by one as if by accident when alone in the bush? True, he had promised Stanley that he would supply us with six hundred men, but that was with the understanding that a steamer with his loads, Salim's woman, and the white officers should come at once to the Falls. This had not been done. Why should he keep his part of the promise until the white man had proved his faith? And why should he show any promptness in keeping his promises?

As to the wild stories of Salim-bin-Mohammed's intention to destroy our camp and obtain the much-coveted ammunition, that is simply answered by a query, Why under the sun did he not do it? What was to prevent him? If he had all along been desirous of attacking us, why had he not done so? There must have been some reason why his wrath should have been specially roused just now, or otherwise why did he wait until this time to threaten us, when he knew that Ward had gone down the river to warn the white men of our danger, and that the steamers would be with us shortly, bringing he knew not how many armed Houssas? Why should he have waited all this time? Why did he not either declare war and lead an attack upon our ill-defended camp, surrounded as it was by nothing stronger than a fence of saplings, garrisoned almost entirely by men of his own religion, whom he could easily have bribed either to betray us, or to desert, or to attack us? Or had he preferred less warlike means of ending us and obtaining the stores, why did he not encircle us with a cordon of hostile Arabs and cut us off from all food supplies? He could have speedily reduced us by starvation, and there would have been no safe way of escape from our terrible position. Had Salim-bin-Mohammed or Tippu-Tib been treacherous, our lives were not worth a day's purchase. Yet up to the last moment of our stay in camp they showed no real sign of treachery. They supplied us, after a long time to be sure, with 400 men, and had kept the letter of the agreement made by Barttelot, which modified Stanley's original instructions. Were they then treacherous because, when they thought their chief was insulted, they rose in their wrath and made threatening speeches and signs? Were they then treacherous because they agreed to allow one of their sheiks, Muni Somai, to go with Barttelot to keep the Manyuemas in order? It is not at all clear to my mind where their treachery comes in; and as far as I can see, Tippu-Tib exercised great control over his followers under most trying circumstances. It should not be forgotten that the delay which *looked* like treachery was, probably, largely due to Tippu-Tib's difficulty in getting his followers to act as our porters. These Manyuemas are wild and undisciplined, never being trained to carry loads. I ought here to mention that the anxiously-awaited steamer did not bring Tippu-Tib's loads or Salim-bin-Mohammed's woman. All the loads of Tippu-Tib's that were entered in the list sent me by Stanley's order were all right; every one of these passed through my hands and reached Leopoldville, and were taken up by Stanley from that point. The missing loads I had nothing to do with, and heard nothing of them until I arrived in camp.

When the steamer *Stanley* left the camp, Major Barttelot entrusted to Captain Shagerström a letter to Ward, instructing him to stop at Bangala with the loads which were being sent down there. It was anticipated that we should meet Ward on his way up with the reply to Barttelot's cablegram to the Committee; but he was to be stopped at Bangala to take care of the surplus loads. It should be remembered that there would naturally be more than enough loads for the four hundred Manyuemas to carry, as the number of surplus loads was increased by those made up of what was taken out in order to reduce each of the four hundred loads by one-third. Among the articles sent down was, we now hear, Mr. Stanley's own personal kit![1] It must also be borne

[1] There were twelve loads, he tells us, of his private baggage left at Yambuya, and I have heard that all these were not sent to Bangala.

in mind that these loads were all sent down here entirely upon Major Barttelot's own responsibility; I have reason to believe that he considered they would be unsafe at Stanley Falls, as he had heard a rumour that, owing to difficulties between Tippu-Tib and one of the State officers, the Arabs threatened to raid the country down to the vicinity of Bangala, and he wished therefore to send the loads beyond their reach.

I was surprised to find that the State officers were so well informed concerning the condition of affairs at Yambuya; the information they had received did not come from me—in fact, they were able to tell me a great deal I had not known before, because, owing to my serious illness during the last six weeks of my stay in camp, I was not in touch with all that occurred; but it should be remembered that they had heard much while at Yambuya, and as Tippu-Tib had come round from Stanley Falls with Vankerckhoven, he had been able to tell his version of the course of events, and had described fully his position as regarded Major Barttelot. Moreover, Assad Farran, I found, was very ready to tell his views of the case, and being an interpreter, he had obtained a large amount of information. He talked much with the Belgian officers, telling them many circumstances of which I was totally ignorant. They questioned me on several points, but I could not and would not give them any satisfaction, saying I knew nothing about them, and would not express my opinion on the probability of their truth.

CHAPTER XVII.

DOWN THE CONGO AND HOME.

On board the *Stanley* was Assad Farran, the Soudanese interpreter, and another Soudanese belonging to the Expedition. They were both incapacitated for further service. The former had bad ulcers on his feet that prevented him from walking, and the other was suffering from a diseased ankle, so it was impossible for either of these to go on with the column.

We continued our voyage down river but slowly, owing to our not having many men on board to cut wood for fuel. I had fever nearly every day, but though gradually improving little by little, I was still extremely weak. We reached Bangala on June 29th, where Captain Shagerström told me he left about 150 Expedition loads. I was told by them of the death of my old friend Deane. This fact had not been mentioned to me before, as no doubt they were afraid that I was too weak to hear such sad news. Poor Deane, the hero of Stanley Falls, had returned to the Congo for elephant shooting in company with Captain Bailey. He had escaped from dangers marvellously heretofore, but he had now been killed by a wounded elephant that turned on him when he was unable to defend himself, owing to being insufficiently armed. From the particulars I heard of his death it must have been a painful one. We left Bangala on July 1st. I was feeling much better for the change and the kind attention that had been shown me by the Belgian State officers. We stopped the next day for a few hours at the Dutch House station, at the mouth of the Loulongo, picking up the agent there, who went with us down to the Equator station, which we reached that night about sunset. We found here Mr. Glave and M. Boulanger, of the Sandford Exploring Association. On Tuesday, July 3rd, we left the Equator station, and about four o'clock in the afternoon we met the small State steamer *En Avant* coming up. Captain Shagerström had been on the look-out, as we were anxious to see Ward, whom we naturally expected to find on the first State steamer we met. Both steamers pulled up at the shore for the night, as we knew that here we could get a good supply of wood. Ward was, as we had anticipated, on the *En Avant*, and soon came on board the *Stanley*. He was, needless to say, much surprised at seeing me, and imagined at first that some great disaster had happened. After I had somewhat satisfied his anxious inquiries, my first question was whether he had heard at the coast any news of Stanley. When I got over the excitement of meeting him, he gave me an account of his recent journey. It will be remembered that I saw him off at the Lomami on his canoe voyage on April 3rd, and he was at Bangala five days later, viz. April 8th. He went on in canoes to the Equator station, where he overtook the s.s. *Stanley*, and going on in her he had reached Leopoldville on Stanley

Pool on April 15th. From there his journey was overland through the cataract region to Matadi, and then by steamer to Banana, at the mouth of the Congo, at which place he arrived on April 29th, in time to catch the Portuguese mail for St. Paul de Loanda. He had occupied only twenty-six days for this journey from Lomami, or twenty-nine days from Yambuya to the coast, a good 1300 miles —a wonderfully rapid journey. He was obliged to wait at St. Paul de Loanda five days for the arrival of the reply to his telegram. I gathered that the substance of this was that Stanley's instructions should be carried out, and that Major Barttelot was to engage no fighting men. I learned, too, that the latest news received from Emin was under date of November 24th, 1887; that Stanley had not joined him then. Emin was going out to meet him, but said that the road he (Stanley) had taken was almost impassable—any amount of swamps, streams, floating vegetation, &c. This is exactly the answer we had anticipated. As regards the road Stanley took, it seems that the information given us by the deserters was not incorrect. I long ago surmised that he had met with great difficulties.

Ward was handed, by Captain Shagerström, the note from Major Barttelot, and he was much disappointed to learn from it that he was on no account to go further than Bangala, and was to look after the loads there, placing himself under the chief of that station. It was a terrible blow to him, after all that he had recently undergone in his rapid journey to and from the coast, not losing a moment's time, travelling night and day for three months, to be told that his exertions were all in vain, and that he must give up all idea of following the Expedition.

I was much delighted to receive from Ward a budget of letters, full of kind messages from hosts of old friends, State officers, missionaries, and numerous people connected with the trading firms. Many of them sent me gifts, including medical comforts, such as champagne, &c. The missionaries had made extraordinary exertions in behalf of all the members of the Rear-guard, on hearing from Ward on his way down of the great straits we had been put to for food. Mr. Ingham, Mr. Hoste, and others had prepared several cases of provisions, medicines, and medical comforts from their private stores, which were ready for Ward to carry on his return up river. Mr. Casement, of the Sandford Exploring Association, had kindly himself accompanied the carriers with these loads from Lutete to Leopoldville. But now all these would be left with Ward at Bangala.

On board the *En Avant* beside Ward was M. Hanneuse, who was on his way up to take over the charge of the new State station at Stanley Falls, along with two other officers, MM. Danen and Müller for Bangala station. Ward sat beside me late into the night talking over all the affairs of the Expedition, &c. The excitement kept me up, but I felt thoroughly tired out later on. We parted company the next morning, the *Stanley* resuming her journey down river, while the *En Avant* went up to Bangala and the Falls. I left poor Ward very dejected over his hard fate, and I felt much the parting with the last comrade of our ill-starred portion of the Expedition.

We proceeded slowly down the Congo. I felt comparatively better; the peace and quiet of the steamer, and the nourishing food and medicines which I was now able to obtain, no doubt did much to improve my condition.

We reached Kinchassa, on Stanley Pool, on July 11th. At this place was my old friend Major Parminter, who had left the State service and joined the Sandford Exploring Association, and also Mr. Swinburne. They, too, were much surprised to see me, and to find me so reduced in health and strength. They insisted on my staying with them, and I was only too glad to accept their invitation. Captain Vangèle, who was on his way home to Europe, went in the *Stanley* to Leopoldville, taking with him Assad Farran and the other Soudanese who had been placed under his charge; they had not been in any way placed under my orders, as it was generally supposed that I could not survive to reach the coast. Major Parminter, on the very day of my arrival, sent for Dr. Sims, who was at Leopoldville, asking him to come over to see me, which he did at once, giving me a thorough medical examination, prescribing for me, and urging me to go home without delay. I was told of numerous additions to the sad death-roll, which had increased to such an extent while I was away at Yambuya. Among those who had gone was Mr. T. Comber, one of the pioneer missionaries on the Congo, and who was much beloved by all those who knew him, and was admired for the good work he had done. He had been for many years connected with the English Baptist Mission. Messrs. Darling and Biggs, of the same mission, had also gone, as well as the wife of Mr. Harvey, of the American Baptist Mission Union. The Sandford Exploring Association had lost Baron von Rothkirch, and the State several of their officers. The news of the deaths of so many friends and acquaintances was very depressing to me in my present condition. However, there were a number of familiar faces among the missionaries left; for instance, Mr. Grenfell, of the English Mission, who is well known not only as a missionary, but for the explorations he has made on the Upper Congo. He was a very highly esteemed friend of mine, and during my stay at Kinchassa he, with some other missionaries, called on me.

I spent nearly ten days very pleasantly among my friends at Kinchassa, where I was the guest of Major Parminter. All continued to do everything they could for my comfort, and every arrangement for my transport through the cataract region to Matadi was kindly made. Carriers were constantly coming in from Manyanga, and they generally returned laden with ivory. There always was some difficulty in getting these men to act as hammock carriers. However, one caravan that came in contained a number of men who had served under me when I was some years before at Leopoldville, and on Major Parminter asking them if they would carry me down, pointing me out as I was on the verandah, there was some hesitation among them; but when I spoke to them, calling one or two by name, they recognized me and consented willingly, so I selected eight of them and prepared to leave next day. We made an early start on July 20th, after saying good-bye to my good friends Parminter and Swinburne, the former up to the last moment impressing on my carriers how ill I was, and how much care they must take of me. I made a short stop at Leopoldville, to call on M. Liebrechts and the missionaries, but soon got away on the road to Lutete. I learned that Captain Vangèle had started down country some days previous, taking with him Assad Farran, but the other Soudanese was too ill to go on. My carriers went well, considering that they were not all accustomed to this sort of work; but all went well, and we got into Lutete on July 24th, where I found M. Dannfeld and Olsen, the "sonny boy." I was delighted to see both of them, and personally thanked the latter for the thoughtful gifts he had sent up to the camp in the *Stanley*. I rested here for two days. I found Assad Farran had been left behind at this place by Captain Vangèle, owing to the great difficulty he had in walking. Messrs. Davies and Slade, of the English Baptist Mission, came in to see me while at Lutete. I left on the 26th, feeling much benefited by the rest, for though my carriers were doing their best, travelling in my present condition was exceedingly fatiguing. I arrived at Manyanga that same afternoon, where Mr. Alfred Parminter, of the Sandford Exploring Association, was expecting me. At this place I had to get fresh carriers for the rest of my journey, and I remained here until August 3rd. During all this time I was well looked after by my kind friend, who did his best for my comfort.

At last I was able to get new carriers, Manyanga boys, among whom were a number of familiar faces, as I had, as I have said, spent some time in these parts. They carried me quite as well as my former ones, though the road in this district is much worse; but I found hammock-travelling dreary work, and if I could only have walked I should have much preferred it. These Manyanga boys were an amusing lot, and tired as I was, I could not help enjoying their jolly *badinage*. My hammock was slung on a pole between two men, who carried it by resting the ends either on their shoulders or head, by a given word-signal shifting it from shoulder to head, or head to shoulder as the case might be, trudging along at a good pace; the other six carriers would take their turn two by two in rotation, each pair carrying me a certain distance agreed on between themselves. On arriving at this point the chances were that the next pair would have lagged behind, and the shouts and cries of my men were very amusing as they stood shifting the carrying-pole, impatiently waiting for their comrades. "Ho, Malanga!" they would shout in stentorian tones. But Malanga, or whatever his name chanced to be, would not be forthcoming for some little time; but when he and his companions did arrive there arose a terrible hubbub, the old carriers reproving them for their delay.

Meanwhile I lay perfectly helpless, but laughing in spite of myself at the way they joked and teased each other, till the new lot grasped the pole and started off at express speed, till I wished they would stop a bit, as the shaking was almost too much for me. These would arrive as usual before their successors, and the same scene would be enacted, perhaps this time with a long conversation between Malanga and myself. As he could speak no English, and I could only just make out his language, it was very exciting. I usually discovered that he was proudly calling my attention to the skill and rapidity that he and his companion had displayed, and was begging me to give them each an extra lot of handkerchiefs by way of reward. All this was urged most solemnly, with a twinkle in the eye and a smile creeping about the corner of his mouth, which when I shook my head and laughingly said, "No, no, go 'long, you Bushman!" would increase to uproarious laughter at the joke, in which all his comrades would join. The humour of it all would strike me, and I could not help the infection, and often felt refreshed by their lively ways on my wearisome journey. At night we would all camp together by the side of some little stream, where the kettle would speedily be boiling, and if I did not take care I would find they had all drawn in close to my tent in almost too cosy a fashion.

So we went on our way to Matadi. M. Vankerckhoven, who had come down behind me in the *A. I. A.*, overtook me on the road. At Matadi I had to wait until the little steamer arrived from Boma some days later. I was glad enough to have a rest, as the daily travelling had been more than tiring. On August 14th the steamer arrived, and I went on that night to Boma, where I found quite a gathering of old friends. Captain Vangèle had arrived before me, and had had prepared very comfortable quarters for my two days' stay. My strength was maintained, but I felt the fatigue of the journey very much.

I went on to Banana in time to catch the homeward bound Portuguese mail-steamer on August 17th. Here I saw again my kind friend Ingham, who had come down to see his wife off to Europe. She was to be a fellow-passenger of mine, and Captain Vangèle was also returning in the same steamer. We had a good passage to Lisbon, arriving there on September 12th. At this port Captain Vangèle and I disembarked, but Assad Farran, who was on board, remaining on her, went on to Hull. I felt that it was most important that I should make my way as rapidly as possible to England; and ill as I was I undertook the fatiguing overland journey. I had telegraphed to the Emin Pasha Relief Committee on my arrival at Lisbon, and then took the express that same evening for Paris, travelling night and day, arriving there on the morning of the 14th. Here I had to part company with Captain Vangèle, who had throughout the long journey from Yambuya in Central Africa till my arrival in the French capital paid the greatest personal attention to my comfort, assisting me in every way in his power down to the very last moment. I owed much to his kindness as I had been suffering from fever all the way on my overland journey; but here our ways divided, he going to Brussels, while I went on to England. He had proved a true "friend in need" to me. On reaching Paris we at once drove together across the city from one station to the other, and I took the tidal train to Calais, crossing to Dover, and at last, after more than eighteen months' absence, found myself again in old England. On the passage over I bought the first English paper I had seen for many a day. On opening it the first thing that caught my eye was the heading of the telegram announcing the murder of Major Barttelot. Though I had feared such an event, it was a terible shock to me to read the fact in print, and I was completely stunned. It was too terrible, alas! to think of it now that it had really happened. The long anticipation of it had not after all prepared me for it; it was so sad to think that a young life had been lost in such a way in Africa, when there was really no necessity for it. Another life had been sacrificed in that country, and the relief of Emin Pasha was again delayed.

I was exceedingly thankful to find my brother waiting for me on the pier at Dover. Had he not met me I do not know how I should have managed to reach London. I suffered severely all the way up, and when I reached Charing Cross Hotel, I was right glad to have ended my journey, as I had an attack of raging fever, from which in my excited state it was hard for me to rally.

I was somewhat surprised that I neither saw nor heard from any member of the Committee for two or three days, although I had kept them fully informed of my movements by telegraph. But at last the Hon. Secretary paid me a visit, and I informed him that I was ready to place myself in the hands of the Committee, and was prepared to make my report to them. The next day I called upon him, and in the presence of a shorthand writer, I made a report concerning the events at Yambuya Camp. This I was told was to be laid before the Committee, and I was given to understand that when the next meeting was held I should be asked to be present and give them full particulars, but I never received any notification or official communication from them.

CHAPTER XVIII.

CORRESPONDENCE WITH MR. STANLEY.

SOME suggestion of the course of events after my return is given in the preface to this volume; but, aside from my personal experience, I would briefly state that on August 17th, 1888, the very day I left the mouth of the Congo, Mr. Jameson died in Mr. Ward's arms, and Mr. Stanley rejoined the wreck of the Rear Column at Bonalya. From that place our leader wrote letters, in which he condemned the officers he had left at Yambuya, on the hurriedly-told report of Mr. Bonny; and these letters on their arrival in England, the following April, were published by the Committee. Meanwhile Mr. Stanley had again vanished into the forest, leaving the public a prey to the many and varying rumours that were circulated from time to time during the ensuing year.

It was not until November, 1889, that authentic news of his approach to the East Coast was made known. Before that month had ended numerous letters of his were printed in the newspapers, in which he repeated his ungenerous attacks upon the Rear Column, without awaiting fuller information concerning it. Early in December he and his comrades arrived at Zanzibar. As soon as I heard that he had reached the coast, I wrote him a long letter, in which I gave him a report of what had occurred at Yambuya during his absence. I forwarded one copy to the care of the Consul at Zanzibar, and another to Sir E. Baring's care at Cairo.

It had happened as I had foreseen, viz., not being in possession of all the facts, Mr. Stanley had drawn wrong inferences, made incorrect surmises, and given the world an inaccurate idea of the course of events at Yambuya. But he had surpassed my expectations of him by publicly condemning, on insufficient data, the officers who had served him faithfully. Although I did not really anticipate that what I wrote would cause him to retract the misstatements he had publicly made, yet I felt it was my duty to present to him, fairly and fully, the true facts, that, in case he should repeat his errors, he should not do so in ignorance. For this reason, and to report to him as my commanding officer, I wrote the following letters, in which I gave him information such as I thought should be made known to him personally; but some of it I do not think necessary to publish at length here, for, as I before suggested, it is not obligatory upon me to burden the public with a detailed account of private affairs, or full particulars of my own opinions.

To H. M. Stanley, Esq., Cairo.

"*4th December*, 1889.

"Dear Sir,

"I have been looking forward with confidence for some time to receive news of your arrival on the East Coast, for I felt sure with your experience in Africa you would emerge from the Dark Continent, even though your appearance was so long delayed that it gave us all cause for grave anxiety. It was, therefore, a great relief, though not wholly a surprise, to learn the good news recently published, that you had so nearly reached the coast. Of course, after so long an absence from the civilized world you must be greatly occupied, not only in making known your own experiences, but in learning what has been going on in the outer world during your absence. At such a time it would seem very egotistical for me to bring myself forward, but I believe you will have already learnt from Sir Francis de Winton that he has brought an action against me in your name, and I therefore think it only right to bring to your notice certain facts of the case, which I shall endeavour to set forth as briefly and concisely as possible.

"When I was compelled to leave Yambuya Camp, owing to ill-health, which prevented me from being of any further service to the Expedition, and on my arrival at Lisbon, I, with all speed, notwithstanding my enfeebled condition, made my way to England to lay the report of events at Yambuya before the Committee. News of Barttelot's assassination reached England the very day I landed at Dover (about 15th September, 1888). Shortly after this the newspapers were filled with laudatory eulogies on Barttelot . . . . . my name was made use of in a most unwarrantable manner, both by those who were attacking and those who were defending him. I did not wish to enter pro or con into the controversy, and avoided making any statements publicly in the Press, confining myself to handing my report to the Committee, which I considered simply to be my duty. I called the attention of Sir Francis de Winton to these public false statements about us all, and requested him to exonerate me in the matter, and for the rest referred him to my report for the truth of events at Yambuya. While admitting we were not to blame, he refused to make any public statement. These reports continued to appear in an exaggerated form. Then it came to my knowledge that a book by (an engineer late in the service of the Congo Free State, who paid a visit to Yambuya when the steamer came up in June 1888, and remained four days), was about to appear, which would contain inaccurate statements about Yambuya. . . . Just at this time, last April 3rd, appeared your letters from Bonalya, dated the August previous. In one of these you condemned all the officers of the Rear-guard, upon what appeared to be incomplete evidence. You called attention to the fact that Major Barttelot was to consult the other officers in certain cases, but as he failed to carry out this instruction, we, the subordinate officers, could not be held responsible for the results. On the appearance of the above statements, I thought it right to call the attention of the public, by a letter to the Press, to the error you had fallen into, and took the opportunity of saying that I had made my report to the Committee, and that they had either discredited it or allowed me to suffer an injustice. I mentioned, as they did not see fit to set matters straight, it became my duty, in order to justify myself and my companions to publish a full account of the course of events. They took no notice of my statement, and it therefore was necessary, in my opinion, to make the truth known, and for this purpose the book, which I entitled *With Stanley's Rear Column,* was prepared. . . . Every arrangement had been made for the publication, it was printed and ready to appear, a great deal of expense having already been incurred, when Sir F. de Winton, in your name, brought an action in Chancery to restrain the publication as a breach of contract. Although I brought to Sir F. de Winton's attention the fact that the contract had not been fulfilled by your subordinates, I had no intention of making public the serious complaint of negligence, until it became necessary in self-defence for me to bring up the matter, to show that the contract seemed to be nullified, because the provision made in that contract, viz. that I should receive tent, bed, &c., &c., had not been carried out. Had Sir F. de Winton not forced this action upon me, the matter of not receiving the specified articles, which every other officer received, would never have been brought forward by me. The result of the action, commenced in June last, was that an interim injunction restraining my publication was granted, which is to be followed by further proceedings in connection with the same, and which will come off, probably early in the coming year. In this way the matter now stands, that I am prevented from making known the truth, by the injunction obtained by Sir F. de Winton.

"I regret to notice that the statement condemning the officers at Yambuya, which I supposed had been made hastily in the first instance when you learned the fate of the Rear Column, was repeated even more circumstantially in the more recent letters which have been published. It seems very unjust that I and the other subordinate officers should be publicly blamed for the conduct of Major Barttelot, who was left in supreme command of the Camp, and was responsible for the course of events. You will admit, I think, that it is unusual for the subordinates to be made responsible for the injudicious action of the superior officer.

"It would seem only right that you, the leader of the Expedition, should hear the report of the senior surviving officer of the Rear-guard, before condemning publicly all the subordinates. I feel that in every way I fulfilled the explicit instructions given me by you; one of which was, on the completion of my special work, to report to Major Barttelot and place myself under his orders. Having a clear conscience in all these matters, I appeal to you—one who knows me and has acknowledged my previous services in Africa—to suspend further public comment until the true course of events, both in Africa and England, have been fully laid before you. Sir F. de Winton himself states it is no personal enmity towards me that has influenced his action; but that he considered the publication of my book would interfere with the sale of your forthcoming work. This was his justification for bringing an action against me. I think this will strike you, as it did me, as a very ridiculous idea, for no book of mine could compete with yours, and I cannot arrogate to myself the power to injure you in this matter, as suggested by Sir F. de Winton. If I believed my publication would have caused you pecuniary loss, in spite of all that has occurred, I should have hesitated before carrying out my intention, for I would have been most unwilling to be the cause of injury to one under whom I have served. . . .

"I trust under all these circumstances you will pardon the length of this explanation. I beg you to remember me with all kindness and cordiality to my late comrades, the other officers of the Expedition. Looking forward to seeing you all in England before long,

"I remain, yours faithfully,
(Signed) "J. Rose Troup."

As I failed to receive any reply to the above letter, I wrote again enclosing a copy of it. This time I took the precaution of registering it. It was as follows :—

To H. M. Stanley, Esq., Shepherd's Hotel, Cairo.

"*4th March*, 1890.

"Dear Sir,

"I wrote to you under date of December 4th last, and directed my letter to you c/o H. B. M.'s Consul at Zanzibar, and sent a duplicate on the same date to you c/o Sir E. Baring, Cairo. As I have received no reply I enclose, herewith, a copy of the same letter, and trust that you will be good enough to acknowledge receipt of it. I would presume that you have not received the copies sent to Zanzibar and Cairo, not only because you have not acknowledged either of them, but have issued your official report, and have therein condemned all the officers of the Rear-guard without in any way giving me—one of them—a chance to place before you the condition I found things in on my arrival at Yambuya and the cause of the disasters that followed. You have publicly made the charge that your written instructions were not carried out by the officer you placed in supreme command of Yambuya, and you blame Jameson, Ward and myself also, giving as your reason that you had in these instructions directed that we were to be consulted before any vital step was taken. . . . Ward and I were not consulted in all vital steps. . . . But your public utterances are confined to that portion of your written instructions regarding our advance when I should arrive at Yambuya, on the second trip of the steamer, with Messrs. Ward and Bonny and the stores left under my charge at Leopoldville. I now propose to show that these instructions were most carefully read and studied by the other officers and myself, and that we tried to carry out the arrangements for our advance, but under the existing circumstances we were forced to make choice of one of the alternatives left to us in your instructions. On my arrival in advance of the Expedition on the Congo, I set about carrying out the special duties for which I had been detailed, taking up my quarters at Manyanga and engaging carriers to transport the loads of the Expedition, on their arrival there, to Leopoldville. When I had forwarded all these loads, acting in accordance with your orders, I pushed on with all speed to Leopoldville, with Mr. Ward, to join you and the main column of the Expedition. This was on the 30th April, 1887, and was the first time I had seen you on your arrival on the Congo, and also the first time I had seen any of the other officers with the exception of Lieut. Stairs, whom I had previously met in your rooms in London. I did not even then meet Major Barttelot or Surgeon Parke, as they had been sent on to Kwamouth. I saw you only for a few hours on the afternoon of my arrival, when you informed me that you had decided to leave me to look after the stores at Leopoldville, as the steamers could not take up all the loads on one trip. In the short conversation we had before dinner, you impressed on me the importance of these stores to the success of the Expedition, and added that when the steamers returned you would send further instructions, but that I was to lose no time in coming up, and would find Major Barttelot at Bolobo and report myself to him. Next day, May 1st, you, with the Expedition, left for the Upper Congo, but as far as I knew it might have been for Stanley Falls or any place else. On the 3rd July, the s.s. *Stanley* returned to Leopoldville, and I received the promised further instructions. You informed me that you had left Ward and Bonny at Bolobo, where I was to pick them up and come on to Yambuya Camp on the Aruwimi. You add: 'On arriving here place yourself under the Major, for he will, no doubt, when you join, push on after me with you all.' After having carried out all your instructions as to the disposal of the surplus stores, and having loaded up with those I was ordered to bring up to Yambuya, I left Leopoldville on the 5th July, picked up Ward and Bonny at Bolobo, and reached Yambuya on the 14th August. Having to the minutest detail carried out all your instructions and landed every packet that you ordered, handing them over to the Commander of the Camp, I also placed myself under his orders as instructed. Your written instructions were given me to read by Major Barttelot, and I need hardly say that I perused them with great care, reading them more than once. Then I had a conversation with Major Barttelot regarding the men promised by Tippu-Tib, and he informed me that he had heard nothing of them; and that when he went up to the Falls in the Mission steamer in the previous June, Tippu-Tib had made constant inquiries about the powder which each man he was to procure was to receive; he was somewhat put out when told that it would not arrive until the second trip of the steamers. Major Barttelot went on to say that you had left on the 28th of June, and that you expected to be back in November, and that if we did not get the men there would be nothing for it but to await your return, as we could not march out with our loads with only 200 men. As near as possible his exact words as to what took place between you and him (repeated again and again in Jameson's presence and confirmed by him) were—'Almost the last words Stanley said as he whisked himself out of that gate were: "Good-bye, Major, I shouldn't be a bit surprised to see you here when I return. I shall be away five months, giving myself three months to go up to Kavalli and two to return, so I will be back in November."' Now to show that he had this firmly rooted in his mind, he writes to the Committee in obedience to your commands—['With the *Stanley's* final departure from Yambuya, you should not fail to send a report to Mr. William Mackinnon, of Gray, Dawes & Co., 13 Austin Friars, London, of what has happened at your camp in my absence, of when I started away eastwards, whether you have heard of or from me at all, when you expect to hear and what you purpose doing. You should also send him a true copy of this order, that the Relief Committee may judge for themselves whether you have acted or propose to act judiciously.']—a letter, dated August 14th, the very day I arrived in the second trip of the steamer, and this letter was taken home by Walker and published by the Committee in the newspapers of November 28th, 1887. A copy of which I enclose marked B. What does he say? I particularly call your attention to the paragraph at the end of his letter. 'If, however, they (the men promised by Tippu-Tib) do not (come) *I shall be compelled to stay here till November, when Mr. Stanley expects to return, or till such time afterwards as he may return.*' Studying your written instructions, with these verbal additions, as furnished by the officer you left in command, what was to be gathered from them? Simply this: you mention certain contingencies and give us certain alternatives, which I condense into the following formula:—

"(1) Advance when the steamers return if Tippu-Tib has sent the promised 600 (six hundred) men.

"(2) Advance with the men you have, if Tippu-Tib sends some but not all of the men.

"(3) If Tippu-Tib does not send any men make marches twice over, or

"(4) If this obliges you to throw away too many loads, stay where you are till I return.

"Twist the paragraphs in any way you like, I cannot see what other meaning they could convey to any one, and although I endeavoured as months went on, by reading and re-reading them over and over again and discussing them with the other officers, I was unable to put any other interpretation on your words than what I have stated above. Such, I say, was the meaning to me of your instructions, and as the others concurred, it was decided to use every endeavour to get the men from Tippu-Tib; as they had not, by this time, put in an appearance, Major Barttelot therefore sent Jameson and Ward to the Falls, a few days after my arrival, to inform Tippu-Tib that the steamer had come up with all the stores, including the powder for his men, and he begged him to send the 600 (six hundred) men as he promised to do in the agreement made at Zanzibar. Jameson and Ward saw Tippu-Tib, who said he had made an attempt, by coming up the Aruwimi, to find our camp, he had with him 600 (six hundred) men; having failed to reach Yambuya, they all returned to the Falls. He then promised to send at once as many men as he could, but was afraid he could not get 600 (six hundred). As you are aware, these men were not forthcoming; many reasons might be mentioned for this, but I will not enter into them here. To consider another of your propositions, I would say, as regards making marches twice over, in my humble opinion, it would have ended in nothing but the loss of all the valuable stores of the Expedition. When we arrived in camp on August 14th with the detachment from Bolobo, the total of our force was 246, consisting of 200 Zanzibaris, 44 Soudanese, and 2 Somalis. I brought up in the *Stanley* 479 loads, and with what you had left in the camp there was a total of over 700 (seven hundred) loads. Of the 200 (two hundred) Zanzibaris which we had, I would say here that in my experience with this class of men, I never came across such a lot of *goy-goys* before. The bulk of them surpassed even the bad lot known as the July detachment of 1885, in the Congo Free State. Here we are entrusted with the care of some 700 (seven hundred) loads, which have been brought all the way from

Zanzibar to the mouth of the Congo, and then transported at great labour and cost to the Upper Congo. for the purpose of being carried to Emin Pasha. We are told these stores are everything to the Expedition, 'without them we should be in the same position as Emin Pasha, wanting relief ourselves.' Bearing this always in mind, impressed on me by your letters and final instructions at Leopoldville concerning these precious stores, I cannot see how with 200 (two hundred) men, or rather less, we were to make marches, even twice over, without throwing away at least 300 (three hundred) loads, nearly half of those in our charge. This matter was never seriously discussed, the commander turning his attention to getting the men from Tippu-Tib; without them he determined he was not competent to march out, and failing their appearance, he decided on the option, left him by your instructions, to remain for your return in November. If the getting of Tippu-Tib's men was of so little importance—'not absolutely necessary,' as you say in your letter to Sir F. de Winton, dated August 31st, 1889, and consequently the transportation of the bulk of the loads to Kavalli equally so, surely it would have been better to let this be clearly stated in the instructions. Why not have written to the effect that when the remainder of the stores and the Bolobo contingent had arrived in August, and if Tippu-Tib had not then found the men, we were to march at once with those we had of our own, make marches twice over which would have enabled us to take about 400 (four hundred) loads, specify these, and tell us to throw the remainder into the Aruwimi or give them to Tippu-Tib! Then there could have been no hesitation as to these precious relief stores, you would have yourself indicated how little you valued them. We should then have had no option, no discretion left as to what we considered best to be done under certain circumstances. You may maintain that your verbal statements to Barttelot, which came to my knowledge for the first time when I read your letter to Sir F. de Winton, were to this effect, but had I heard such from him, I should have ignored them as contrary to your written instructions, which bore your signature, and I should have considered anything else unauthorized. If we of the Rear-guard were simply an exploring expedition, and had only been left behind to await the arrival of a few supplies of cloth, &c., to enable us to barter with the natives, we would have advanced with whatever men we had, but this was not the case. We were the Rear-guard of a *Relief Expedition,* and were entrusted with the safe keeping and transport of a large quantity of stores. Our instructions were to bring up these stores if we could to Kavalli; but if unable so to do, to guard them safely till your return. It was not a matter of five men rushing bald-headed after you and letting the loads go hang, for you may be sure that had I had this option, I should not have hesitated to follow on your track sooner than have sat down in Yambuya Camp * * * It was clear to me that under all these circumstances, that is not having obtained the men promised by Tippu-Tib, and sooner than throw too many loads away, there was nothing for us to do but to choose the alternative left to us by you, viz. to remain for your return, which we were given to understand by Barttelot would be in November, five months, and which time tallied with your postscript to the written instructions:—'Give one brass rod per week to each man to buy fish, &c., in five months these amount to 2580. Give also six cowries per man per week, in five months these will number 15,480.'

* * * * * * *

"As you placed him (Barttelot) in supreme command, making me, like all the others, his subordinate, I decline to accept your statement that I was in any way responsible for the disaster to the Rear-guard. Yet, while repudiating responsibilities for consequences for which I was not liable, I maintain your instructions gave us the option of remaining at Yambuya until we got the men or you returned, as you positively stated your intention of doing. When Tippu-Tib sent some of the men Major Barttelot marched as instructed. As to the Committee in London understanding the instructions differently, that is, that we ought to have marched in August; if so, they should have sent instructions to Major Barttelot when he stated, in his letter to them, his intention of awaiting your return. I have written thus at length, and stated matters as plainly as possible; but I may here remark, that I am much surprised that you have neglected to ask me, the senior surviving officer of the Rear-guard, for any statement of the course of events at Yambuya, especially as I may presume that it is your intention to write further on the subject in your forthcoming book; but I do not wish you to have the power of saying that these facts herein mentioned have never been brought to your notice. There are several other points on which, to judge from your statements, you have been misinformed, viz. treachery of Tippu-Tib; sending of Ward to the coast; sending of loads to Bangala; two scenes of severe punishments in camp, &c. &c. And if you desire any information on these or other subjects of which I have cognizance, I need hardly say I am ready to lay it before you, *if you simply ask me to do so.*

"Trusting you will give the matters I have mentioned your careful consideration,— "I remain, yours faithfully,

"J. ROSE TROUP."

The above letter received almost immediate attention, for the short note from Mr. Stanley given below reached me soon after. He does not even here ask for any full report from me as his officer. Mr. Ward's statement referred to in this is given in his book, *In Darkest Africa*, vol. i. p. 496, and covers two closely printed pages.

"VILLA VICTORIA, CAIRO,
"*March* 15*th*, 1890.

"DEAR MR. TROUP,

"Your letter is very interesting, but I am sorry to say that it is so intemperate that I cannot use it in my book as it stands, whereas nothing would have pleased me more than to have allowed you to state your own case. There are some things in it that would do very well, but unless you give me permission to expunge some intemperate expressions, of course I can do nothing with it. If you will let me know, I will try and manage to let you state your own case without injuring yourself by attacks on other people, or if you yourself will write such a statement that I can make use of in the same way that Ward has done, nothing will please me better. I thank you all the same for your letter.

"I am, yours very faithfully,
(Signed) "HENRY M. STANLEY.

"J. R. TROUP, ESQ."

I had not written my letter for publication, but to give information to him, my chief. However, to be courteous to him, and in order to discover just what he meant by my "intemperate expressions," I wrote him a short note, which I give here in full.

"*March 27th*, 1890.

"DEAR SIR,

"I am in receipt of your letter dated Cairo, March 15th, and hasten to acknowledge it, and thank you for the prompt reply to mine of March 4th. I was much pleased to hear directly from you, and to learn that you had read my statement concerning certain events at Yambuya. I am also glad to know that you contemplate inserting in your book what I said, provided I am willing to omit the intemperate language you refer to. I regret that you consider the language in any instance intemperate, and that you think that in my attempt to place the matter clearly before you, I have, in stating the case, attacked others. But I am quite ready to modify my remarks as far as I can do so consistently with the truth. You do not, unfortunately, say what statement or words you desire eliminated, so I beg you to underline in red ink the phrases to which you object, and return me the letter, that I may then be able to judge whether they can be struck out or altered. I suggest this as the method least likely to occupy your valuable time, but of course I should prefer that you should draw out the sort of statement you would like me to make for insertion in your forthcoming volume. I would try then to adopt the moderate language you suggest, and it would be more in consonance, probably, with the rest of your book.

"I regret that, owing to your absence from England, it has been impossible for me to talk over this matter with you; it would be so much easier in a conversation to lay the subject clearly before you, and you could then give me an outline of what it would be agreeable to have me state for publication. As I understand that you are to be in England very shortly, perhaps you will give me the opportunity of talking it over.

"Trusting you will favour me with an early reply, I remain

"Yours faithfully,

(Signed) "J. ROSE TROUP.

"H. M. STANLEY, ESQ.,
"VILLA VICTORIA, CAIRO, EGYPT."

Mr. Stanley thereupon wrote again, but failed to comply with my request, that he would point out my intemperate language.

"VILLA VICTORIA, CAIRO,
"*April 5th*, 1890.

"DEAR MR. TROUP,

"I send you your letter back, and I beg you will be good enough, if you wish it put into my book, to reduce your arguments into *two* of your sheets, not more, and to write a plain, temperate narrative, without any argument whatever, as to why, having received instructions to follow on our track on arrival of the Bolobo contingent, and concentration of the Rear Column at Yambuya, you preferred to rest inactive at Yambuya for eleven months. Why, having contracted loyal and devoted service, you should render your chief naked. Why, instead of distributing the provisions, as is mentioned in the letter of instructions, among the sick men, and among the thirty-three abandoned in camp, you sent the provisions down river. Why, after knowing that Tippu-Tib had promised to be there in nine days, and then in a second term of ten days, and then in a third term of twenty days, you perpetually pinned your faith to him instead of pinning it to your duty. Why, all of you, burning, eager, interested in the Relief Expedition, decided to sit down when action would have been preferable, and would have taken you further towards the Albert. Why you spent 1200 miles running between Yambuya and Stanley Falls, instead of putting that distance between Yambuya and the Albert. Why, with a plain letter of instructions before you, you should send Ward down to the sea, and so have me lose the services of a young man who professed devotion. No use for you to charge Barttelot or Jameson as having been the cause of it, because you are mentioned in the instructions that if any vital step is to be taken, your voice and that of Ward must be taken too. Major Barttelot was chief, but the letter of instructions expressly says that he cannot take any vital step without you. It is a vital step to deprive your employer of his baggage and clothing. It is a vital step to leave thirty-three men starving in the camp. It is a vital step to go contrary to the spirit of the instructions, which are that you are to follow the track. It is a vital step to change the spirit of the instructions in the slightest, even though they were drawn in conformity with your wishes.

"Now if you can condense your narrative to two sheets, and not make the mistake of thinking that you defend yourself by attacking Barttelot, I may possibly find room in the book, but I really am doubtful.

"My address will be, Royal Palace, Brussels. That is the first stopping-place where I could possibly receive your letter.

"Yours faithfully,

(Signed) "HENRY M. STANLEY.

"J. ROSE TROUP, ESQ."

He wishes me to condense my arguments, which must not be arguments, into less space than he allows Mr. Ward—in fact, two sheets would have been little more than that occupied by the above letter of his. He also puts forward a series of questions, brings against me grave charges, and holds me responsible for events that occurred after I had left the camp. These questions are answered as follows :—

1. *I* never received these instructions, but I was shown those given to Major Barttelot, who was left in command of Yambuya. *I* did not prefer to rest inactive there, but was forced to do so by circumstances, and in obedience to my commanding officer.

2. I had nothing to do either with clothing or "rendering naked" my chief—the matter was not referred to in my contract. It is presumable that this points to the fact that some of Mr. Stanley's personal effects were sent to Bangala. By reference to page 20 of this volume, it will be seen that I knew nothing about these loads; they had never been entrusted to my care, and I was ignorant of their contents. I was too ill to have anything to do with, and was unable to prevent, their despatch.

3. There is not a word in Mr. Stanley's instructions (which are given at length on page 139 *et seq.* of this book) about distributing provisions to sick or other men; and moreover, we had none to distribute. It will also be seen that he gave no express orders that Messrs. Ward, Bonny, and myself should receive provisions.

I had nothing whatever to do with these provisions. Until mentioned in this letter of Mr. Stanley's, I never heard that thirty-three men were left in camp. I was carried out of Yambuya in a hammock on June 8th, and it was not until June 11th that the Rear Column set out from that place; at that time it may have been that sick men were left behind, but I had no knowledge of such fact, neither was I responsible for it.

4. It was not a matter of pinning one's faith to Tippu-Tib. The commander of Yambuya, by Mr. Stanley's instructions, was given the option of remaining where he was if Tippu-Tib failed him, and he wrote to the Committee on his own account the day after I arrived in camp (August 15th) as follows:—"Up to the time of writing this Tippu-Tib's men have not arrived, though these marauding Arabs before mentioned may be them. Should it turn out to be so, or should they eventually come, I should march at once in Mr. Stanley's track. If, however, they do not, I shall be compelled to stay here till November, when Mr. Stanley expects to return, or till such time afterwards as he may return."

5. Action, by which he seems to mean advance, was, in the circumstances, impossible unless we forsook our loads, and would have been an impracticable undertaking even at the outset, for our 200 men were picked out by Mr. Stanley as the worst of the 600. Where he succeeded, as he says, through the intervention of "a gracious Providence," in marching 400 of the best men, carrying few loads, how can any one expect that we with 175 inferior men could carry full loads four times over the same course? As time went on our numbers were reduced, and it became more than ever impossible. Mr. Stanley demands of his subordinates what even he would have found a task taxing his powers to the uttermost. If, however, the loads were unimportant, as he indicates in one of his letters, why did he not take with him all the officers and men with such loads as could be carried?

6. It is difficult to understand how 1200 miles were spent in running between Yambuya and Stanley Falls—a distance of less than 100 miles. The country between these places was traversed only a few times by individual officers of the Expedition, and they did not carry 60 lb. loads with them. Even if the distance traversed equalled the figures Mr. Stanley gives, the journeying of unladen men was a decidedly different thing from forcing 175 men, carrying

700, or even 500, loads, to march four times over through the jungle that Stanley could scarcely penetrate with his 400 picked men in light marching order, and where they nearly starved.

7. *I* had nothing to do with the scheme of sending Ward to the coast. I was a subordinate officer. All the arrangements for the trip were made by Major Barttelot while he was absent from Yambuya, and I was ignorant of his intention until he returned. By reference to page 225 it will be seen that I distinctly disapproved of it, and it was done without my sanction.

8. I have never associated Mr. Jameson's name with such charges as Mr. Stanley suggests.

9. Compare the words "your voice and that of Ward's must be taken too," with these of the original instructions which he is supposed to be quoting—"When Messrs. Troup and Ward are here, pray admit them to your confidence, and let them speak freely their opinions." In these there was no imperative *must*, and Barttelot was only requested to admit us to his confidence, and let us speak freely our opinions. This cannot possibly be construed into placing us on equal terms with Major Barttelot, or even with Mr. Jameson. It does *not* say that Major Barttelot "*cannot* take any vital step without you."

10. The vital step of depriving Mr. Stanley of his clothes was done without my knowledge, when I was too helpless to prevent it.

11. The vital step of leaving 33 men starving in camp occurred, if it did occur, three days after I was carried out of Yambuya.

12. The spirit of Mr. Stanley's instructions was not made clear *if* this was it; as any reader can see for himself, we had the option of remaining.

I felt that if I forwarded this *reductio ad absurdum* of his questions to Mr. Stanley it would only annoy him, so I contented myself with the following general reply. I saw at this late date, when his volume was in the hands of the printers, that it was useless to expect that he would publish any statement I wrote out, as he distinctly says—"*I may possibly find room in the book, but I really am doubtful.*" It was too apparent that he wished to ignore the report of one of his surviving officers. I took care to register my reply.

"*April* 17*th*, 1890.

"DEAR SIR,

"I am in receipt of your letter of the 5th inst. from Cairo. It was not my desire to furnish you with any statement intended for publication when I first wrote to you; my letter of March 4th was only written to lay before you, for your personal guidance, what I considered to be a true explanation of circumstances attending our year's delay at Yambuya, which would correct any misapprehension of those facts which you might entertain. I see it is impossible to modify those statements in the direction that you now indicate. I had no cognizance of several of the matters referred to by you, and the rest are totally contradicted by what I have already written. I do not wish to shift all blame of what occurred at Yambuya on to Major Barttelot's shoulders, but I contend that by your letter of instructions he was left in sole command, and was merely given the option by you of consulting the others if he chose, and was never forced to accept our views. By this it follows, that instead of having an equal right with Major Barttelot, as you assume, of directing the affairs of the camp, I and the others only had such right as Major Barttelot chose to allow us. I feel that I have placed before you a sufficiently clear outline of what happened at Yambuya in your absence to enable you to form a just estimate of the true position of affairs there, and having thus taken the necessary steps to prevent you from falling into error, I consider that I have done my duty. Whatever decision you may come to as to the judgment to be meted out to those at Yambuya, I feel that I have done all in my power to enable you to arrive at a right one. Having regard to what I consider the explicit nature of the statements contained in my letter of March 4th, I do not see my way to meet your suggestion to condense those statements into two sheets, while repudiating all intention of endeavouring to *defend* myself by unduly blaming Major Barttelot.

"Yours faithfully,
"J. ROSE TROUP.

"H. M. STANLEY, ESQ.,
"*Royal Palace, Brussels.*"

I trusted that he could gather from this, that I thought he was partly to blame for the disaster to the Rear-guard, and also that I did not consider that I had done wrong, and required to defend myself.

Our correspondence now ceased for a few months, and I had by this time given up all hope of receiving justice in his book; but when I was in London early in May, when he was making his first speeches, I called upon him. He received me well, and listened apparently with keen interest to my narrative, which was an elaboration of my letter of March 4th. Being unable to finish all that he wished to hear in a conversation that lasted three hours, he asked me to call again at 4 o'clock on Wednesday, May 7th, and I then continued my story for another three hours. He asked question after question, but seemed to admit that my arguments were unanswerable, only he continued to insist that we should have tried to march out with the loads, making extra journeys, although I pointed out that my belief was, that if we had done so it would have caused an even more serious loss of life and loads than did occur when the march was at last undertaken. He assured me that *he* could have managed to make the men march back and forth three or four times over, but I replied that it was one thing what *he*, with his experience and influence, would be able to accomplish, and quite another what Major Barttelot, who had not had his experience, could have done.

Up to this time, it should be noted, his book had not been published, and it was quite within his power to make additional notes to indicate what information he had received from me, or to correct his inaccurate statement that there was no answer to the simple question propounded on page 6 of the first volume of *In Darkest Africa*. But he did not do either.

It was after this that he entered into the agreement with me mentioned in my preface. When I read the prefatory letter published in *Scribner's Magazine* for June, before his book appeared, I wrote a letter to the newspapers, at the same time sending him a copy, pointing out some of his mis-statements. Of this he took no notice. When his book appeared I felt it incumbent upon me to write again, pointing out the very serious errors and grave charges made therein, and contradicting them. This I did in the following words:—

"*July 3rd*, 1890.

"DEAR SIR,

"I have just received your volume, *In Darkest Africa*, in which I find you have repeated certain statements in spite of my clear denial and correction of the same. Aside from unnecessary aspersions cast upon me and others, you, on page 472, blame me in these words: 'Three young Englishmen come last, who are attached to the Major's staff, two of whom—Mr. Herbert Ward and Mr. Troup—are to be associated with the commander and his second in the discussion of every vital step, and no important decision can be taken unless a council of the four has been convened to consider it as to its bearing upon the enterprise for which they have assembled on the verge of the unknown region of woods. They are therefore implicated in the consequences of any resolution and every sequent act.'

"I have stated in writing and by word of mouth that I was *not consulted*, and *was not* called upon to take part in the discussion of many of the acts decided upon by our commanding officer, and I therefore repudiate the blame cast upon me, and ask you to retract this incorrect statement, made with full knowledge of the facts. Also on page 474, speaking of the loads sent to Bangala by Major Barttelot, you state: 'But Messrs. Jameson, Troup, and

Bonny are present concurring and assisting, and the two last-named receive salaries, and both present their accounts and are paid, not a penny deducted, and a liberal *largesse* besides in first-class passages home is granted them' (!).

"You know equally well, by statements made by me, that I was not present, and was incapable of assisting in the despatch of the loads referred to: this is borne out by the statement of Major Barttelot in his report, and by Mr. Bonny's medical certificate. All the time I was on my back, unable to move hand or foot, and had been in this condition for six weeks previous to my departure from the camp, my condition being largely due to the absence of proper medicines. I was not out of my hut during that time, and was unable to participate in any councils that may have been held. Being well aware that I was not present or consulted, yet you blame me for this action, whereof I am entirely innocent.

"Your further reference to the fact that I received the miserable pittance of £150 a year, voluntarily offered me by you, and that I had the marvellous *largesse* of a first-class passage home, sounds strangely when we read elsewhere in your book of the payment of expenses, and the *first* donation of £400 each to your four officers, two of whom were drawing full pay in her Majesty's Service during the three years they were with you. But this unnecessary slur on me is hardly worthy of notice. There are very many other statements of yours that I can prove to be totally inaccurate, but I will not touch upon here.

"I would request you to publicly retract you mis-statements, or else I shall find it necessary to take such steps as I think fit to clear my character of the aspersions you have cast upon it. I would presume that you would have no objection to the publication of this letter, and any reply I may receive from you.

"Yours truly,

(Signed) "J. ROSE TROUP.

"H. M. STANLEY, ESQ.,
"De Vere Gardens, London, W."

Mr. Stanley then wrote a letter of nine folio pages, quite too long for insertion in the newspapers, and as it did not contain a word of reply to my last letter, it did not seem necessary to publish it at that time; so I have reserved it to appear in this place:—

"*De Vere Gardens, W.*
"*July 5th*, 1890.

"DEAR SIR,

"I am sorry that the great demands on my time from various people and varying circumstances will not permit me to reply at any great length to your letter. I have before this received your written statement, but—as I wrote to you—the intemperate language and the reflections on the Major required eliminating before it could enter my book. Mr. Herbert Ward wrote a statement also, but, as you observe, it is a clear straightforward narrative, and it was published by me in justice to him. You declined to alter your statement, so as to relate the narrative from your point of view, without needlessly attacking one who is dead, and cannot answer for himself, and you could show no letter or written communication from the Major which would prove that you protested against certain resolutions which involved the Rear Column in disaster, and subjected its members to appalling suffering.

"A mere glance at your contract ought to convince you, on reviewing all, that you have failed in the performance of your duty.

"'1. I agree to place myself under the command of Mr. H. M. Stanley, and to accept any post or position in that Expedition to which he may appoint me.

"'2. I agree to serve him loyally and devotedly.

"'3. To obey all orders.

"'4. To follow him by whatever route he may choose.

"'5. To use my utmost endeavours to bring the Expedition unto a successful issue.

"'6. Should I leave the Expedition without his orders, I agree to forfeit all claim to pay due to me, to return passage-money, and to become liable to a refund of all monies advanced to me for passage and outfit.

"'7. I undertake not to publish anything connected with the Expedition, or to send any account to the newspapers, for six months after the issue of the official publication of the Expedition by the leader or his representative.'

"Can you declare in all seriousness that you have been faithful to any one of these seven articles enumerated above? If you can persuade yourself that you have done so, I cannot but admire your easy views of duty, and I hope that you will always find an employer disposed to take the same view as yourself of your duties. As to Article 1, my command has been issued in the letter of instructions to you and your comrades alike. Can you solemnly say that you observed the command or followed the suggestions? If so, give me proof—not your diary—but a letter or note from Major Barttelot acknowledging the receipt of a protest from you against staying at Yambuya, against sending my baggage down river, against deporting the vital effects of the Expedition, such as medicines, provisions, wine, &c., against sending yourself home, against shooting the Soudanese[1] for stealing a piece of meat, against the reported severities we have heard of, against one act to which I—were I in Yambuya—could have taken exception to.

"2. Do you consider you served me loyally and devotedly by stripping me naked, or remaining silent while I was being reduced to poverty?

"3. Do you consider you obeyed my orders when you went home?

"4. Did you follow my route?

"5. Did you use your utmost endeavours to bring the Expedition into a successful issue, by skulking in your house for some weeks, by brooding over fancied wrongs (these I only know of through your own verbal confessions to me), and finally accepting your sick leave from Major Barttlelot? Well, Stairs was sick too, but he pleaded to be allowed to follow us; the Somalis were sick also, but none pleaded to be sent back. We have all been sick, but you are the only one I have ever had on an African Expedition who accepted his leave from a subordinate.

"6. Having left the Expedition by permission of the Major, did you agree to forfeit all claims to pay due you, &c., &c.?

"7. About the last clause, can you deliberately say you denied yourself the pleasure of publishing, or attempting to publish anything?

[1] This Soudanese was only flogged for this offence (see *Diary*, Dec. 4th). The Soudanese who was shot had deserted, taking with him a rifle (see *Diary*, Feb. 9th).

"Before responding to these questions, try to understand that Major Barttelot is my representative with the Rear Column for the 'ordinary duties of the defence and conduct of the camp, or of the march,' but should any vital step be proposed to be taken, he is to take the voice of Mr. Jameson, and of Messrs. Troup and Ward when they are in camp, and these gentlemen are to be admitted to his confidence and to be allowed to speak freely their opinions. You say you were not consulted or called upon to discuss acts decided upon by your commanding officer. I ask, why not? The same commission that recognized him as the senior officer of the Rear Column, gave you also a right to speak freely your opinions, and to be consulted upon all vital steps proposed. Why was this right not exercised as much as the other? You are, it is true, to submit to his authority, but you are also to observe that his authority is exercised judiciously and wisely, because you are bound to be loyal and devoted to the Expedition as per contract—and your reputation depends on it, your friend and friends rely on you, and if you will reflect a little, for a score and more of reasons besides. The commanding officer has no right to exercise any authority unwisely, injudiciously, tyrannically, or cruelly, and you are not expected to obey him if he attempts to do so, because you are all expected to render an account of what you have done later on, and naturally you will expect recognition from your friend and principal in the Expedition, of the good that you have done. This letter of yours to me, I take it, is a plea for this recognition. Well, state the good that you have done; prove it, and I am satisfied. I desire to know it, to see it with your eyes and appreciate it.

"As far as Yambuya I observe that you are the same Mr. Troup who won from me the following tribute in *The Congo and the Founding of its Free State.* 'Among the later British arrivals who have been conspicuous for their capacity and their all round fitness for special fields of labour are John Rose Troup, &c.' 'It is my belief that in Mr. John Rose Troup we have a thoroughly good officer,' &c. The stores which you have had in your charge are delivered safely at Yambuya. The lists of them are made out in a methodical and business-like manner. There is absolutely nothing to complain of but much to admire, and to bear me out in my statement to the advance officers, which is, 'Whatever the Major may be, I rely upon

Troup.' But on reaching the wrecked Rear Column I cannot hear of one sensible act done by you or any of your comrades. I hear much after reaching Yambuya of brisk journeying and long travel, of sickness and suffering—and alas! I hear of many things which had better never be mentioned—and what I see dare not be described—because it will serve no useful purpose to do so, and I wish to forget what cannot be amended. My limited knowledge of the actual facts will not permit me to judge of who is to blame for this fearful condition of things. What is the use of enumerating anything of what strikes me dumb with horror, and aghast with wretchedness. It is enough that day after day I hear of nothing else—and never of what I should call one sane act. I am not competent to decide for the public who is right or wrong—Bonny presents his written statement, Ward gives his, and you give yours—but in the absence of documents, letters, or notes, to substantiate your statements, each must be revised to be fit for publication, that is, you may say what you will in your own defence, but I cannot consent to publish any violent or intemperate remarks upon the Major. You may see what I mean by reading Ward's narrative or Bonny's statement, and when you send me such a story I will publish it with pleasure. But if you think that is not sufficient, you have the privilege of stating anything you please in your own book after October 15th next.—My public and private opinions differ somewhat. My publicly expressed opinion of who is to blame must be based upon statements, documents, or other evidence than my own belief, and none of you can furnish anything to prove that the Major was in the wrong—that you did what you ought to have done, protested against those acts which caused so many deaths and misery, demanded that a council should be called for consultation, and to be heard for the good of the Expedition—and failing such demand being listened to, compelling a declaration from the Major in writing that he relieved you and your comrades from all responsibility connected with the fate of the Expedition, inasmuch as he, and he alone, was responsible for it. My private opinion is reserved for myself alone. Some benefit of it, however, I give in my book to you and your friends. I can give no more without evidence from both sides. You three survivors of the Rear Column can give sufficient testimony to prove conclusively that the Major took upon himself the disposition of the property of the Expedition, and the entire command of you and the Column; but there is not one tittle forthcoming to prove that either of you, though bound to do loyal and devoted service to me as Commander of the Expedition, remonstrated, expostulated, or protested formally, as intelligent devoted officers, and as humane Englishmen, against the tyrannous exercise of that control and command. Do not ask me—for your own sake—what instance I mean, lest in so doing you confess yourself as being morally blind. Accept from me that I know of many acts committed at Yambuya, the hearing of which has filled me with horror and grief. Did you attempt to stop any of these acts? Did you write a letter of protest against any one of them? If so, give me a copy of it, and the Major's answer. Did you lift your voice in condemnation of such acts? If so, give me proof certified by a written admission from the Major that you did so. The Major has gone to his final account, he cannot speak for himself, but a note or a letter from him during his life with you, admitting that he heard your protest, that he disregarded it, and acted on his own responsibility, would exonerate you greatly, though not completely. Even that would not prove that you were a loyal, trustworthy, intelligent officer, because to merit such a character no loyal, trustworthy, and intelligent officer would have permitted himself to be deprived of his rights and privileges, secured to him by contract. I have not met anything like this in the whole course of my life. You want medicines, and you say you cannot get them. You need medical comforts—you say they are denied to you. You desire to be of use to me, and yet not one of you seemed to know how to set about it. You know your faithful Zanzibaris cannot be replaced in Mid-Africa, and yet you look on and see them die without making any effort to save them. You say that the Major is the cause of all this—and not one of you seem to know how to persuade the Major to do what you profess to believe, and to know ought to be done; you say that the Major is going beyond his powers, but yet you do not know how to protect yourselves from the consequences of his acts.

"Before you leave Yambuya for England you write a most amiable letter to me, and yet no sooner do you reach England than you commenced transgressing the terms of your contract, and compel my agent to go to law to protect a man to whom you professed devotion. You may probably get the entire press of the world—for you are eloquent and smart in your writing, and clever and learned—to accept your view of your case; but I adhere to my own opinion still, which is that my conscience will not acquit you and your companions from blame in the case of John Henry, the Soudanese soldier, the boy Soudi, and a dozen other cases; for surrendering without a decent struggle for your reputation as officers and Englishmen, and freemen of equal breed and intelligence, rights and privileges to which you were fairly entitled—for retiring to brood sullenly in your huts, when a calm, temperate, polite, but resolute attitude, such as duty and common-sense dictated, would have saved the Rear Column from the miserable and helpless state in which I found the dying and despairing remnant of it.

"If you can supply me with that which is necessary to correct this opinion, no one will be readier to acknowledge his error than I, your friend of Congo days,

"Yours truly,

(Signed) "HENRY M. STANLEY.

"JOHN ROSE TROUP, ESQ., &c."

In the foregoing letter he brings forward disgraceful charges, that really do not refer to me at all, although he blames me for what happened. The injustice of his accusations, made as they are without documentary or, as far as I can learn, any evidence, can hardly be made clear to the public, but they must be aware, when they read what has preceded this correspondence, that he has acted as no one in his position should have acted. It is vain to try to express the indignation I felt as I penned the following lines:—

"11*th July*, 1890.

"DEAR SIR,

"I have received your letter of July 5th, which is not a reply to mine of July 3rd, for if you will re-read that you will perceive that I called your attention to two incorrect statements of yours made with full knowledge of their inaccuracy, and in the nine folio pages of your letter no reference is made to them. Your silence on the subjects condemns yourself, as you thus admit that I am right. Your letter is instead filled with more misstatements and unjust accusations, which are entirely unsupported by documentary evidence, and you charge me with being responsible for events that occurred after I had been sent home by the medical officer, and you overlook the documentary evidence of his medical certificate, which refers to my serious illness, and the entire absence of proper medicines. You call upon me to produce documentary evidence that would incriminate my superior officer. As I told you, I was not in Africa to collect evidence against Major Barttelot, and did not devote my time to getting up a case at law against him. You call my attention to certain paragraphs of the contract made between yourself and me; I would, therefore, ask you to recall the last paragraph but one, which reads as follows:—'In addition to the outfit, Mr. Stanley will supply the following—Tent, bed, 1 Winchester rifle, 1 revolver, ammunition for the same, canteen, a due share of Europeon provisions taken for the party, besides such provisions as the country can supply.' I would ask you if you consider that you had fulfilled this contract of yours, as the only article mentioned therein that I received was a revolver, and I would call upon you to produce documentary evidence that I received these articles, or even that these articles were taken to the Camp and left there for me. I have stated on oath that I did not receive these promised articles. Speaking of documentary evidence, you have mentioned a 'death-warrant signed by the officers,'—if this document can be produced, it will be found that it does not have my signature attached, as I never heard of its existence till you mentioned it; you bring various unsupported charges against me, a matter much more serious to have made without documentary evidence than my failure to produce documentary evidence against my commanding officer. How I could have forced him to take me into his counsels and confidence I don't understand, and I never heard of such a thing as a subordinate officer thus forcing himself upon his superior.

"In my letter I did not plead for recognition of my good services, which you have denied me, but I note that you state in your letter that if you knew them you would with pleasure acknowledge them. You thereupon give a page in praise of me, saying what good service I rendered the Expedition up to the time I reached Yambuya, but even this you have not given me public credit for; but as you promise to acknowledge what good I have done, kindly fulfil this promise by acknowledging these services even now. You state that your public opinion and your private opinion differ. This is an extraordinary statement, but it might be interpreted that as you publicly condemn me, privately you would praise me. You ask if I believe I have done my duty,—if my duty was, as you said, to tie up Major Barttelot and send him to the Coast, I have not done it, for that would have been mutiny, and if I had been shot for doing so, I should have deserved my fate. You have asked me many questions, let me ask you one. In your book you state you had a number of loads, and it is to be presumed that those 12 left at Yambuya contained clothing, as you constantly refer, in a ludicrously pathetic manner, to your nakedness. I was not responsible for these being at Bangala, but had I been, should I have committed an action likely to cause such serious consequences as that for which you were responsible, viz. that I did not receive tent or bed to protect me from the climate, a Winchester rifle to defend myself from the natives, or proper food or medicines to preserve my health? all these articles having been promised to me by you. You speak as if medicines had been obtainable in the camp, but the documentary evidence of Mr. Bonny states that there were no proper medicines, and as he was medical officer and commanding *pro tem.* in the camp, he knew more of what was there than you did when you were hundreds of miles away. You speak ungraciously of me as 'skulking in my hut,' when I was so ill that I could not move hand or foot out of my bed, the seriousness of my illness being testified to by the documentary evidence of Mr. Bonny's certificate. I continue to decline to alter my narrative, as I tell only the truth. I served you loyally and faithfully, and only when I was told that by continuing with the Expedition I should be an incumbrance, did I yield to Major Barttelot's repeated suggestion that I should be sent home. I never voluntarily requested permission to return. As he was your representative, as you yourself say, and you were hundreds of miles away, I had no option of asking your permission to be carried, for I could not move because of my weakness, out of the camp.

"It would be interesting to learn whether the four officers with you were allowed to interfere when they thought you acted unwisely.

"I did attempt again and again to stop many acts I believed to be injurious to the Expedition, and reiterated my remonstrances. You begin by stating that I had no letter from Major Barttelot, and on page after page you demand such a letter. You object to me because I do not produce such documentary evidence to support my statements, some of which I have made on oath, yet you are totally destitute of documentary evidence to support your *ex parte* statements about events that occurred while you were absent from Yambuya, and of which you were unable to have personal cognizance.

"Many of your questions contradict each other, and some statements you make are contradicted by yourself.

"Yours truly,
(Signed) "J. Rose Troup.

"H. M. Stanley, Esq."

APPENDIX A.

[COPY OF J. ROSE TROUP'S REPORT TO MR. H. M. STANLEY.]

"*Yambuya Camp,*
August 17, 1887.

"H. M. STANLEY, ESQ.,
"COMMANDING EMIN RELIEF EXPEDITION.

"SIR,

"I have the honour to report that I arrived here on the 14th instant, in the s.s. *Stanley*, with Messrs. Ward and Bonny, 130 men and boys, three donkeys, and all the stores and ammunition, as ordered by you in your letter to me of the 18th June last—all well.

"The s.s. *Stanley*, with the *Florida*, returned to Leó from the Aruwimi on the afternoon of Sunday, 3rd July. We loaded up on Monday, and left next day, the 5th. None of the other steamers had arrived when we started. M. Leibrechts, chief of Leó, had requested me as a matter of urgency to take back forty time-expired Bangalas to their homes, and also to forward some stores for the chief of that station. I consented, as I found we would have sufficient space for these men and stores with the aid of the whaleboat, which I had arranged to have when I met it coming down. On the day after leaving Leopoldville we met the s.s. *Peace* with Mr. Walker on board, who informed us that the s.s. *Henry Reed* had been seized by Captain Vangèle at Bangala, along with the *En Avant* and whaleboats. I had then to consider whether I ought not to send back the Bangalas and loads for that station to Leó in the *Peace*, as we would be crowded without the whaleboat after picking up the men at Bolobo. After consulting with the captain of the s.s. *Stanley*, and on his assuring me that he would be able to take the lot, I decided to go on as we were, after taking Mr. Walker on board. On the 10th we reached Bolobo, leaving again next morning, after taking Messrs. Ward and Bonny, 100 Zanzibaris (twenty were already on board, having gone down in the *Stanley* to help in cutting wood), and one donkey on board.

"We stopped at Lukolela for wood half a day, calling at the mission to give them their mail and two or three cases, leaving again about 1 o'clock in the afternoon.

"About 4 o'clock, going at full speed, we struck violently on something, rebounded, and then stuck fast. Found the fore compartment filled with water, and after throwing all the wood overboard, tried to steam off, but with no success. Soundings were taken, and over six fathoms of water found all round. A sail was drawn underneath the bows, and we found we were fast on the compartment under the captain's cabin, where the plates were indented, but fortunately making no water. We came to the conclusion that we had struck a big snag, tearing a large hole in the fore compartment, and then sticking on fast aft. We sighted some canoes, and shouting to them, after considerable delay, they came alongside, and consented to take the men off on payment of mitakos. This was done without a hitch before dark. We then tried again to steam off, but in vain, so we decided to wait till morning, and send some of the cargo on shore. Mr. Ward went to the village to arrange for the canoes to return there for that purpose. Steam was kept up all night, with a sharp look-out. In the morning the canoes came out, Messrs. Ward and Bonny going on shore to look after the reception of the cargo, whilst I remained on board to dispatch the same. After taking out over 100 loads we managed at last to steam off, going to the shore to pick up men and loads, and then making for a sandbank opposite as a more suitable place for beaching. We then took out all the cargo, stacking it, covered with tarpaulins; Messrs. Ward, Bonny, and myself keeping watch and watch every night over it. We had to remain here three days, whilst Mr. Walker and the engineer of the *Stanley* fixed a new plate on the hole made in the fore compartment. At Bangala we found the steamers and whaleboats; I requisitioned one of the latter, as although from there we had ample space after disembarking the forty Bangalas and stores, I thought it advisable to have a boat of some kind to communicate with the shore, in case of any further accident to the s.s. *Stanley*, considering her condition. However, I was informed I could not have one, as all the craft were required for the service of the State. We reached the camp here without any further serious mishap, with the exception of the donkey-engine getting out of gear. In accordance with your instructions I sold the surplus rice, between the State and Sandford Expeditions, and the surplus brass rods to the Dutch house; all subject to one pound sterling per load for transport from Matadi to Leó. I was unable to get a due bill from the different parties, both M. Leibrechts and Mr. Greshoff stating that as the expedition had an account with them, the value of these goods would be placed to the credit of the expedition. So I did not press the matter, taking their receipts and forwarding the originals to Mr. W. Mackinnon, Austin Friars, London, as requested by you. I beg to enclose copies of these marked A. B. C.

"The forty-one drums of engine oil was made over to M. Leibrechts as directed. There were twenty-three cases of biscuits over and above the number ordered to be brought up. I tried to sell them, but no one would have them, so I left them with M. Leibrechts. I beg also to enclose statement of loads left under my charge at Leó, distribution of same, and number landed here. By this statement you will see that all your instructions were carried out, the various payments made along the route, and the requisite number of loads brought to Yambuya.

"I had to leave one donkey boy very sick at Leó, also one donkey in the same condition. The other donkey boy, as well as the three Zanzibaris left behind at Kimpoko, came up with me. I was fortunate enough to be able to purchase eight sacks salt from M. Leibrechts, but I regret I could not get the soap anywhere. Two cases of Madeira came up for you to Leó after your departure; these I brought on along with the other loads, as well as your mail and the letters of the other officers of the expedition. You were good enough to say at Kinchassa that I might open and have a look at your newspaper, so I availed myself of your permission by opening the packet of newspaper cuttings. In conclusion, I would beg to mention that I received every assistance from Mr. A. B. Swinburne, whose guest I was at Kinchassa, as well as from M. Leibrechts, chief of Leó; both these gentlemen having wood ready at their stations at my disposal, thus enabling me to lose no time in starting on the second trip of the *Stanley*.

"I have the honour, &c.

(Signed) "J. ROSE TROUP.

"I enclose P.S. on a separate piecé of paper.

"P.S.—On 4th April, 1887, you wrote to me from Kwilu Ferry, requesting me to send down 200 men from Manyanga to Matadi to bring up surplus rice left there. This was over and above the original number of loads on the list. The men were sent down, but did not return in time. Before leaving Leó I received a letter from the chief of Manyanga, informing me that some of the rice had arrived at his station. I arranged with M. Leibrechts that he should purchase all that came to Manyanga, he undertaking to bring it up to Leó. You will note that your order to deliver at Bangala was 30 (thirty) cases of mitakos and 300 mitakos. I reckoned that the cases would be of 500 each, and this I found to be the case. M. Baert asked me how many those I had in sacking contained, and on my informing him 400, demanded repayment at the rate of 500 per case; so I handed him 38 bundles of 400 each = 15,200 mitakos, and 100 mitakos, making a total of 15,300 mitakos, equal to your order of 30 cases, reckoning at 500 per case, 300 mitakos. I had brought up some extra loads of mitakos on this account, and also because a number of the loads had nothing like 400 in them. In the original list brass and iron wire, beads, and cowries are put down as a total of 63 (sixty-three) loads. There are 73 loads, made up of 27 brass and iron wire, 33 beads, and 13 cowries, all of which I brought up.

"Some of the compartments of the *Stanley* let in water. When the cargo was taken out on the sandbank, the scene of accident, a number of the loads were wet. On arrival at the camp here I had them all dried and patched up. Some of the loads of beads and cowries I put into empty canvas-bags for protection, without taking them out of original packing. Some of the loads of handkerchiefs burst through, having got wet, portions of the sacking rotten. Used these first as required, keeping the better conditioned ones, as regards packing, for transport on road. Mitakos in sacking had to be sewn up again. Tippoo Tib has been complaining that the State have not sent up some of his loads left behind. The only loads of his noted in the list sent to me, arrived at Manyanga, 3rd April, 1887, and were forwarded to Leó next day, 4th April, 1887. These were 18 loads, and consisted of—

12	loads done up in sacking.	
2	„ „	sacks.
3	„ „	large tins.
1	case.	
18	total.	

These loads were taken up on the first trip of the *Stanley*, as stated on list handed by you to me, copy herewith marked W.

(Signed) "J. R. T."

A copy of this report I handed to Major Barttelot to give Mr. Stanley, and a duplicate to Mr. Bonny.

APPENDIX B.

CORRESPONDENCE RESPECTING THE EXPEDITION FOR THE RELIEF OF EMIN PASHA.[1]

No. 1.

Acting Consul-General Holmwood to the Earl of Iddesleigh. (*Received October* 25.)

Zanzibar, September 23, 1886.

(Extract.)

I telegraphed to your Lordship to-day, giving the heads of the latest intelligence from Emin Bey at Wadelai. I also transmitted to Sir Evelyn Baring at Cairo a brief account of Emin's situation.

The sources from which this information was derived were: letters from Mr. Mackay, of the Church Missionary Society at Uganda, dated the 14th and 26th May and 28th June; letters from Emin Bey, dated Lado, 16th November, 1884, and Wadelai, 1st January and 27th February, 1886.

Besides these, I have been shown a letter from Mr. Mackay to the Society's agent here, dated the 12th July, and two letters from Dr. Junker and one from Emin Bey to the German Consul-General. The last of these letters, dated the 16th August, was written from Msalala, the Church Mission depôt on the south of the Victoria Lake. I have also seen a number of Arabic letters written by Emin Bey and officers of his Staff.

[1] This Parliamentary Paper appears by the kind permission of the Comptroller of Her Majesty's Stationery Office.

Mr. Mackay writes that they are in great danger and practically prisoners, and that their every action is reported to the King, unless in instances where they can secure silence by bribery. After Dr. Junker's departure the King stopped the porters he had engaged to carry supplies to Emin, and it was uncertain whether he would eventually permit their leaving.

The other letters confirm generally all that Mr. Mackay writes.

Dr. Junker was proceeding at once to Uyui, and anticipated arriving at Zanzibar towards the end of the year. He speaks in the highest terms of Emin's capacity and behaviour, and of the heroism and loyalty of his troops and the 4000 natives who remain faithful to the Egyptian Government. He states, however, that Emin is urgently in need of ammunition and stores, and that though he purchased of Mr. Mackay cotton goods to the value of £400, and engaged porters to convey them to Wadelai, he hopes prompt measures may be taken for his more effectual relief. Most of the loyal Egyptian subjects in Emin's province, being natives of the district, had refused to accompany him to the coast, and Emin had written to say he would never abandon them unless they could be left in a safe position. Neither he nor Mr. Mackay mention the latest dates from Wadelai, though it is clear they have both received more recent letters than those inclosed. Dr. Junker is under great misgivings regarding the situation of our missionaries in Uganda, whom he describes as in great danger and virtually in slavery.

No. 2.

Mr. Portal to the Earl of Iddesleigh.—(Received October 25.)

Cairo, October 19, 1886.

(Extract.)

I have the honour to inclose herewith copies of two dispatches which I have received from Her Majesty's Acting Consul-General at Zanzibar, giving the latest intelligence which has been received on the subject, and containing his views as to the possibility of helping Emin Bey and the Egyptian troops with him.

The documents and letters inclosed in Mr. Holmwood's dispatches are all in Arabic, and are addressed, some to Egyptian

Ministers and some to private individuals. There is also among them a long Report addressed to the Minister of War. This and the other official correspondence is being communicated to the Egyptian Ministry.

I have also shown Mr. Holmwood's dispatches both to Nubar Pasha and to General Grenfell, Sirdar of the Egyptian army, who are considering whether any means exist within the power of the Egyptian Government by which the difficulties and dangers of Emin Bey's position may be lessened.

Inclosure 1 in No. 2.

Acting Consul-General Holmwood to Sir E. Baring.

Zanzibar, September 25, 1886.

(Extract.)

At the same time that I received the information regarding the situation of Emin Bey which I telegraphed to you on the 23rd instant, I also learned from the English missionaries at Uganda and from Dr. Junker, who quitted that country on the 12th July, that King Mwanga, the son and successor of Mtesa, had instituted a violent persecution against Christianity, that he had put to death or barbarously mutilated all natives suspected of being in communication with the Mission, and forcibly detained the missionaries themselves who wished to quit the country.

Dr. Junker had purchased stores to the value of £400, and engaged porters, with the King's permission, to convey them to Wadelai, but after his departure the King had refused to allow their transmission, and the latest intelligence received is that all Europeans in Uganda were in extreme personal danger.

I forward herewith an Arabic dispatch and Report from Emin Bey to the President of the Council, also other letters for various officials in Egypt, but fear there are others of more recent date which, up to the present time, have failed to reach Zanzibar.

In Emin's letters to me he only reports his situation up to the 27th February last, when he proposed arranging for the evacuation of his province by detachments, the first of which he hoped to dispatch at the close of the rains, towards the end of July; but both Dr. Junker and Mr. Mackay, the Head of the Church Mission in Uganda, inform me that they have since heard from Emin that the majority of the 4000 loyal Egyptian subjects who have remained faithful to Egypt throughout, and have supported him in the face of constant attacks from the Mahdi's adherents, aggravated by an imminent danger of starvation, refuse to leave their country; and he had therefore determined, if he could possibly do so, to remain at his post and continue to protect Egyptian interests till relief arrived, and he might be able to leave his followers in a safe position. Neither of my informants mentions the date of Emin's latest letters, but I know that Dr. Junker was in Unyoro, two days' march from the Albert Lake, with which Emin had ready communication by means of two steam-vessels, in the latter part of May, since which he had to abandon the protection of Kabarega, an ally of Emin's, in consequence of his being driven into the forest by the fierce attacks of the Uganda forces. Both Dr. Junker and Mr. Mackay strongly urge the necessity of the immediate relief of Wadelai, if it is not to be abandoned to the same fate as that which overtook Khartoum, the transmission of full supplies of ammunition and cotton cloth being its chief needs.

But for the dangerous attitude of Uganda, this would have been a mere question of expenditure, regarding which I should have telegraphed you an estimate, leaving you to forward ammunition, the pattern of which must be known at the War Office in Cairo; but under present circumstances, other important considerations are involved, which it is beyond my province to pass an opinion on. I would, however, suggest that, should it be decided to relieve Emin, this would be a good opportunity for dealing at the same time with Uganda, the infamous conduct of whose King has for many years been prejudicial to the development of the interior, and has, indeed, been the direct cause of Emin's isolation and of his present straitened situation. Were Uganda free from this tyrant, the Equatorial Province, even should the present elementary system of communication remain unmodified, would be within eight weeks' post of Zanzibar, and a safe depôt on the Albert Nyanza would provide a base from which any further operations that might be decided upon for the retention of the Upper Nile could be undertaken effectively and without anxiety.

Dr. Junker states that the country to the east of the Ripon Falls has proved impracticable, and that Emin has lost many troops in endeavouring to open communications through it. If such be the case, the alternative line by which Dr. Fischer tried to relieve Junker, and which, I believe, he still recommends, could not be relied on for turning Uganda and its eastern dependency, and the well-known route *viâ* Uganda would be the only one available for an expedition of moderate size.

On the eve of the departure of the mail, it is impossible to give you any exact estimate regarding the *personnel* of the relief expedition, but I am instituting inquiries as to the nature and value of the goods best suited for Emin's province. As far as I am able to judge, without making any special calculation, I consider that 1200 porters would be the smallest number that would suffice, and a well-armed guard of at least 500 natives would be necessary. It would be a great advantage were a body of seasoned Nubian or Egyptian troops under experienced leaders available, and it may be worth consideration whether the Sultan of Zanzibar might not be induced to place a few companies of his well-armed native force, under Brigadier Mathews, a retired officer of our Royal Navy, at disposal.

In the meantime, I am writing to Emin and to Mr. Mackay, warning them privately that, though nothing is decided on, they should be prepared for hearing of some action being taken, and should therefore spare no pains in forwarding all possible detailed information which may serve to guide the leader of any relief party; and I have indicated means for insuring the delivery of their communications to the proper persons, should the messengers they may dispatch hear of any such expedition on their way to the coast.

I have also assumed the responsibility for any expenses Mr. Mackay may incur in inducing the King of Uganda's officers to facilitate the transmission of his letters.

Should any further measures occur to me as being advisable, pending a reply to my telegram, I shall not fail to take them.

Inclosure 2 in No. 2.

Acting Consul-General Holmwood to Sir E. Baring.

Zanzibar, September 27, 1886.

Sir,

In reply to your telegram of the 25th instant, I telegraphed yesterday the general lines on which I would recommend an expedition for Emin Bey's relief to be organized.

Some other suggestions regarding the measures that would be necessary are given in my dispatch of the 25th instant, but until I have made further inquiries as to the weight and nature of the goods which must be forwarded, I can supply no exact details.

General Mathews, whom I had consulted as to the force necessary for the safety of the expedition, is of opinion that I have formed far too low an estimate; but after weighing the testimony of many experienced persons acquainted with Uganda, I must adhere to my opinion, that 500 native troops, armed with modern rifles and under experienced leaders, would, if supplemented by an irregular force, fully suffice.

On the supposition that the well-travelled route *viâ* Uganda would be adopted, the plan I would recommend is as follows:—

Temporary possession of all the Arab dhows and other large craft on the south of the Victoria Lake having been obtained, by peaceable means if possible, the advance guard shall proceed to the landing-place a few hours' march from the King of Uganda's town, and, being furnished with presents and letters from the Sultan of Zanzibar and this Agency requesting every assistance in passing through his country, they shall communicate with the missionaries and avail themselves of their knowledge of the country and its people.

A little tact would, with ordinary precaution, secure their safety until the arrival of the main body, and unless I am greatly misinformed, the Waganda would never venture to meddle with such a body as would then be assembled.

Arrangements could then be made either to march to Wadelai, or to meet Emin's steamers at the Albert Lake, a five days' journey; and after his relief any measures for dealing with Uganda could be undertaken by an overwhelming force made up from the expedition and Emin's people.

I have sent through the Sultan for an Arab who bears a high character and has his Highness' confidence. He has just reached the coast from Uganda, where he has resided many years. I expect him here to-morrow, and after carefully advising with him regarding points on which I feel somewhat doubtful, it may be necessary to amend my present recommendations in some respects.

I am in communication with the Foreign Office regarding assistance for the missionaries in Uganda.

I have, &c.

(Signed) FREDC. HOLMWOOD.

No. 3.

The British and Foreign Anti-Slavery Society to the Earl of Iddesleigh.—(Received November 12.)

55, *New Broad Street, London, November* 8, 1886.

MY LORD,

At a meeting of the Committee of this Society held on the 5th instant, I had the honour to lay before them a letter addressed to myself by Emin Bey, and one from him to Dr. Felkin, a former traveller through the Equatorial provinces governed by Dr. Emin Bey, and a member of this Committee.

Dr. Emin Bey, who was appointed by General Gordon to administer, on behalf of the Egyptian Government, the Equatorial provinces on the Upper White Nile, has long been cut off from all help from the civilized world, and is now, as shown by his letters, in a very precarious position. After considering the subject-matter of the two letters (of which I have the honour to inclose proof copy), the following Resolution was passed unanimously, and I was directed to forward a copy to your Lordship.

I have, &c.

(Signed) CHAS. H. ALLEN, *Secretary.*

INCLOSURE IN No. 3.

Resolution.

That in view of the services rendered by Dr. Emin Bey, both in the suppression of the Slave Trade, and in administering for a considerable period a settled and peaceful Government in the Equatorial provinces of Egypt, the Committee consider that the position of Dr. Emin Bey presents a very strong claim upon her Majesty's Government.

While not suggesting any measure of a military character for his relief, the Committee hold that both her Majesty's Government and that of Egypt are bound to be sparing of neither exertion nor expense in order to rescue him from the destruction which seems to await him, or by the supply of money and goods to enable him to hold a friendly position amongst the natives of his province.

No. 4.

Mr. Mackinnon to Foreign Office.—(Received November 16.)

Trentham, Stoke-upon-Trent, November 15, 1886.

DEAR SIR JAMES,

I have no doubt it is well known to you and to her Majesty's Government that there is a general feeling in the country, especially in London, that some effort should be made to open up communication with Emin Bey, and that the Government should in some way or other take immediate action in the matter.

It is not for me to say how it should be done; but I venture to suggest that if her Majesty's Government desire to respond to the general feeling I have indicated, it might be worthy of their consideration whether the services of the well-known eminent explorer, Mr. H. M. Stanley, should not be secured.

I have reason to believe that he would promptly and unhesitatingly undertake such a service as this, provided the means for doing so were placed at his disposal, although the service would involve great personal and pecuniary sacrifices on his part.

I believe he would be quite willing, with such support, and without demanding any official recognition for his mission from her Majesty's Government, to undertake the service and relieve them of all responsibility for his personal safety.

Mr. Stanley proceeds to America on the 18th to fulfil an engagement there, but should it be desired, I believe he would be ready, if a decision is come to promptly by her Majesty's Government, to delay his departure or to return from New York on receipt of a telegram.

Might I ask you (provided you deem it expedient) to bring the matter under the notice of the Secretary of State for Foreign Affairs? And if you think my services, whether by communicating with Mr. Stanley or otherwise, can be of any use, I beg you will not hesitate to inform me.

Believe me, &c.

(Signed) W. MACKINNON.

No. 5.

Sir E. Baring to the Earl of Iddesleigh.—(Received by telegraph, November 16.)

Cairo, November 16, 1886.

(Extract.)

I have the honour to enclose herewith copy of a dispatch which I have to-day received from her Majesty's Acting Agent and Consul-General at Zanzibar.

The letters from Emin Bey inclosed therein were all in Arabic, and were addressed to various private individuals and officials in this country, including one Report addressed to the President of the Council.

I would draw your Lordship's attention to the last paragraph in Mr. Holmwood's dispatch, and venture to ask your Lordship to instruct me as to what answer I should send him as to the possibility of taking any steps with a view to sending stores or relief to Emin Bey.

The Egyptian Government is not in a position which will enable it to send an expedition to Wadelai.

Inclosure in No. 5.

Acting Consul-General Holmwood to Sir E. Baring.

Zanzibar, October 23, 1886.

(Extract.)

In continuation of my Report of the 27th September, I have the honour to inform you that letters from Emin Bey have been received, dated Wadelai, 7th July, when he was still at his post, and hoped his stores would hold out until relief arrived. It is evident from his letters that Mr. Mackay, the missionary at Uganda, had greatly encouraged him and the subjects of Egypt to believe that relief would shortly reach him.

I inclose herewith letters for the President of the Council and other Egyptian officials. Emin forwarded copy of a letter from the Mahdi, giving a description of the capture of Khartoum and Gordon's death. This and other letters have been transmitted to the Foreign Office; also letters from Uganda up to the 24th August.

In reference to the relief of Wadelai, I have ascertained that 500 porters, carrying goods to the value of about £2500, would suffice, and I conclude that 300 porters would carry the necessary supply of ammunition. For your guidance I may mention that each load would be about 60 lbs. in weight.

I shall await your reply to my recent dispatch before reporting further.

No. 6.

Foreign Office to the British and Foreign Anti-Slavery Society.

Foreign Office, November 19, 1886.

Sir,

I am directed by the Earl of Iddesleigh to acknowledge the receipt of your letter of the 8th instant with regard to the present position of Emin Bey, and I am to request that you will assure your Society that the matter is receiving attention from her Majesty's Government.

I am, &c.

(Signed) T. V. Lister.

No. 7.

Acting Consul-General Holmwood to the Earl of Iddesleigh.
(*Received November* 22.)

Zanzibar, October 18, 1886.

(Extract.)

I have the honour to report that letters from Mr. Mackay and Emin Bey, dated respectively the 24th August and 7th July, were received on the 16th instant, when a short telegraphic summary of the intelligence they contained was forwarded to your Lordship.

Emin Bey's letter indicates his ability to remain at his post for some time longer; his followers evidently anticipated the organization of some form of relief, but this is no doubt a question which is occupying the attention of her Majesty's Government as well as that of Egypt. I am forwarding this information to our Minister at Cairo, together with letters from Emin to the President of the Council and other Egyptian officials.

The inclosed letter from the Mahdi to the Governor he had appointed to the Bahr-Ghazal provinces, describing the capture of Khartoum and the death of Gordon, is a most interesting document, and may prove a valuable historical record.

Inclosure in No. 7.

The Mahdi to Sheikh Karamalla.

12*th Rajab*, 1302 (*April* 28, 1885).

(Translation.)

From the miserable Mahomed Ahmed, who is called El Mahdi, to his miserable Wali, Karamalla-el-Sheikh.

After praises to Almighty God—

From the poor slave of God, Mahomed El Mahdi-bin-Abdullah, to his friend and his Governor, Karamalla-bin-Sheikh Mahomed, God grant him, &c. I present to you excellent salaams, &c. Then I inform you, oh, my dear friend, that according to the fulfilled promise of God, the city of Khartoum was entered by the help of

God on the 9th Rabeen-el-Akhur (26th January, 1885), at the instant of daybreak, through the helpers of our religion, who were ready, and jumped over the ditch acting upon the command of the Lord of the whole world; it was a quarter of an hour or less that they came upon the enemies of God, their cutting off even from the beginning to the end of them, and the surrounding of them, notwithstanding they were strong in their arms of strength. From the commencement they fled away from before the troops of God, and though thinking to obtain safety by entering inclosures and shutting the doors, they were met face to face, and hewn with swords or stabbed with spears until their cries were terrible, and they were cut to pieces at once there upon the ground, and then the troops of God fell upon the rest of the people who had shut the doors fearing the same fate, who were taken up and killed properly, and there were none left remaining of them but little children and free slaves.

But as to the enemy of God, Gordon, though we had warned him and talked kindly to him that he might return to God, yet he never did so, because his miserable state had been fore-ordained, and also the access of his foolishness, before his fate was ended, and he was sorry for the sins he had sown in this world; and he was removed by God to the place of His wrath, which is a bad place to remain in.

The end of this guilty people is that they were cut off, which, thanks to God, befalls those who are to receive fire as their reward, whilst light is reserved for those who shall receive Heaven as their dwelling-place.

Then he began to bless himself and his people in prayer. There were ten persons only of our people who were killed in this holy strife for the conquest of Khartoum, and the rest of our people neither received wound nor hurt.

All has happened by the Providence of God, &c. And we bow our heads in thanks to God for the help we have received from Him. May you also do so. Bow the head to God, and thank His Holy Name.

No. 8.

Mr. H. M. Stanley to Mr. Mackinnon.—(Communicated to the Foreign Office, November 22.)

160, *New Bond Street, London, November* 15, 1886.

MY DEAR MR. MACKINNON,

Reverting to the subject which we have been verbally discussing, viz. the relief of Emin Bey, let me say that if there is real intention and purpose to carry the proposal into effect, you may fully command my services. To that object I would devote them freely and gratuitously, without hope of fee or reward. I am abundantly occupied and most profitably employed, as you know, in lecturing; but if such a mission as this is offered to me, I will gladly put myself under your orders at a moment's notice. Personally, I am prepared to start within an hour; but an expedition's necessities would detain me about two weeks here.

After casting up the various items of expenditure, I venture to promise you that the cost of such a relief expedition would not exceed £20,000; and if you could obtain transport by sea by a man-of-war or a Government store-ship the cost would be still less. It would also be necessary to obtain the good offices of her Majesty's Government, without which the expedition would be expensive and tedious. Presuming that you could obtain the latter, I beg to point out to you that there are four routes by which Emin Bey could be relieved and extricated from his dangerous position. That which offers the least obstacles for the march of the expedition is also that on which the moral help of her Majesty's Government would be most effective. By a few strokes of the pen in the shape of a recommendatory letter, without attracting notice or comment from curious people, her Majesty's Government can expedite me on my way, and contribute materially, though unostentatiously, to the success of the expedition. By this route, which I consider safest and most feasible, Emin Bey can be reached safely by the middle of June 1887, and Emin Bey's as well as the relief expedition could reach the sea by the middle of December 1887.

The last week I have been industriously sketching out the routes, with the various costs attending each route. You must remember that for fourteen years I have been in Africa; that I know personally the region to be traversed within 350 miles of Emin Bey's position at Wadelai; that the three expeditions I have led into Africa have been perfectly successful each time; that my recommendation to King Leopold was based upon the celerity of my movements and the economy attending the equipment of the previous expeditions.

I do not know Emin Bey personally, but if I am ready to sacrifice an income of £10,000 to his relief, surely it would not be a difficult task to find a few noble-hearted gentlemen willing to subscribe such a small sum as £20,000 to equip an expedition, under my voluntary guidance, to effect this desirable object. Think of the years this man has patiently waited at the post of duty. All that I have sounded upon the subject join with me in wishing that his rescue may be accomplished.

I propose sailing for New York, *viâ* Southampton, on steam-ship *Allen*. I submit this letter to you in the hope that such a trip will not be necessary, and that you may be able to arrive at a definite decision before I sail, as every day is valuable. Should I sail, however, you know my address; and you may cable to me in the most peremptory manner, feeling sure that by first steamer I shall return, and abandon everything to carry out this honourable mission.

Yours, &c.

(Signed) HENRY M. STANLEY.

No. 9.

Sir E. Baring to the Earl of Iddesleigh.—(Received by Telegraph, November 23.)

Cairo, November 23, 1886.

(Extract.)

I have the honour to report that I have had an interview with Dr. Schweinfurth on the subject of Emin Bey's position, and the possibility of affording him relief.

Dr. Schweinfurth suggests that an expedition should be organized for the relief of Emin by Dr. Junker, who is well acquainted with those countries, and that the Egyptian Government should contribute £10,000 towards the cost. He does not think it would cost more than £10,000, and anticipates that, even if it did, the Missionary Societies in England would subscribe a certain amount towards paying the expenses. His plan would include dethroning the present King of Uganda by giving help to the party hostile to him.

Dr. Schweinfurth attaches great importance to Dr. Junker being made the leader of the expedition, and is much adverse to any officer of the Sultan of Zanzibar being given the command.

The Egyptian Government would not be unwilling to contribute £10,000 for the relief of Emin Bey, but they have no troops whom they could send; Dr. Schweinfurth does not think that Egyptian troops would be necessary. Before any steps are taken I venture to request your Lordship to inform me whether her Majesty's Government has any objection to the Egyptian Government taking this proposed action.

I have to-day telegraphed to Consul Holmwood, and have asked him to give me his opinion on the subject, and it would, I submit, be desirable to obtain Sir John Kirk's opinion as to the practicability of the plan.

At the same time, I would recall to your Lordship's notice the opinions expressed by Consul Holmwood.

No. 10.

The Earl of Iddesleigh to Sir E. Baring.

Foreign Office, November 25, 1886.

SIR,

I have to inform you that Mr. Stanley has proposed to organize and conduct a peaceful relief expedition, on the condition that her Majesty's Government will supply the necessary funds.

It has, however, been decided by her Majesty's Government that this is a responsibility which they cannot undertake.

An endeavour is being made by some private individuals to

provide the funds necessary to supplement the £10,000 which might, in your opinion, be possibly contributed by the Egyptian Government towards an expedition.

Her Majesty's Government will address a further communication to you on the subject.

I am, &c.

(Signed) IDDESLEIGH.

No. 11.

Mr. Mackinnon to the Earl of Iddesleigh.—(Received November 27.)

Burlington Hotel, 30, *Old Burlington Street,*
November 27, 1886.

DEAR LORD IDDESLEIGH,

I have considered the matter of our conversation of Wednesday evening last, and now submit the inclosed Memorandum for your consideration.

If the plan suggested in the Memorandum meets with your Lordship's approval, I shall at once set to work to get together the £10,000 required from my friends and myself.

I would hope that if the £10,000 required from Government should be provided, as would seem to be reasonable, out of the resources of Egypt, the Committee should not on that account be placed under obligation to have any direct communication with the Egyptian Government. They would doubtless prefer to communicate with or through her Majesty's Government only.

It appears, from information in the newspapers, that Emin Bey is believed to have considerable quantities of ivory, which might be utilized for repayment of outlays connected with any scheme of relief. The Committee would naturally expect that if this expectation should be realized, a just proportion should be made over to them in repayment of their outlay.

I remain, &c.

(Signed) W. MACKINNON.

INCLOSURE IN NO. 11.

Memorandum on the subject of the Relief of Emin Bey.

It is suggested, for the consideration of her Majesty's Government, by Mr. W. Mackinnon, acting for himself and others, that a small Committee be formed to organize and send out a private expedition to open communications with, and carry relief to, Emin Bey.

It has been estimated by the most competent authorities that for this purpose a sum of £20,000 will be needed, and it is proposed that, the sum being provided, the Committee should intrust the leadership and sole conduct of the expedition to Mr. H. M. Stanley, who offers his services gratuitously, and proposes to engage only Zanzibaris and other East Africans. He is ready to accept all the personal risks involved, and to relieve every one else of responsibility.

It would be necessary, in order to carry out this proposal, that Government should assist the expedition by instructing its agents and naval officers to render every assistance and exert themselves in its favour, and that Government should facilitate the supply of the arms and ammunition necessary to Mr. Stanley's native escort.

Mr. Mackinnon and friends will endeavour to provide £10,000 towards the expenses of the expedition, on condition that a similar amount be placed at the disposal of the Committee through her Majesty's Government.

Mr. Stanley would be left to decide what would be the best and quickest mode of reaching Emin Bey. He would, in consideration of the Government grant, be instructed to convey any communications with which he might be intrusted by her Majesty's Government or by the Egyptian Government, and to assist the withdrawal to the coast of the Egyptian garrison.

(Signed) W. MACKINNON.

No. 12.

The Earl of Iddesleigh to Sir E. Baring.

Foreign Office, December 3, 1886.

SIR,

I have informed you to-day, by telegraph, that some private persons personally interested in Emin Bey are organizing a small Committee with the view of setting on foot an expedition for his relief of a purely pacific character. The sum of £10,000 will be raised by them for the purpose, and they will take on themselves all responsibility, provided that a similar sum is given by the Egyptian Government.

The expedition will be under the direction of Mr. H. M. Stanley.

Her Majesty's Government approve the proposal, and the persons by whom the offer is made have been informed that it is understood by us that the agreement of the Egyptian Government is assured.

I am, &c.

(Signed) IDDESLEIGH.

No. 13.

Sir E. Baring to the Earl of Iddesleigh.—(Received December 4, 10.30 A.M.)

Cairo, December 4, 1886.

(Telegraphic.)

Egyptian Government fully concur in the arrangement, and agree to contribute £10,000 towards the expenses of the expedition.

No. 14.

Sir E. Baring to the Earl of Iddesleigh.—(Received December 6.)

Cairo, November 28, 1886.

MY LORD,

I have the honour to inclose herewith to your Lordship copy of a telegram which, at the request of the Egyptian Government, I have forwarded to her Majesty's Consul at Zanzibar for transmission to Emin Bey.

I have, &c.

(Signed) E. BARING.

INCLOSURE IN NO. 14.

Sir E. Baring to Acting Consul-General Holmwood.

Cairo, November 28, 1886.

(Telegraphic.)

Send by sure and early opportunity following message from President Council of Ministers to Emin Bey.

Message begins:—

"Your letters of July received. His Highness the Khedive, in order to show his appreciation of the good services rendered, desires you to express his thanks to the loyal Egyptian troops, and in recognition of your services has promoted you to the rank of Pasha.

"You are authorized to retire from the Mudirieh to the best position for ultimate withdrawal *viâ* Zanzibar when circumstances allow.

"Any troops and refugees who desire to remain where they are should be permitted to do so. Fuller instructions and details will follow by post. A letter has been written to Sultan Kabarega desiring his assistance.

"Every effort will be made for the dispatch and safe-conduct of letters. Endeavour to keep us well informed."

Message ends.

NO. 15.

Foreign Office to Acting Consul-General Holmwood.

Foreign Office. December 7, 1886.

SIR,

You were this day informed, by telegram, that some private gentlemen, of whom Mr. Mackinnon, the Chairman of the British India Steam Navigation Company, is the head, are, with the approval of her Majesty's Government, making arrangements with Mr. Stanley to lead an expedition for the relief of Emin Pasha. To the expenses of this scheme the Egyptian Government have agreed to contribute £10,000.

I am, &c.

(Signed) JULIAN PAUNCEFOTE.

No. 16.

Mr. Mackinnon to Mr. H. M. Stanley.—(Communicated to the Foreign Office, December 11.)

London, December 11, 1886.

(Telegraphic.)

Your plan and offer accepted. Authorities approve. Funds provided. Business urgent. Come promptly. Reply.

No. 17.

Mr. H. M. Stanley to Mr. Mackinnon.—(Communicated to the Foreign Office, December 13.)

New York, December 13, 1886.

(Telegraphic.)

Just received Monday's cablegrams. Many thanks. Everything all right. Will sail per *Eider*, 8 o'clock Wednesday morning. If good weather and barring accidents arrive 22nd December Southampton. It is only one month's delay after all. Tell authorities prepare Holmwood, Zanzibar, and Seyyid Barghash. Best compliments to you.

No. 18.

Sir F. de Winton to Foreign Office.—(Received January 3, 1887.)

160, *New Bond Street, December* 30, 1886.

SIR,

I have the honour to inform you, for the information of the Right Honourable the Secretary of State for Foreign Affairs, that a Committee has been formed of the following gentlemen, to conduct the affairs of the Emin Bey Relief Expedition:—

Mr. W. Mackinnon, Chairman; Honourable G. Dawnay; Mr. H. M. Stanley; Sir L. Pelly; Mr. A. F. Kinnaird; Colonel Grant; Reverend H. Waller; Colonel Sir F. de Winton, Acting Secretary.

A meeting of this Committee was held on the 29th instant, when the action of Mr. Mackinnon, and the appointment of Mr. Stanley to command the expedition, were confirmed.

I have, &c.

(Signed) F. DE WINTON.

No. 19.

Emin Relief Committee to the Marquis of Salisbury.
(*Received January* 13.)

160, *New Bond Street, London, January* 12, 1887.

MY LORD,

I have been instructed by the Committee of the Emin Pasha Relief Expedition to request the favourable consideration of her Majesty's Government to the question of transport of the expedition from Zanzibar to Banana, the mouth of the Congo, for the following reasons:—

That his Majesty the Sovereign of the New Congo Free State has placed a sufficiency of transport on the waters of the Haut Congo at the disposal of Mr. Stanley, for the use of the expedition, gratuitously.

That the adoption of the Congo route would prevent any misfortune happening to the missionaries, French and English, now in the power of Mwanga, the King of Uganda.

That the funds of the Committee are not sufficient to enable them to carry out the whole object of the expedition, *viâ* the Congo route, in consequence of the cost of transport from Zanzibar to Banana Point.

The Congo route has the further advantages, that it offers far greater facilities for the withdrawal of the women and children at present with Emin Bey. It would also open up a sure and certain route for any further relief to Emin Pasha, and for his own ultimate retreat.

The Committee have therefore to request, as the Congo route is for so many reasons the more suitable, that her Majesty's Government may see fit to place a steam-vessel at the disposition of the expedition to carry out the service of transport of 500 Zanzibaris, and the material belonging to the expedition, from Zanzibar to Banana; and, if the request be granted, that the steamer may be at Zanzibar not later than the 21st February to await Mr. Stanley's orders.

I have, &c.

(Signed) F. DE WINTON, *Colonel*,
Acting Secretary, Emin Pasha Relief Expedition.

No. 20.

Foreign Office to Emin Relief Committee.

Foreign Office, January 13, 1887.

SIR,

I am directed by the Secretary of State to acknowledge the receipt of your letter of the 12th instant, expressing the hope of your Committee that her Majesty's Government may see fit to place a steam-vessel at the disposition of the expedition for the relief of Emin Pasha to carry out the service of transport of 500 Zanzibaris, and the material belonging to the expedition, from Zanzibar to Banana, the mouth of the Congo, and that, if the request be granted, the steamer may be at Zanzibar not later than the 21st February to await Mr. Stanley's orders.

In reply, I am to state to you that, although, as your Committee are aware, her Majesty's Government assume no responsibility for the conduct of the expedition, still the interest taken by them in its success has already induced them to make inquiries at the Admiralty as to whether one of her Majesty's ships could be spared to undertake the duty referred to in your letter of yesterday's date. The outcome, however, of these inquiries has been that the Admiralty report that there is no ship of war, troop or store ship, available, that is capable of carrying 350 to 400 porters and stores from Zanzibar to the Congo mouth.

Under these circumstances, the Secretary of State regrets that it is not in the power of her Majesty's Government to assist the Committee in the manner requested, but inquiries are being made —the result of which will be immediately communicated to your Committee—as to whether substantial assistance, in some other shape, cannot be unofficially extended to the expedition.

I am, &c.

(Signed) JAMES FERGUSSON.

No. 21.

The Marquis of Salisbury to Sir E. Baring.

Foreign Office, January 22, 1887.

SIR,

I have requested you, by telegraph, to inform the Egyptian Government that the Emin Relief Committee are desirous that the Egyptian contribution of £10,000 should be immediately placed to their credit at Messrs. Ransom, Bouverie, and Co.'s, No. 1, Pall Mall, London.

I have added that Mr. Stanley is of opinion that Emin Pasha will probably be in need of clothes, and he therefore suggests that one or two good uniforms should be sent for his personal use.

I am, &c.

(Signed) SALISBURY.

No. 22.

The Marquis of Salisbury to Sir E. Baring.

Foreign Office, January 24, 1887.

SIR,

I informed you to-day, by telegraph, that you are authorized to give Mr. Stanley all the assistance you can in regard to the expedition for the relief of Emin Pasha, bearing in mind that her Majesty's Government have thought it right to avoid giving to the expedition any countenance, which, in possible contingencies, might impose on Great Britain large military responsibilities.

I am, &c.

(Signed) SALISBURY.

No. 23.

Foreign Office to Emin Relief Committee.

Foreign Office, January 31, 1887.

SIR,

I am directed by the Marquis of Salisbury to acquaint you that a letter has been received from the Admiralty, stating that orders have been telegraphed to the Commander-in-chief on the East India Station and to the Commander-in-chief at the Cape of Good Hope and the Senior Naval Officer at Zanzibar to supply the steam-ship *Madura* with coal free of charge, to enable her to proceed from Bombay to Zanzibar to embark the Relief Expedition and convey them to the Congo River, calling at the Cape of Good Hope.

I am, &c.

(Signed) P. CURRIE.

No. 24.

Foreign Office to Emin Relief Committee.

Foreign Office, February 3, 1887.

SIR,

With reference to your letter of the 21st ultimo respecting the payment of the sum which the Egyptian Government are to contribute to the Emin Relief Expedition, I am directed by the Marquis of Salisbury to inform you that a telegram has been received from her Majesty's Agent and Consul-General at Cairo stating that the money will be paid at once to the account of the Relief Committee.

I am, &c.

(Signed) T. V. LISTER.

No. 25.

Emin Relief Committee to Foreign Office.—(*Received February* 11.)

30, *Old Burlington Street, W.*, *February* 10, 1887.

SIR,

I have the honour to report, for the information of the Marquis of Salisbury, that the Emin Relief Expedition has departed, that it will reach Zanzibar on or about the 21st February, and unless Mr. H. M. Stanley, on his arrival there, receives intelligence from the interior necessitating an advance from the East Coast, he will proceed with his expedition from Zanzibar in the British India Steam Company's steam-ship *Madura*, 2000 tons, to Banana, off the mouth of the Congo, where he expects to arrive about the 23rd March. The expedition will then push as rapidly as possible up the Congo to a point about 23° east longitude; from this point it will take about thirty-five days to march to the Lake Albert Nyanza, when the object of the expedition will have been accomplished.

The earliest date at which Mr. Stanley can arrive at Wadelai is about the middle or end of July.

With reference to this Congo route, it is very probable, after Mr. Stanley has opened up a road from the Congo to Wadelai, that Emin Pasha will send the women and children, and other Egyptians who may be desirous of returning to their country, by this route, which will have a land journey of about 600 miles as against 1200 miles by the route to the East Coast.

The Committee are therefore desirous that such a possibility may be brought to the notice of the Egyptian Government in order that the co-operation of the authorities of the Congo Free State may be invited to assist these people down the Congo, and also that suitable stores of food may be established at certain parts of the route where drought extending for a period of three years has caused a scarcity of provisions.

They have deemed it their duty to bring the fact to the notice of her Majesty's Government, for although Mr. H. M. Stanley will afford Emin Pasha all the assistance in his power, arrangements for transport and food for refugees, whether men or helpless women and children, must be made by the Egyptian Government, the Committee undertaking only the expense of the expedition, and they have made no provision for the passage down the Congo and back to Egypt of any refugees. Neither would it be, in Mr. Stanley's opinion, advisable to convey such by the land route to the East Coast, on account of the length of the land journey, as the women and children would have to be carried all the way, with the probability of meeting fierce and hostile tribes between Wadelai and Karagwe.

Should it be deemed desirable, the Committee will be happy to submit a short Memorandum showing the probable requirements of such a service as they have indicated.

I have, &c.

(Signed) W. MACKINNON,
Chairman of Committee.

No. 26.

Acting Consul-General Holmwood to the Marquis of Salisbury.
(*Received February* 14.)

Zanzibar, January 17, 1887.

MY LORD,

I have the honour to acknowledge the receipt of your Lordship's telegram instructing me to inform the Sultan that the expedition for the relief of Emin Pasha would proceed by the Congo route instead of *viâ* Zanzibar, as originally proposed, and to request his Highness to allow the porters, already retained for the land journey from this coast, to amend their engagements in accordance with the change.

To this his Highness at once assented, promising that, should any of these men refuse to serve under the new conditions, he would make every effort to procure substitutes who should be equally qualified.

In consequence of General Mathews' detachment on other service, his Highness deputed two of his clerks, who are fully acquainted with the work, to select the best porters procurable, and 500 of these had already been retained by Mr. Mackinnon's agent, here, all of whom have passed a medical examination.

Large quantities of provisions and stores have already been transmitted to various points along the route which had been indicated by telegram to Mr. Mackinnon's agent, and the rifles and ammunition which Mr. Stanley had desired the Egyptian Government to forward will arrive here in a few days.

I have, &c.

(Signed) FREDC. HOLMWOOD.

No. 27.

Sir E. Baring to the Marquis of Salisbury.—(Received February 14.)

Cairo, February 2, 1887.

(Extract.)

I have the honour to state that I have accompanied Mr. Stanley on visits both to the Khedive and to Nubar Pasha, and that, after hearing an explanation of his views on the subject, they have both expressed themselves thoroughly satisfied of the wisdom of Mr. Stanley's choice in adopting the Congo route in preference to that from Zanzibar.

I have also the honour to inclose herewith copy of a letter which Nubar Pasha has written, in French, to Emin Pasha, and which will be taken by Mr. Stanley; the Khedive has also written one much to the same effect in Arabic.

INCLOSURE IN No. 27.

Nubar Pasha to Emin Pasha.

Le Caire, le 2 *Février*, 1887.

MON CHER EMIN PACHA,

Je vous avais expédié, par les soins obligeants du Consulat de Sa Majesté Britannique à Zanzibar, une lettre que son Altesse vous avait adressée pour vous complimenter sur votre conduite et vous féliciter, vous, vos officiers, et vos soldats d'avoir triomphé des difficultés que vous avez eu à combattre.

Son Altesse vous faisait savoir dans cette lettre qu'elle vous accordait votre grade de Général et conférerait tous les grades et récompenses que vous avez accordés aux officiers et autres. Moi-même je vous annonçais une expédition se préparant à aller à votre secours. Cette expédition est formée; elle est commandée par Mr. Stanley et c'est lui qui vous remettra ma lettre avec celle que vous écrit son Altesse et une autre que je vous écris en Arabe.

L'expédition que commande Mr. Stanley a été formée et constituée pour se rendre auprès de vous avec les provisions et les munitions dont vous devez avoir nécessairement besoin. Elle est chargée de vous ramener vous, vos officiers, et vos soldats en Égypte par la voie que Mr. Stanley jugera la plus convenable. Je n'ai rien à ajouter à ce que je viens de dire relativement au but de l'expédition. Seulement son Altesse vous laisse complètement libres, vous, vos officiers, vos soldats, et autres de rester dans les régions ou vous êtes, ou de profiter des secours qu'elle vous envoie pour votre retour.

Mais il est bien entendu, et il faudrait que vous le fassiez bien comprendre à vos officiers, soldats, au autres, au cas ou quelques-uns désireraient ne pas revenir, qu'ils sont libres de rester, mais qu'ils restent à leurs risques et périls et par leur propre volonté et qu'ils n'ont conséquemment à attendre aucun secours subséquent de la part du Gouvernement. C'est ce que je désire que vous fassiez bien comprendre à ceux qui pourraient vouloir rester.

Je n'ais pas besoin d'ajouter que vous, vos officiers, vos soldats et les autres fonctionnaires vous serez réglés et payés a votre arrivée en Égypte et ce qui peut vous être dû sur vos traitements et autres allocations, vos grades ayant été déjà confirmés par son Altesse.

J'espère, mon cher Pacha, que Mr. Stanley vous trouvera tous sains et saufs et jouissant d'une pleine santé. C'est ce que nous souhaitons tous de tout cœur et c'est avec ces souhaits que je vous prie d'agréer tous mes sentiments de dévouement et mes sincères félicitations sur votre belle conduite.

(Signé) N. Nubar.

(Translation.)

Cairo, February 2, 1887.

My dear Emin Pasha,

I had sent you, by the kind favour of her Britannic Majesty's Consulate at Zanzibar, a letter addressed to you by his Highness complimenting you on your conduct, and congratulating you, your officers, and your soldiers on having overcome the difficulties with which you had to cope.

His Excellency made you aware in that letter that he had promoted you to the rank of General, and would confirm all promotions and rewards given by you to your officers and others. I informed you myself of the preparation of an expedition for your relief. The expedition is now formed; it is commanded by Mr. Stanley, who will himself hand you my letter with that which his Highness is writing to you, and another which I am writing to you in Arabic.

The expedition commanded by Mr. Stanley has been formed and organized in order to go to you with the provisions and stores, of which you must certainly be in want. Its object is to bring you, your officers, and soldiers back to Egypt by the way which Mr. Stanley shall think most suitable. I have nothing to add to what I have just said of the objects of the expedition. Only his Highness leaves you, your officers, your soldiers and others entirely free to stay where you are, or to make use of the help he sends for your return.

But, of course, and this must be made clear to your officers, soldiers, or others, if some do not wish to return they are free to remain, but at their own risk and by their own desire, and that they cannot expect any other help from the Government. That is what I wish you to make clear to those who may wish to remain.

I have only to add that you, your officers, your soldiers and other officials will have your accounts settled and be paid on your arrival in Egypt what is due to you on account of salaries and other allowances, as all your promotions have been already confirmed by his Highness.

I hope, my dear Pasha, that Mr. Stanley will find you all safe and sound, and enjoying good health. That is what we all wish with all our hearts, and it is with these wishes, &c.

(Signed) N. NUBAR.

No. 28.

The Marquis of Salisbury to Sir E. Baring.

Foreign Office, February 17, 1887.

SIR,

I inclose herewith copy of a letter which has been received from the Chairman of the Emin Bey Relief Committee with reference to the withdrawal of Emin's party through the Congo Territory;[1] and I have to request you to ascertain whether the Egyptian Government would wish the Committee to be asked to supply any further information. You will see that the Committee is prepared to furnish a Memorandum if required.

I am, &c.

(Signed) SALISBURY.

No. 29.

Sir E. Baring to the Marquis of Salisbury.—(Received February 21.)

Cairo, February 12, 1887.

MY LORD,

I have the honour to inform your Lordship that £8600 out of the £10,000 which the Egyptian Government are contributing to the Emin Pasha Relief Expedition was remitted to Messrs. Ransom, Bouverie, and Co. on the 9th instant.

Of the remaining £1400, £400 has been paid to Dr. Junker for certain expenses incurred by him in connection with the expedition, as to the proper disbursement of which sum Nubar Pasha has expressed himself fully satisfied, and of which the accounts will be obtained as soon as possible, while £1000 is retained here as a reserve fund in payment of any expenses that may be incurred in this country on behalf of the expedition.

I should add that, as the arms and ammunition will be given free of payment, it is almost certain that the expenses incurred will be much less than £1000.

I have, &c.

(Signed) E. BARING.

[1] No. 25.

No. 30.

Foreign Office to Emin Relief Committee.

Foreign Office, February 25, 1887.

SIR,

I am directed by the Marquis of Salisbury to inform you that a dispatch has been received from Sir E. Baring, her Majesty's Agent and Consul-General at Cairo, stating that of the £10,000 which the Egyptian Government are contributing towards the Emin Relief Expedition, £8600 have been remitted to Messrs. Ransom, Bouverie, and Co., £400 have been paid to Dr. Junker for expenses incurred by him in connection with the expedition, and £1000 has been retained at Cairo as a reserve fund in payment of any expenses that may be incurred in Egypt on behalf of the expedition.

Sir E. Baring adds that, as the arms and ammunition will be given free of payment, it is almost certain that the expenses incurred will be much less than £1000.

I am, &c.

(Signed) T. V. LISTER.

No. 31.

Sir E. Baring to the Marquis of Salisbury.—(Received March 14.)

Cairo, March 7, 1887.

MY LORD,

With reference to your Lordship's dispatch of the 17th February, inclosing copy of a letter from the Chairman of the Emin Pasha Relief Committee with reference to the withdrawal of Emin's party through the Congo Territory, I have the honour to inclose herewith copy of a note which I have received from Nubar Pasha, from which your Lordship will perceive that the Egyptian Government accepts the proposal of the Committee to ask the assistance of the Congo Free State for the storage of provisions at different points on the route to be followed by the fugitives.

But, in the first place, the Egyptian Government would be glad to be supplied with further information, and ask that they may be shown the Memorandum which Mr. Mackinnon said the Committee was prepared to furnish.

I would call your Lordship's attention to Nubar Pasha's remarks as to the necessity for observing the strictest economy in all preparations for the withdrawal of Emin's party, and also to the remark that the Egyptian Government expects to be, to some extent, remunerated by the sale of the ivory which it possesses at Wadelai.

I have, &c.

(Signed) E. BARING.

INCLOSURE IN NO. 31.

Nubar Pasha to Sir E. Baring.

Le Caire, le 7 *Mars*, 1887.

M. LE MINISTRE,

J'ai reçu avec la dépêche que vous avez bien voulu m'écrire, le 27 du mois dernier, la lettre adressée par le Comité institué pour la délivrance d'Emin Pacha à Lord Salisbury concernant la coopération du Gouvernement de son Altesse, en vue des mesures à prendre pour aider au rapatriement des réfugiés Egyptiens par la voie du Congo.

Le Gouvernement de son Altesse apprécie, avec toute la sympathie qu'elle mérite, la situation intéressante de ces réfugiés et ne peut qu'être disposé à prêter son concours à leur rapatriement.

Aussi, à cet effet, accepte-t-il la proposition faite par la Comité de demander l'assistance de l'Etat Libre du Congo pour que des provisions de bouche soient préparées sur divers points de la route que les réfugiés auront à parcourir.

Mais le Gouvernement de son Altesse serait désireux, au préalable, d'avoir connaissance du Mémoire que le Comité est disposé à préparer et indiquant d'une façon au moins approximative les frais éventuels qui pourraient en résulter pour le Trésor Égyptien. La situation difficile des finances des pays étant donnée, il serait à désirer que ce projet soit rédigé avec un esprit de grande économie, d'autant plus que le Gouvernement de son Altesse compte, pour se rembourser de ces dépenses, sur le produit de la vente de l'ivoire qu'il possède à Wadelai.

Permettez-moi, M. le Ministre, de compter sur votre obligeance pour soumettre cette manière de voir du Gouvernement de son Altesse à Lord Salisbury, et veuillez, &c.

(Signé) NUBAR.

(TRANSLATION.)

Cairo, March 7, 1887.

M. LE MINISTRE,

With the dispatch, which you were so good as to address to me on the 27th ultimo, I received the letter from the Committee formed to effect the relief of Emin Pasha to Lord Salisbury about the co-operation of the Government of his Highness in measures to be taken for helping the repatriation of the Egyptian refugees by way of the Congo.

The Government of his Highness, with all due sympathy for the interesting position of these refugees, cannot but be willing to lend assistance for their repatriation.

With this view, therefore, it agrees to the proposal made by the Committee to ask the help of the Congo Free State in providing provisions at various places on the route which the refugees will have to follow.

But the Government of his Highness would first wish to see the Memorandum, which the Committee is willing to prepare, and which is to indicate, at least approximately, the ultimate burden which may eventually be entailed on the Egyptian Exchequer. Considering the difficult state of the finances of the country, it is to be wished that the scheme should be formed with the most careful economy, the more so, that the Government of his Highness counts on the proceeds of the sale of the ivory belonging to it at Wadelai to reimburse it for these expenses.

Permit me, M. le Ministre, to trust that you will be so good as to bring the views of his Highness' Government before Lord Salisbury.

Accept, &c.

(Signed) NUBAR.

No. 32.

Foreign Office to Emin Relief Committee.

Foreign Office, March 16, 1887.

SIR,

With reference to your letter of the 10th ultimo relative to the withdrawal of Emin's party through the Congo Territory, I am directed by the Marquis of Salisbury to acquaint you that Sir E. Baring has received a note from Nubar Pasha accepting the proposal of the Committee to ask the assistance of the Congo Free State for the storage of provisions at different points on the route to be followed by the fugitives. In the first place, however, the Egyptian Government would be glad to be supplied with further information, and they desire to be furnished with the Memorandum which Mr. Mackinnon stated the Committee was ready to prepare; and I am to request that it may now be forwarded to this Department for communication to the Egyptian Government.

Nubar Pasha lays stress upon the necessity for observing the strictest economy in all the preparations for the withdrawal of Emin's party, and states that the Egyptian Government expect to recover their expenses by the sale of the ivory which they possess at Wadelai.

I am, &c.

(Signed) JULIAN PAUNCEFOTE.

No. 33.

Emin Relief Committee to Foreign Office.—(Received March 26.)

28, *Wynnstay Gardens, Kensington, March* 25, 1887.

SIR,

In the concluding part of your letter of the 16th instant allusion is made to the question of ivory, and the hope expressed by Nubar Pasha that the expenses incurred by the Egyptian Government would be covered by the sale of the ivory they possess at Wadelai.

Referring to a letter from Mr. W. Mackinnon to the late Lord Iddesleigh, dated the 27th November, 1886, the following paragraph occurs:—

"It appears from information in the newspapers that Emin Bey is believed to have considerable quantities of ivory which might be utilized for repayment of outlays connected with any scheme of relief. The Committee would naturally expect that, if this anticipation be realized, a just proportion should be made over to them."

I have now the honour to request that her Majesty's Secretary of State for Foreign Affairs may see fit to have the just claims of the Committee for their proportion of the ivory at Wadelai acknowledged by the Egyptian Government, in order that no misunderstanding may arise in the future, and that the utmost harmony of purpose may exist between the objects of the Committee and that of the Egyptian Government.

I have, &c.

(Signed) F. DE WINTON, *Hon. Secretary,*
Emin Relief Committee.

No. 34.

Emin Relief Committee to Foreign Office.—(Received March 27.)

28, *Wynnstay Gardens, Kensington, March* 25, 1887.

SIR,

In further reply to your letter of the 16th March relative to the withdrawal of Emin's party through the territory of the Congo Free State, I have the honour to transmit a Memorandum, adopted by the Committee, to which is appended an estimate showing probable expense for 100 persons, for the information of the Egyptian Government.

In framing this estimate, it is presumed that a sufficient escort will be provided by Emin Pasha to transport all the loads required by the party, as well as to carry the women and children.

The estimates are based upon the prices paid by the Congo Free State to feed their employés.

With reference to the assistance of the Congo Free State, the

Committee, as they have no official position, would suggest, if the Egyptian Government approve of this route, that her Majesty's Government communicate with the Sovereign of the Congo Free State, in order to obtain the necessary authority for the carrying out of the following services:—

1. The engagement of the steamer *Stanley* to proceed to Stanley Falls about the last week in September.

2. To send to the Congo in June, or to arrange with the authorities of the Congo Free State to have in their stores, six tons of rice, to be distributed as follows:—

2 tons at Stanley Pool.
2 tons at Lukungu.
2 tons at Matadi.

3. To obtain authority for the purchase from the stores of the Congo Free State of such cloth and goods as the Emin party may reasonably require.

It is probable that about next June or July the Committee will be in possession of further intelligence from Mr. Stanley, which might modify these arrangements as regards time, but they earnestly advise that the general scheme they have suggested be carried out.

In conclusion, the Committee append a comparative table of the distances of the different routes from Wadelai to the East Coast, and to the West Coast *viâ* the Congo. These distances only refer to the land portion of the different roads, and are given according to the number of days each would take.

I have, &c.

(Signed) F. DE WINTON, *Hon. Secretary, Emin Relief Committee.*

INCLOSURE 1 IN NO. 34.

Memorandum for the Information of the Egyptian Government on the Congo as a Route for the Return of Emin Pasha's Party from Wadelai.

The Congo route should be divided into four portions:—

1. From Wadelai to Stanley Falls (land).
2. From Stanley Falls to Stanley Pool (water).

3. From Stanley Pool to Matadi (land).

4. Matadi to Banana (water).

From Wadelai to Stanley Falls the necessary supplies would be arranged by Emin Pasha, and the Committee have no means of ascertaining what these expenses would be.

From Stanley Falls to Stanley Pool each person would cost about 1*d.* per diem. There would also be the expense of the steamers and the cost of miscellaneous expenditure.

From Stanley Pool to Banana rice would have to be laid down at Stanley Pool, Lukungu, and Matadi.

The Committee are of opinion that if the route from Wadelai to Stanley Falls be found practicable and free from danger, Emin Pasha and Mr. Stanley will arrange for the return of the women and children with a suitable escort by this route, for the reason that as most of the women will have to be carried, it will save a great deal of labour, time, and expense.

Much of the ivory now in possession of Emin Pasha can also be sent by this route.

By sending the women and children by this route, it will enable the force under Mr. Stanley and the men belonging to Emin Pasha whom it is decided to return to Egypt, to make rapid marches to the East Coast, thus saving expense, and avoiding the possibility of conflict with Mwanga of Uganda and other tribes on the route which might be hostile to them.

It has been calculated that Mr. Stanley's expedition will reach Wadelai about middle of July, and, if this date proves correct, the return caravan from Wadelai with women, children, and ivory may be expected at Stanley Falls about end of October.

If steamer should be there, the party would reach Stanley Pool late in November, and Banana the end of December. From Banana they could be transported *viâ* England or Antwerp by the regular line of steamers, or *viâ* the Cape by a steamer sent to the Congo specially for this service.

The following are the comparative distances by land:—

	Days.
To Zanzibar—	
Viâ Kavironda	115
„ Msalala	100
„ Karagwe	145
„ Ujiji	116
To Banana—	
By Congo	55

Inclosure 2 in No. 34.

Estimate of the Cost of 100 Persons from Stanley Falls to Banana.

Number of Days.		£ s. d.
	Stanley Falls to Stanley Pool.	
17	100 persons at $1\frac{1}{2}$ mitakos per diem = 2550 mitakos: 500 mitakos at £3	15 0 0
	Stanley Pool to Banana.	
25	100 persons at $1\frac{1}{2}$ lbs. rice per diem = 3750 rations at 2*d.* per ration	31 5 0
	2 bales cloth, at £5 per bale	10 0 0
	Hire of steamer *Stanley*	100 0 0
	Miscellaneous, to include transport of goods and provisions, and other incidental expenses ...	43 15 0
	Total expenses, Stanley Falls to Banana, for 100 persons, about	200 0 0

No. 35.

Emin Relief Committee to Foreign Office.—(Received March 29.)

28, *Wynnstay Gardens, Kensington, March* 25, 1887.

Sir,

I have the honour to acknowledge the receipt of your letter having reference to the grant of £10,000 by the Egyptian Government towards the expenses of the Emin Relief Expedition,[1] and I am directed by the Committee to request you to convey to the Marquis of Salisbury their thanks for the information it contains.

I am further to ask that the particulars of the application of the £1000 retained in Egypt may be furnished to the Committee as soon as it is convenient.

I am, &c.

(Signed) F. de Winton, *Hon. Secretary, Emin Relief Committee.*

[1] No. 30.

No. 36.

Foreign Office to Emin Relief Committee.

Foreign Office, April 1, 1887.

SIR,

I am directed by the Marquis of Salisbury to request that you will inform the Committee that his Lordship has caused copies of your three letters of the 25th ultimo to be forwarded to Sir E. Baring, who has been requested to ascertain Nubar Pasha's views respecting the proportion in which the proceeds of ivory in Emin Pasha's possession should be shared between the Egyptian Government and the Committee.

As regards your suggestion, that her Majesty's Government should negotiate with the Congo Free State for the supply of transport and food to any party sent by Mr. Stanley down the Congo to Banana, I am to point out that it was understood that all the African arrangements were undertaken by the Committee, under whose directions the persons rescued would be brought either to the East or West Coast as might be more convenient, their conveyance from the coast to Egypt being undertaken by the Egyptian Government.

I am to suggest, therefore, that the Committee should arrange the question of the transport down the Congo as they have already arranged that of the transport up the river.

I am, &c.

(Signed) JULIAN PAUNCEFOTE.

No. 37.

The Marquis of Salisbury to Sir E. Baring.

Foreign Office, April 1, 1887.

SIR,

I inclose copy of a Memorandum, giving an estimate of the cost of conveying to Banana the party from Wadelai which Mr. Stanley proposes to send home by the Congo route, which has been furnished by the Emin Relief Committee, in compliance with the request contained in your dispatch of the 7th ultimo.[1]

I inclose also copy of an explanatory letter from the Committee.[2] You will see that the question of the expense of conveyance from Banana to Egypt is untouched, though various routes are indicated.

I inclose also copy of a letter which has been addressed to the Committee by my directions, informing them that it is for them rather than for her Majesty's Government to undertake the negotiations with the Congo State respecting transport.[3]

I inclose copy of a further letter from the Committee respecting the ivory believed to be in Emin Pasha's possession.[4]

Nubar Pasha claims that the Egyptian Government should be repaid out of the proceeds, while the Committee, referring to Mr. Mackinnon's letter of the 27th November, 1886, copy of which is in your possession, claim that they also are entitled to a just proportion of any sum realized.

It is presumed that the Egyptian Government do not contest the Committee's claim, but it may be advisable that the question of the proportion in which the proceeds should be shared should be settled before Wadelai is reached by the expedition. It might therefore be advisable that you should endeavour to obtain some suggestion on the subject from Nubar Pasha.

I add copy of another letter from the Committee respecting the £1000 retained by the Egyptian Government out of their contribution of £10,000 to the expenses of the expedition.[5]

I am, &c.

(Signed) SALISBURY.

No. 38.

Acting Consul-General Holmwood to the Marquis of Salisbury. (*Received April* 11.)

Zanzibar, February 25, 1887.

MY LORD,

In continuation of my dispatch of the 17th January regarding the Emin Relief Expedition, I have the honour to report the

[1] Inclosure 1 in No. 34. [2] No. 34. [3] No. 36. [4] No. 33. [5] No. 35.

departure this morning of Mr. H. M. Stanley and his party in the British India Company's steamer *Madura,* which will proceed to the Congo after touching at the Cape.

On Mr. Stanley's arrival here in the mail-steamer on the 22nd he found all the preparations for his expedition completed, and a body of 650 picked men awaiting him.

I presented Mr. Stanley to the Sultan at a public reception the next morning, when he thanked his Highness most heartily for the prompt and valuable assistance he had rendered to the Relief Expedition. In the afternoon he paid a long private visit to his Highness, who communicated to him without any reserve all the details of his present and recent difficulties.

Admiral Knorr and all the foreign Consuls, with the exception of the German Consul-General, called upon Mr. Stanley during his stay here.

Mr. Stanley personally attended at the muster of the Zanzibar porters which took place before embarkation: a considerable number of these had served under him on former occasions, and his immediate recognition of them, accompanied in every instance with the recollection of their names, and generally of some little incident in connection with their former service, evidently afforded especial gratification; indeed, this trait, or gift, which he eminently possesses, together with a practical knowledge of the Swahili language, would alone account for his undoubted popularity and great influence among his native followers.

After embarkation on board the *Madura* the roll was called, when not one man of the large party was missing.

Mr. Stanley was accompanied by a staff of six English gentlemen, all of whom have had the benefit of military experience in the Soudan or South Africa: the discipline, energy, and efficiency of these and a few other Europeans who have volunteered to serve, I believe at their own expense, under Mr. Stanley, augurs well for the success of the expedition.

I was able to assist in effecting an arrangement between Tippoo Tib and Mr. Stanley, by which the former has engaged to place 600 of his followers at the latter's disposal on arrival of the expedition at Stanley Falls. I have referred more fully to this arrangement, which I consider indispensable for the prompt carrying out of the present undertaking, in a separate Report.

The relief party which has just left Zanzibar is in every way the most perfectly organized expedition that has hitherto entered tropical Africa, and it is organized on a scale exactly suited to the work in hand.

I have, &c.

(Signed) FREDC. HOLMWOOD.

No. 39.

Acting Consul-General Holmwood to the Marquis of Salisbury. (*Received April* 19.)

Zanzibar, March 3, 1887.

MY LORD,

I have the honour to explain the considerations which led me on the occasion of Mr. Stanley's visit to the place to take an active part in bringing about a definite understanding between the King of the Belgians and Tippoo Tib in regard to the latter's future relations with the Congo Free State.

In the dispatch above-mentioned, I have already reported that I had induced Tippoo Tib to send special messengers overland to Stanley Falls, with orders to his *locum tenens* there not to allow his people to be drawn into further hostilities with the officers of the new State pending his arrival; but in the absence of any instructions from your Lordship, I had refrained from any further interference in the differences between the Europeans and Arabs of the Congo.

I had, however, taken note of the dispatch by Tippoo Tib of 1500 percussion muskets and 200 breech-loading rifles, with a full complement of ammunition, to his Agent on the Congo, and I knew that his followers to the number of 3000, armed with guns, had mustered at his station on the Falls.

On learning that Mr. Stanley had decided to lead the Emin Relief Expedition by the Congo route, and that he was visiting the King of the Belgians on his way out to Zanzibar, I foresaw the absolute necessity of coming to a distinct understanding with Tippoo Tib before the Relief Expedition was finally committed to the route that had been chosen, as, the larger the force under Mr.

Stanley's command, the more effectually could the Arabs of the Upper Congo arrange to cut off their food supplies during the most difficult part of this march between their depôt on the Congo and the Monbuttu region, during which, owing to the nature of the country, they would be wholly dependent on the friendly attitude of the natives for any further supplies than those they could carry with them.

Tippoo Tib had himself already taken alarm in consequence of Mr. Stanley's visit to Brussels having been reported by telegram, and I discussed the subject with him in a perfectly straightforward manner, in no way attempting to deny his right to defend his stations below the Stanley Falls, which he had occupied prior to the arrival of the International Association, and whose inhabitants acquiesced in his claims to Chieftainship, but basing my recommendation, that he should make sacrifices in order to arrive at a peaceful solution of the dispute, on the fact that, with uninterrupted river communication up to Stanley Falls, his position on the banks of the Congo must be untenable, should the armed steamers of the Congo State decide to ignore his claims.

He seemed, however, confident in his power to maintain his position below the Falls, and on Mr. Stanley's arrival the latter at once frankly informed me that he had advised the King of the Belgians that Tippoo Tib could prevent the Congo State from establishing permanent stations in the country claimed by him. Mr. Stanley added that he had offered to bring about, if possible, an amicable arrangement with Tippoo Tib, but that the King had only authorized him to do so on terms which this Chief's attitude seemed likely to render impracticable. He asked me, however, to assist him in endeavouring to bring the question to a favourable issue, as it was most important for the welfare of his expedition that he should obtain Tippoo Tib's co-operation.

I therefore again entered into communication with Tippoo Tib, and eventually he promised to accept any settlement I arranged. Finding that the occupation of the Congo up to the Falls was a point the King had insisted upon, I called Tippoo Tib to an interview with Mr. Stanley, who showed him, during a very long and clear explanation, that the result of his persistence would be the permanent abandonment by the Congo State of this enterprise, in event of which they would immediately be supplanted by the

French. Mr. Stanley described the eventual result in a very forcible exposition, which seemed to convince his hearer: at any rate, after prolonged discussion, the basis of an understanding was arrived at, and Tippoo Tib left me to settle all details and draw up an Agreement.

This Agreement was duly prepared, but the terms being beyond the discretion with which Mr. Stanley had been invested, he telegraphed these to the King of the Belgians, who replied in the course of a few hours, fully ratifying them, and duplicate deeds were immediately signed in my presence.

This Agreement, of which I inclose copy, is in my opinion most advantageous for both parties, and my action will, I trust, be approved by her Majesty's Government. By it Tippoo Tib, as Governor of the Congo State District of Stanley Falls, has bound himself to suppress slave raiding and Slave Trade throughout the whole of this large district. He acknowledges the sovereignty of the Congo State up to Stanley Falls, though, so long as the Agreement stands, it will virtually remain under his rule. Provided the officers of the Congo State do not interfere with the legitimate trade of the Arabs of the adjacent affluents of the Congo, I anticipate that Tippoo Tib will carry out all that he has undertaken.

I have, &c.

(Signed) FREDC. HOLMWOOD.

INCLOSURE IN NO. 39.

Agreement.

Mr. Henry Morton Stanley, on behalf of his Majesty the King of the Belgians, and Sovereign of the Congo State, appoints Hamed-bin-Mahommed al Marjebi, Tippoo Tib, to be Wali of the independent State of the Congo, at Stanley Falls district, at a salary of £30 per month, payable to his Agent at Zanzibar, on the following conditions:—

1. Tippoo Tib is to hoist the flag of the Congo State at its station near Stanley Falls, and to maintain the authority of the State on the Congo and all its affluents at the said station down-

wards to the Bujine or Aruwimi river, and to prevent the tribes thereon, as well as Arabs and others, from engaging in the Slave Trade.

2. Tippoo Tib is to receive a resident officer of the Congo State, who will act as his Secretary in all his communications with the Administrator-General.

3. Tippoo Tib is to be at full liberty to carry on his legitimate private trade in any direction, and to send his caravans to and from any places he may desire.

4. Tippoo Tib shall nominate a *locum tenens,* to whom in case of his temporary absence his powers shall be delegated, and who, in event of his death, shall become his successor in the Waliship; but his Majesty the King of the Belgians shall have the power of veto should there be any serious objection to Tippoo Tib's nominee.

5. This arrangement shall only be binding so long as Tippoo Tib or his representative fulfils the conditions embodied in this Agreement.

(Signed) HENRY M. STANLEY.
TIPPOO TIB (in Arabic).

(Signed) FREDC. HOLMWOOD.
KANJI RAJPAR (in Hindi).

Zanzibar, February 24, 1887.

No. 40.

Foreign Office to Acting Consul-General Holmwood.

Foreign Office, April 19, 1887.

SIR,

I am directed by the Marquis of Salisbury to acknowledge the receipt of your dispatch of the 3rd ultimo, and I am to state to you, in reply, that his Lordship approves your efforts to promote the success of Mr. Stanley's expedition for the relief of Emin Bey. It is to be hoped that the negotiations to which you refer between the Congo State and the Arabs will have the desired effect of contributing to the suppression of Slave Trade on the Congo.

I am, &c.

(Signed) JULIAN PAUNCEFOTE.

No. 41.

Emin Relief Committee to Foreign Office.—(*Received April* 26.)

28, *Wynnstay Gardens, Kensington, April* 25, 1887.

SIR,

I have the honour to acknowledge the receipt of your letter of the 1st April, having reference to the subjects contained in my three letters of the 25th ultimo, and I am requested by the Emin Relief Committee to convey their grateful acknowledgments to the Right Honourable the Secretary of State for Foreign Affairs for the reference he has caused to be made to the Egyptian Government concerning the ivory stated to be in the possession of Emin Pasha.

With regard to the repatriation of the Egyptian troops, and their women and children, *viâ* the Congo route, the Committee beg to refer you to the letter of Mr. W. Mackinnon, dated the 10th February, 1887. In this letter the position of, and the duties undertaken by, the Emin Relief Expedition are clearly defined.

The Egyptian Government, as we are informed by Sir John Kirk, have already made arrangements, through her Majesty's Consul-General at Zanzibar, for the payment of all expenses incurred by Emin Pasha's forces should they return *viâ* the East Coast.

It would therefore appear proper they should make arrangements such as are suggested by the Committee, and forwarded, under cover, in my letter of the 26th ultimo, for the West Coast, so as to insure the return of the refugees without fear of possible disasters which might arise from want of proper arrangements for their transport from the Congo to Egypt, and of a proper food supply. As regards the transport on the Haut Congo of the relieving force by the boats of the Congo Free State, this service was offered to the expedition by his Majesty the King of the Belgians, and was the chief inducement to adopting that route; but for the transport down the Congo, required by the Egyptian Government for refugees, the Committee suggest that the Government of his Highness the Khedive should apply for such service direct to the Sovereign of the Congo Free State, which is under no engagement to the Committee to provide any transport downwards.

In conclusion, the Committee desire to repeat that the only duty they have undertaken is that of attempting to relieve Emin Pasha, and to open up communications with him, as will be seen by the inclosed copies of a letter written to the late Lord Iddesleigh, and the Memorandum which accompanied it.[1]

Mr. H. M. Stanley will, of course, co-operate with Emin Pasha and help him to the utmost of his ability; but the Committee are unable to undertake the repatriation of any refugees, or to provide for the expenses of the journey either by East or West Coast, though they are desirous to give every information and afford every assistance that may lie in their power.

I have, &c.

(Signed) F. De Winton, *Hon. Secretary,*
Emin Relief Expedition.

No. 42.

Foreign Office to Emin Relief Committee.

Foreign Office, May 11, 1887.

Sir,

I am directed by the Marquis of Salisbury to observe that, in order that he may submit to the Egyptian Government the proposition contained in your letter of the 25th ultimo, some considerations appear to his Lordship to be necessary.

You state that your Committee arranged with the Congo Free State to transport up the river for the Relief Expedition, but that no arrangement has been made for the transport of refugees down the river to the coast; and you suggest that it is the duty of the Egyptian Government to make arrangements for this transport with the Congo Free State, or otherwise.

I am to inquire whether your meaning is that, in your opinion, the Egyptian Government should be prepared to pay for the transport of refugees from the time when they may reach the banks of the Congo, the conveyance being arranged by the leaders

[1] No. 11 and Inclosure.

of the expedition; or whether it is your idea that that Government should organize a further expedition to proceed to Stanley Falls or the neighbourhood, with sufficient transport to bring down refugees.

If the latter is your proposal, it would seem necessary that the fullest information should be given to the Egyptian Government as to the nature of the existing arrangements with the Congo Free State, and as to the instructions given to Mr. Stanley; and it would seem advisable that some plan should be settled before Mr. Stanley ascends the river, as, if the Egyptian Government should not be able to undertake the task which it is proposed to assign to it, or should fail in making arrangements, the consequences might be disastrous.

His Lordship has no wish to interefere in the plans of the expedition; but he feels that if he is asked to place an application of this nature before the Egyptian Government on behalf of the Committee, he should be enabled to accompany it with some scheme for their acceptance; and this can only be framed by those who are acquainted with the local resources which have been employed for the outward journey and voyage.

The repatriation of refugees, after they shall have reached either the East or West Coast, was understood to be undertaken by the Egyptian Government.

I am, &c.

(Signed) James Fergusson.

No. 43.

Emin Relief Committee to Foreign Office.—(Received May 20)

28, *Wynnstay Gardens, Kensington, May* 19, 1887.

Sir,

With reference to your letter of the 11th instant, I have the honour, by direction of the Committee, to reply as follows:—

The repatriation of refugees *viâ* the Congo would require steam transport on the upper waters of that river. The only available steamers for such a service belong to the Congo Free State. Arrangements would therefore have to be made with the authorities of that State, by which the use of these steamers could be obtained if required; and the hire or cost of the same was calculated in the estimate transmitted in my letter of the 25th March.

In that letter a scheme was suggested with, the Committee trust, sufficient detail to give the information asked for by the Right Honourable the Secretary of State for Foreign Affairs in the concluding portion of your letter. This scheme suggested the engagement of the steamer *Stanley* to proceed to Stanley Falls about September next; the purchase of six tons of rice, to be placed in magazines at different named stations of the route between Stanley Pool and the coast; and the obtaining authority from the Congo Free State for the purchase of such other goods as might be required from their stores.

The Committee, in suggesting this scheme, had no idea of organizing a further expedition, as the employés of the Free State are quite capable of carrying out the services above-mentioned.

No definite instructions were given to Mr. Stanley, but it was Mr. Stanley's idea to adopt the Congo route for the repatriation of the women and children—though he himself and his party would return *viâ* the East Coast—as being in every way more expeditious and less dangerous, if Emin Pasha concurred.

The Committee have no application to submit to the Egyptian Government, as they consider their duty has been discharged when they bring the facts before her Majesty's Government as they have done, leaving it to them to take such steps in dealing with the Egyptian Government as they may deem expedient; but they are prepared to furnish any other detail if the scheme and estimate and letter of the 25th March, above referred to, are not found to be sufficient.

The Committee learn with satisfaction, as they have always understood, that the repatriation of refugees, after they shall have reached either the East or West Coast, will be undertaken by the Egyptian Government.

I have, &c.

(Signed) F. De Winton, *Hon. Secretary,*
Emin Relief Expedition.

No. 44.

The Marquis of Salisbury to Sir E. Baring.

Foreign Office, May 27, 1887.

SIR,

I inclose for your observation, copies of correspondence with the Emin Relief Committee respecting the arrangements for the transport of refugees down the Congo.[1]

I am, &c.

(Signed) SALISBURY.

[1] Nos. 41 and 43.

Richard Clay & Sons, Limited,
London & Bungay.

11, Henrietta Street, Covent Garden, W.C.

October, 1890.

A

Catalogue of Books

PUBLISHED BY

CHAPMAN & HALL,

LIMITED.

FOR

Drawing Examples, Diagrams, Models, Instruments, etc.,

ISSUED UNDER THE AUTHORITY OF

THE SCIENCE AND ART DEPARTMENT,
SOUTH KENSINGTON,

FOR THE USE OF SCHOOLS AND ART AND SCIENCE CLASSES,

See separate Illustrated Catalogue.

BEATTY-KINGSTON (W.)—Continued—

MONARCHS I HAVE MET. 2 vols. Demy 8vo, 24s.

MUSIC AND MANNERS: Personal Reminiscences and Sketches of Character. 2 vols. Demy 8vo, 30s.

BELL (JAMES, Ph.D., &c.), Principal of the Somerset House Laboratory—

THE CHEMISTRY OF FOODS. With Microscopic Illustrations.

PART I. TEA, COFFEE, COCOA, SUGAR, ETC. Large crown 8vo, 2s. 6d.
PART II. MILK, BUTTER, CHEESE, CEREALS, PREPARED STARCHES, ETC. Large crown 8vo, 3s.

BENSON (W.)—

UNIVERSAL PHONOGRAPHY. To classify sounds of Human Speech, and to denote them by one set of Symbols for easy Writing and Printing. 8vo, sewed, 1s.

MANUAL OF THE SCIENCE OF COLOUR. Coloured Frontispiece and Illustrations. 12mo, cloth, 2s. 6d.

PRINCIPLES OF THE SCIENCE OF COLOUR. Small 4to, cloth, 15s.

BINGHAM (CAPT. THE HON. D.)—

THE MARRIAGES OF THE BOURBONS. With Illustrations. 2 vols. Demy 8vo, 32s.

A SELECTION FROM THE LETTERS AND DESPATCHES OF THE FIRST NAPOLEON. With Explanatory Notes. 3 vols. Demy 8vo, £2 2s.

THE BASTILLE. With Illustrations. 2 vols. Demy 8vo, 32s.

BIRDWOOD (SIR GEORGE C. M.), C.S.I.—

THE INDUSTRIAL ARTS OF INDIA. With Map and 174 Illustrations. New Edition. Demy 8vo, 14s.

BLACKIE (JOHN STUART), F.R.S.E.—

THE SCOTTISH HIGHLANDERS AND THE LAND LAWS. Demy 8vo, 9s.

ALTAVONA: FACT AND FICTION FROM MY LIFE IN THE HIGHLANDS. Third Edition. Crown 8vo, 6s.

BLENNERHASSETT (LADY)—

MADAME DE STAEL: Her Friends, and Her Influence in Politics and Literature. Translated from the German by J. E. GORDON CUMMING. With a Portrait. 3 vols. Demy 8vo, 36s.

BLEUNARD (A.)—

BABYLON ELECTRIFIED: The History of an Expedition undertaken to restore Ancient Babylon by the Power of Electricity, and how it Resulted. Translated from the French by F. L. WHITE, and Illustrated by MONTADER. Royal 8vo, 12s.

BLOOMFIELD'S (BENJAMIN LORD), MEMOIR OF—MISSION TO THE COURT OF BERNADOTTE. With Portraits. 2 vols. Demy 8vo, 28s.

BONVALOT (GABRIEL)—

THROUGH THE HEART OF ASIA OVER THE PAMIR TO INDIA. Translated from the French by C. B. PITMAN. With 250 Illustrations by ALBERT PÉPIN. Royal 8vo, 32s.

BOULGER (DEMETRIUS C.)—

GENERAL GORDON'S LETTERS FROM THE CRIMEA, THE DANUBE, AND ARMENIA. 2nd Edition. Crown 8vo, 5s.

BOWERS (G.)—

HUNTING IN HARD TIMES. With 61 coloured Illustrations. Oblong 4to, 12s.

BRACKENBURY (COL. C. B.)—

FREDERICK THE GREAT. With Maps and Portrait. Large crown 8vo, 4s.

BRADLEY (THOMAS), of the Royal Military Academy, Woolwich—

ELEMENTS OF GEOMETRICAL DRAWING. In Two Parts, with Sixty Plates. Oblong folio, half bound, each Part 16s.

MRS. BRAY'S NOVELS AND ROMANCES.

New and Revised Editions, with Frontispieces. 3s. 6d. each.

THE WHITE HOODS.
DE FOIX; a Romance of Bearn.
FITZ OF FITZFORD; a Tale of Destiny.
HENRY DE POMEROY.
TRELAWNY OF TRELAWNE.
A FATHER'S CURSE AND A DAUGHTER'S SACRIFICE.
TRIALS OF THE HEART.
THE TALBA; or, The Moor of Portugal.
THE PROTESTANT.
WARLEIGH; or, The Fatal Oak.
COURTENAY OF WALREDDON.
HARTLAND FOREST AND ROSE-TEAGUE.

BRIDGMAN (F. A.)—

WINTERS IN ALGERIA. With 62 Illustrations. Royal 8vo, 10s. 6d.

BRITISH ARMY, THE. By the Author of "Greater Britain," "The Present Position of European Politics," etc. Demy 8vo, 12s.

BROADLEY (A. M.)—

HOW WE DEFENDED ARABI AND HIS FRIENDS. A Story of Egypt and the Egyptians. Illustrated by FREDERICK VILLIERS. Demy 8vo, 12s.

BROCK (DR. J. H. E.), Assistant Examiner in Hygiene, Science and Art Department—

ELEMENTS OF HUMAN PHYSIOLOGY FOR THE HYGIENE EXAMINATIONS OF THE SCIENCE AND ART DEPARTMENT. Crown 8vo, 1s. 6d.

BROMLEY-DAVENPORT (the late W.), M.P.—

SPORT: Fox Hunting, Salmon Fishing, Covert Shooting, Deer Stalking. With numerous Illustrations by General CREALOCK, C.B. New Cheap Edition. Post 8vo, 3s. 6d.

——— Small 4to, 21s.

BUCKLAND (FRANK)—

LOG-BOOK OF A FISHERMAN AND ZOOLOGIST. With numerous Illustrations. Fifth Thousand. Crown 8vo, 5s.

BUCKMAN (S. S.), F.G.S.—

JOHN DARKE'S SOJOURN IN THE COTTESWOLDS AND ELSEWHERE: A Series of Sketches. Crown 8vo, 3s. 6d.

BROWN (J. MORAY)—

POWDER, SPEAR, AND SPUR: A Sporting Medley. With Illustrations by G. D. GILES and EDGAR GIBERNE from Sketches by the Author. Crown 8vo, 10s. 6d.

BURCHETT (R.)—

DEFINITIONS OF GEOMETRY. New Edition. 24mo cloth, 5d.

BURCHETT (R.)—Continued—

LINEAR PERSPECTIVE, for the Use of Schools of Art. New Edition. With Illustrations. Post 8vo, cloth, 7s.

PRACTICAL GEOMETRY: The Course of Construction of Plane Geometrical Figures. With 137 Diagrams. Eighteenth Edition. Post 8vo, cloth, 5s.

BURGESS (EDWARD)—

ENGLISH AND AMERICAN YACHTS. Illustrated with 50 Beautiful Photogravure Engravings. Oblong folio, 42s.

BUTLER (A. J.)—

COURT LIFE IN EGYPT. Second Edition. Illustrated. Large crown 8vo, 12s.

CAFFYN (MANNINGTON)—

A POPPY'S TEARS. Crown 8vo, 1s.; in cloth, 1s. 6d.

*CARLYLE (THOMAS), WORKS BY.—**See pages 29 and 30.*

THE CARLYLE BIRTHDAY BOOK. Compiled, with the permission of Mr. Thomas Carlyle, by C. N. WILLIAMSON. Second Edition. Small fcap. 8vo, 3s.

CARSTENSEN (A. RIIS)—

TWO SUMMERS IN GREENLAND: An Artist's Adventures among Ice and Islands in Fjords and Mountains. With numerous Illustrations by the Author. Demy 8vo, 14s.

CHALDÆAN AND ASSYRIAN ART—

A HISTORY OF ART IN CHALDÆA AND ASSYRIA. By GEORGES PERROT and CHARLES CHIPIEZ. Translated by WALTER ARMSTRONG, B.A. Oxon. With 452 Illustrations. 2 vols. Imperial 8vo, 42s.

CHAPMAN & HALL'S SHILLING SERIES.

Bound in Cloth, 1s. 6d.

A POPPY'S TEARS. By MANNINGTON CAFFYN.
NOTCHES ON THE ROUGH EDGE OF LIFE. By LYNN CYRIL D'OYLE.
WE TWO AT MONTE CARLO. By ALBERT D. VANDAM.
WHO IS THE MAN? A Tale of the Scottish Border. By J. S. TAIT.
THE CHILD OF STAFFERTON: A Chapter from a Family Chronicle. By CANON KNOX LITTLE.
THE BROKEN VOW: A Story of Here and Hereafter. By CANON KNOX LITTLE
THE STORY OF AN AFRICAN FARM. By OLIVE SCHREINER.
PADDY AT HOME. By BARON E. DE MANDAT-GRANCEY.

CHARLOTTE ELIZABETH, LIFE AND LETTERS OF, Princess Palatine and Mother of Philippe d'Orléans, Regent of France, 1652–1722. Compiled, translated, and gathered from various Published and Unpublished Sources. With Portraits. Demy 8vo, 10s. 6d.

CHARNAY (DÉSIRÉ)—

THE ANCIENT CITIES OF THE NEW WORLD. Being Travels and Explorations in Mexico and Central America, 1857—1882. Translated from the French by J. Gonino and Helen S. Conant. With upwards of 200 Illustrations. Super Royal 8vo, 31s. 6d.

CHURCH (PROFESSOR A. H.), M.A. Oxon.—

FOOD GRAINS OF INDIA. With numerous Woodcuts. Small 4to, 6s.

ENGLISH PORCELAIN. A Handbook to the China made in England during the Eighteenth Century, as illustrated by Specimens chiefly in the National Collection. With numerous Woodcuts. Large crown 8vo, 3s.

CHURCH (PROFESSOR A. H.)., M.A. Oxon. (Continued)—

ENGLISH EARTHENWARE. A Handbook to the Wares made in England during the 17th and 18th Centuries, as illustrated by Specimens in the National Collections. With numerous Woodcuts. Large crown 8vo, 3s.

PLAIN WORDS ABOUT WATER. Illustrated. Crown 8vo, sewed, 6d.

FOOD: Some Account of its Sources, Constituents, and Uses. A New and Revised Edition. Large crown 8vo, cloth, 3s.

PRECIOUS STONES: considered in their Scientific and Artistic Relations. With a Catalogue of the Townsend Collection of Gems in the South Kensington Museum. With a Coloured Plate and Woodcuts. Large crown 8vo, 2s. 6d.

CLINTON (R. H.)—

A COMPENDIUM OF ENGLISH HISTORY, from the Earliest Times to A.D. 1872. With Copious Quotations on the Leading Events and the Constitutional History, together with Appendices. Post 8vo, 7s. 6d.

COBDEN, RICHARD, LIFE OF. By the RIGHT HON. JOHN MORLEY, M.P. With Portrait. New Edition. Crown 8vo, 7s. 6d.

Popular Edition, with Portrait, 4to, sewed, 1s.; cloth, 2s.

COLLINS (WILKIE) and DICKENS (CHARLES)—

THE LAZY TOUR OF TWO IDLE APPRENTICES; NO THOROUGHFARE; THE PERILS OF CERTAIN ENGLISH PRISONERS. With 8 Illustrations. Crown 8vo, 5s.

*** These Stories, which originally appeared in "Household Words," are now reprinted for the first time complete.

COOKERY—

THE PYTCHLEY BOOK OF REFINED COOKERY AND BILLS OF FARE. By MAJOR L——. Fourth Edition. Large crown 8vo, 8s.

BREAKFASTS, LUNCHEONS, AND BALL SUPPERS. By MAJOR L——. Crown 8vo. 4s.

OFFICIAL HANDBOOK OF THE NATIONAL TRAINING SCHOOL FOR COOKERY. Containing Lessons on Cookery; forming the Course of Instruction in the School. Compiled by "R. O. C." Twentieth Thousand. Large crown 8vo, 6s.

BREAKFAST AND SAVOURY DISHES. By "R. O. C." Ninth Thousand. Crown 8vo, 1s.

HOW TO COOK FISH. Compiled by "R. O. C." Crown 8vo, sewed, 3d.

SICK-ROOM COOKERY. Compiled by "R. O. C." Crown 8vo, sewed, 6d.

THE ROYAL CONFECTIONER: English and Foreign. A Practical Treatise. By C. E. FRANCATELLI. With numerous Illustrations. Fifth Thousand. Crown 8vo, 5s.

THE KINGSWOOD COOKERY BOOK. By H. F. WICKEN. Crown 8vo, 2s.

COOPER-KING (LT.-COL.)—

GEORGE WASHINGTON. Large crown 8vo. With Portrait and Maps. [*In the Press.*

COURTNEY (W. L.), M.A., LL.D., of New College, Oxford—

STUDIES NEW AND OLD. Crown 8vo, 6s.

CONSTRUCTIVE ETHICS: A Review of Modern Philosophy and its Three Stages of Interpretation, Criticism, and Reconstruction. Demy 8vo, 12s.

CRAIK (GEORGE LILLIE)—

ENGLISH OF SHAKESPEARE. Illustrated in a Philological Commentary on his "Julius Cæsar." Eighth Edition. Post 8vo, cloth, 5s.

OUTLINES OF THE HISTORY OF THE ENGLISH LANGUAGE. Eleventh Edition. Post 8vo, cloth, 2s. 6d.

CRAWFURD (OSWALD), Her Majesty's Consul at Oporto—

ROUND THE CALENDAR IN PORTUGAL. With numerous Illustrations by Mrs. H. M. STANLEY (DOROTHY TENNANT), Mrs. ARTHUR WALTER, Mr. TRISTRAM ELLIS, Miss WOODWARD, Miss THOMSON, Mr. A. LEE, and the Author. Royal 8vo. [*In the Press.*

BEYOND THE SEAS; being the surprising Adventures and ingenious Opinions of Ralph, Lord St. Keyne, told by his kinsman, Humphrey St. Keyne. Second Edition. Crown 8vo, 3s. 6d.

CRIPPS (WILFRED JOSEPH), M.A., F.S.A.—

COLLEGE AND CORPORATION PLATE. A Handbook for the Reproduction of Silver Plate. [*In the South Kensington Museum, from celebrated English collections.*] With numerous Illustrations. Large crown 8vo, cloth, 2s. 6d.

DAIRY FARMING—

DAIRY FARMING. To which is added a Description of the Chief Continental Systems. With numerous Illustrations. By JAMES LONG. Crown 8vo, 9s.

DAIRY FARMING, MANAGEMENT OF COWS, etc. By ARTHUR ROLAND. Edited by WILLIAM ABLETT. Crown 8vo, 5s.

DALY (J. B.), LL.D.—

IRELAND IN THE DAYS OF DEAN SWIFT. Crown 8vo, 5s.

DAUBOURG (E.)—

INTERIOR ARCHITECTURE. Doors, Vestibules, Staircases, Anterooms, Drawing, Dining, and Bed Rooms, Libraries, Bank and Newspaper Offices, Shop Fronts and Interiors. Half-imperial, cloth, £2 12s. 6d.

DAVIDSON (ELLIS A.)—

PRETTY ARTS FOR THE EMPLOYMENT OF LEISURE HOURS. A Book for Ladies. With Illustrations. Demy 8vo, 6s.

DAVITT (MICHAEL)—

LEAVES FROM A PRISON DIARY. Crown 8vo, sewed, 1s. 6d.

DAY (WILLIAM)—

THE RACEHORSE IN TRAINING, with Hints on Racing and Racing Reform, to which is added a Chapter on Shoeing. Sixth Edition. Demy 8vo, 9s.

DAS (DEVENDRA N.)—

SKETCHES OF HINDOO LIFE. Crown 8vo, 5s.

DE AINSLIE (GENERAL)—

A HISTORY OF THE ROYAL REGIMENT OF DRAGOONS. From its Formation in 1661 to the Present Day. With Illustrations. Demy 8vo, 21s.

DE CHAMPEAUX (ALFRED)—

TAPESTRY. With numerous Woodcuts. Cloth, 2s. 6d.

DE FALLOUX (THE COUNT)—

MEMOIRS OF A ROYALIST. Edited by C. B. PITMAN. 2 vols. With Portraits. Demy 8vo, 32s.

D'HAUSSONVILLE (VICOMTE)—

SALON OF MADAME NECKER. Translated by H. M. TROLLOPE. 2 vols. Crown 8vo, 18s.

DE KONINCK (L. L.) and DIETZ (E.)—

PRACTICAL MANUAL OF CHEMICAL ASSAYING, as applied to the Manufacture of Iron. Edited, with notes, by ROBERT MALLET. Post 8vo, cloth, 6s.

DE LESSEPS (FERDINAND)—

RECOLLECTIONS OF FORTY YEARS. Translated from the French by C. B. PITMAN. 2 vols. Demy 8vo, 24s.

DE LISLE (MEMOIR OF LIEUTENANT RUDOLPH), R.N., of the Naval Brigade. By the Rev. H. N. OXENHAM, M.A. Third Edition. Crown 8vo, 7s. 6d.

DE MANDAT-GRANCEY (BARON E.)—

PADDY AT HOME; OR, IRELAND AND THE IRISH AT THE PRESENT TIME, AS SEEN BY A FRENCHMAN. Translated from the French. Fifth Edition. Crown 8vo, 1s.; in cloth, 1s. 6d.

D'OYLE (LYNN CYRIL)—

NOTCHES ON THE ROUGH EDGE OF LIFE. Crown 8vo, 1s.; in cloth, 1s. 6d.

DE STAËL (MADAME)—

MADAME DE STAËL: Her Friends, and Her Influence in Politics and Literature. By LADY BLENNERHASSETT. Translated from the German by J. E. GORDON CUMMING. With a Portrait. 3 vols. Demy 8vo, 36s.

DE WINDT (H.)—

FROM PEKIN TO CALAIS BY LAND. With numerous Illustrations by C. E. FRIPP from Sketches by the Author. Demy 8vo, 20s.

*DICKENS (CHARLES), WORKS BY—**See pages* 31—37.

THE LETTERS OF CHARLES DICKENS. Two vols. uniform with "The Charles Dickens Edition" of his Works. Crown 8vo, 8s.

THE LIFE OF CHARLES DICKENS—*See "Forster."*

THE CHARLES DICKENS BIRTHDAY BOOK. With Five Illustrations. In a handsome fcap. 4to volume, 12s.

THE HUMOUR AND PATHOS OF CHARLES DICKENS. By CHARLES KENT. With Portrait. Crown 8vo, 6s.

DICKENS (CHARLES) and COLLINS (WILKIE)—

THE LAZY TOUR OF TWO IDLE APPRENTICES; NO THOROUGHFARE; THE PERILS OF CERTAIN ENGLISH PRISONERS. With Illustrations. Crown 8vo, 5s.

*** These Stories are now reprinted in complete form for the first time.

DILKE (LADY)—

ART IN THE MODERN STATE. With Facsimile. Demy 8vo, 9s.

DINARTE (SYLVIO)—

INNOCENCIA: A Story of the Prairie Regions of Brazil. Translated from the Portuguese and Illustrated by JAMES W. WELLS, F.R.G.S. Crown 8vo, 6s.

DIXON (CHARLES)—

IDLE HOURS WITH NATURE. With Frontispiece. Crown 8vo.

ANNALS OF BIRD LIFE: A Year-Book of British Ornithology. With Illustrations. Crown 8vo, 7s. 6d.

DOUGLAS (JOHN)—

SKETCH OF THE FIRST PRINCIPLES OF PHYSIOGRAPHY. With Maps and numerous Illustrations. Crown 8vo, 6s.

DOWN WITH ENGLAND. Translated from the French. With Maps. Crown 8vo, 1s.

DRAYSON (MAJOR-GENERAL A. W.), Late R.A., F.R.A.S.—

THIRTY THOUSAND YEARS OF THE EARTH'S PAST HISTORY. Large Crown 8vo, 5s.

EXPERIENCES OF A WOOLWICH PROFESSOR during Fifteen Years at the Royal Military Academy. Demy 8vo, 8s.

PRACTICAL MILITARY SURVEYING AND SKETCHING. Fifth Edition. Post 8vo, cloth, 4s. 6d.

DREAMS BY A FRENCH FIRESIDE. Translated from the German by MARY O'CALLAGHAN. Illustrated by Fred Roe. Crown 8vo, 7s. 6d.

DUCOUDRAY (GUSTAVE)—

THE HISTORY OF ANCIENT CIVILISATION. A Handbook based upon M. Gustave Ducoudray's "Histoire Sommaire de la Civilisation." Edited by REV. J. VERSCHOYLE, M.A. With Illustrations. Large crown 8vo, 6s.

DUFFY (SIR CHARLES GAVAN), K.C.M.G.—

THE LEAGUE OF NORTH AND SOUTH. An Episode in Irish History, 1850–1854. Crown 8vo, 8s.

DYCE (WILLIAM), R.A.—

DRAWING-BOOK OF THE GOVERNMENT SCHOOL OF DESIGN; OR, ELEMENTARY OUTLINES OF ORNAMENT. Fifty selected Plates. Folio, sewed, 5s.; mounted, 18s.

ELEMENTARY OUTLINES OF ORNAMENT. Plates I. to XXII., containing 97 Examples, adapted for Practice of Standards I. to IV. Small folio, sewed, 2s 6d.

SELECTION FROM DYCE'S DRAWING BOOK. 15 Plates, sewed, 1s. 6d.; mounted on cardboard, 6s. 6d.

TEXT TO ABOVE. Crown 8vo, sewed, 6d.

EDWARDS (H. SUTHERLAND)—

FAMOUS FIRST REPRESENTATIONS. Crown 8vo, 6s.

EGYPTIAN ART—

A HISTORY OF ART IN ANCIENT EGYPT. By G. PERROT and C. CHIPIEZ. Translated by WALTER ARMSTRONG. With over 600 Illustrations. 2 vols. Imperial 8vo, £2 2s.

ELLIS (A. B., Major 1st West India Regiment)—

THE EWE-SPEAKING PEOPLE OF THE SLAVE COAST OF WEST AFRICA. With Map. Demy 8vo, 10s. 6d

WEST AFRICAN STORIES. Crown 8vo, 6s.

THE TSHI-SPEAKING PEOPLES OF THE GOLD COAST OF WEST AFRICA: their Religion, Manners, Customs, Laws, Language, &c. With Map. Demy 8vo, 10s. 6d.

SOUTH AFRICAN SKETCHES. Crown 8vo, 6s.

THE HISTORY OF THE WEST INDIA REGIMENT. With Maps and Coloured Frontispiece and Title-page. Demy 8vo, 18s.

THE LAND OF FETISH. Demy 8vo, 12s.

ENGEL (CARL)—

MUSICAL INSTRUMENTS. With numerous Woodcuts. Large crown 8vo, cloth, 2s. 6d.

ESCOTT (T. H. S.)—

POLITICS AND LETTERS. Demy 8vo, 9s.

ENGLAND: ITS PEOPLE, POLITY, AND PURSUITS. New and Revised Edition. Demy 8vo, 3s. 6d.

EUROPEAN POLITICS, THE PRESENT POSITION OF.
By the Author of "Greater Britain." Demy 8vo, 12s.

FANE (VIOLET)—

AUTUMN SONGS. Crown 8vo, 6s.

THE STORY OF HELEN DAVENANT. Crown 8vo, 3s. 6d.

QUEEN OF THE FAIRIES (A Village Story), and other Poems. Crown 8vo, 6s.

ANTHONY BABINGTON: a Drama. Crown 8vo, 6s.

FARR (WILLIAM) and THRUPP (GEORGE A.)—

COACH TRIMMING. With 60 Illustrations. Crown 8vo, 2s. 6d.

FIELD (HENRY M.)—

GIBRALTAR. With numerous Illustrations. Demy 8vo, 7s. 6d.

FIFE-COOKSON (LIEUT-COL. J. C.)—

TIGER-SHOOTING IN THE DOON AND ULWAR, AND LIFE IN INDIA. With numerous Illustrations by E. HOBDAY, R.H.A. Large crown 8vo, 10s. 6d.

FITZGERALD (PERCY), F.S.A.—

THE HISTORY OF PICKWICK. Demy 8vo.

THE CHRONICLES OF BOW STREET POLICE OFFICE. With numerous Illustrations. 2 vols. Demy 8vo, 21s.

FLEMING (GEORGE), F.R.C.S.—

ANIMAL PLAGUES: THEIR HISTORY, NATURE, AND PREVENTION. 8vo, cloth, 15s.

PRACTICAL HORSE-SHOEING. With 37 Illustrations. Fifth Edition, enlarged. 8vo, sewed, 2s.

RABIES AND HYDROPHOBIA: THEIR HISTORY, NATURE, CAUSES, SYMPTOMS, AND PREVENTION. With 8 Illustrations. 8vo, cloth, 15s.

FORSTER (JOHN)—

THE LIFE OF CHARLES DICKENS. Uniform with the Illustrated Library Edition of Dickens's Works. 2 vols. Demy 8vo, 20s.

THE LIFE OF CHARLES DICKENS. Uniform with the Library Edition. Post 8vo, 10s. 6d.

THE LIFE OF CHARLES DICKENS. Uniform with the "C. D." Edition. With Numerous Illustrations. 2 vols. 7s.

THE LIFE OF CHARLES DICKENS. Uniform with the Household Edition. With Illustrations by F. BARNARD. Crown 4to, cloth, 5s.

FORSTER, THE LIFE OF THE RIGHT HON. W. E.
By T. WEMYSS REID. With Portraits. Fourth Edition. 2 vols. Demy 8vo, 32s.
FIFTH EDITION, in one volume, with new Portrait. Demy 8vo, 10s. 6d.

FORSYTH (CAPTAIN)—

THE HIGHLANDS OF CENTRAL INDIA: Notes on their Forests and Wild Tribes, Natural History and Sports. With Map and Coloured Illustrations. A New Edition. Demy 8vo, 12s.

FORTESCUE (THE HON. JOHN)—

RECORDS OF STAG-HUNTING ON EXMOOR. With 14 full page Illustrations by EDGAR GIBERNE. Large crown 8vo, 16s.

FORTNIGHTLY REVIEW (*see page* 40)—

FORTNIGHTLY REVIEW.—First Series, May, 1865, to Dec. 1866. 6 vols. Cloth, 13s. each.

New Series, 1867 to 1872. In Half-yearly Volumes. Cloth, 13s. each.

From January, 1873, to the present time, in Half-yearly Volumes. Cloth, 16s. each.

CONTENTS OF FORTNIGHTLY REVIEW. From the commencement to end of 1878. Sewed, 2s.

FORTNUM (C. D. E.), F.S.A.—

MAIOLICA. With numerous Woodcuts. Large crown 8vo, cloth, 2s. 6d.

BRONZES. With numerous Woodcuts. Large crown 8vo, cloth, 2s. 6d.

FOUQUÉ (DE LA MOTTE)—

UNDINE: a Romance translated from the German. With an Introduction by JULIA CARTWRIGHT. Illustrated by HEYWOOD SUMNER. Crown 4to, 5s.

FRANCATELLI (C. E.)—

THE ROYAL CONFECTIONER: English and Foreign. A Practical Treatise. With Illustrations. Fifth Edition. Crown 8vo, 5s.

FRANCIS (FRANCIS), JUNR.

SADDLE AND MOCASSIN. 8vo, 12s.

FRANKS (A. W.)—

JAPANESE POTTERY. Being a Native Report, with an Introduction and Catalogue. With numerous Illustrations and Marks. Large crown 8vo, cloth, 2s. 6d.

FROBEL, FRIEDRICH; a Short Sketch of his Life, including Fröbel's Letters from Dresden and Leipzig to his Wife, now first Translated into English. By EMILY SHIRREFF. Crown 8vo, 2s.

GALILEO AND HIS JUDGES. By F. R. WEGG-PROSSER. Demy 8vo, 5s.

GALLENGA (ANTONIO)—

ITALY: PRESENT AND FUTURE. 2 vols. Dmy. 8vo, 21s.

EPISODES OF MY SECOND LIFE. 2 vols. Dmy. 8vo, 28s.

IBERIAN REMINISCENCES. Fifteen Years' Travelling Impressions of Spain and Portugal. With a Map. 2 vols. Demy 8vo, 32s.

GASNAULT (PAUL) and GARNIER (ED.)—

FRENCH POTTERY. With Illustrations and Marks. Large crown 8vo, 3s.

GILLMORE (PARKER)—

THE HUNTER'S ARCADIA. With numerous Illustrations. Demy 8vo, 10s. 6d.

GIRL'S LIFE EIGHTY YEARS AGO (A). Selections from the Letters of Eliza Southgate Bowne, with an Introduction by Clarence Cook. Illustrated with Portraits and Views. Crown 4to, 12s.

GLEICHEN (COUNT), Grenadier Guards—

WITH THE CAMEL CORPS UP THE NILE. With numerous Sketches by the Author. Third Edition. Large crown 8vo, 9s.

GORDON (GENERAL)—

LETTERS FROM THE CRIMEA, THE DANUBE, AND ARMENIA. Edited by DEMETRIUS C. BOULGER. Second Edition. Crown 8vo, 5s.

GORST (SIR J. E.), Q.C., M.P.—

An ELECTION MANUAL. Containing the Parliamentary Elections (Corrupt and Illegal Practices) Act, 1883, with Notes. Third Edition. Crown 8vo, 1s. 6d.

GOWER (A. R.), Royal School of Mines—

PRACTICAL METALLURGY. With Illustrations. Crown 8vo, 3s.

GRAHAM (SIR GERALD), V.C., K.C.B.—

LAST WORDS WITH GORDON. Crown 8vo, cloth, 1s.

GRESWELL (WILLIAM), M.A., F.R.C.I.—

OUR SOUTH AFRICAN EMPIRE. With Map. 2 vols. Crown 8vo, 21s.

GRIFFIN (SIR LEPEL HENRY), K.C.S.I.—

THE GREAT REPUBLIC. Second Edition. Crown 8vo, 4s. 6d.

GRIFFITHS (MAJOR ARTHUR), H.M. Inspector of Prisons—

FRENCH REVOLUTIONARY GENERALS. Large crown 8vo. [*In the Press.*

CHRONICLES OF NEWGATE. Illustrated. New Edition. Demy 8vo, 16s.

MEMORIALS OF MILLBANK: or, Chapters in Prison History. With Illustrations by R. Goff and Author. New Edition. Demy 8vo, 12s.

HALL (SIDNEY)—

A TRAVELLING ATLAS OF THE ENGLISH COUNTIES. Fifty Maps, coloured. New Edition, including the Railways, corrected up to the present date. Demy 8vo, in roan tuck, 10s. 6d.

HAWKINS (FREDERICK)—

THE FRENCH STAGE IN THE EIGHTEENTH CENTURY. With Portraits. 2 vols. Demy 8vo, 30s.

ANNALS OF THE FRENCH STAGE: FROM ITS ORIGIN TO THE DEATH OF RACINE. 4 Portraits. 2 vols. Demy 8vo, 28s.

HILDEBRAND (HANS), Royal Antiquary of Sweden—

INDUSTRIAL ARTS OF SCANDINAVIA IN THE PAGAN TIME. With numerous Woodcuts. Large crown 8vo, 2s. 6d.

HILL (MISS G.)—

THE PLEASURES AND PROFITS OF OUR LITTLE POULTRY FARM. Small 8vo, 3s.

HOLBEIN—

TWELVE HEADS AFTER HOLBEIN. Selected from Drawings in Her Majesty's Collection at Windsor. Reproduced in Autotype, in portfolio. £1 16s.

HOLMES (GEORGE C. V.), Secretary of the Institution of Naval Architects, Whitworth Scholar—

MARINE ENGINES AND BOILERS. With Sixty-nine Woodcuts. Large crown 8vo, 3s.

HOPE (ANDRÉE)—

CHRONICLES OF AN OLD INN; or, a Few Words about Gray's Inn. Crown 8vo, 5s.

HOUSSAYE (ARSÈNE)—

BEHIND THE SCENES OF THE COMÉDIE FRANCAISE, AND OTHER RECOLLECTIONS. Translated and Edited, with Notes, by ALBERT D. VANDAM. Demy 8vo, 14s.

HOVELACQUE (ABEL)—

THE SCIENCE OF LANGUAGE: LINGUISTICS, PHILOLOGY, AND ETYMOLOGY. With Maps. Large crown 8vo, cloth, 5s.

HOZIER (H. M.)—

TURENNE. With Portrait and Two Maps. Large crown 8vo, 4s.

HUEFFER (F.)—

HALF A CENTURY OF MUSIC IN ENGLAND. 1837—1887. Demy 8vo, 8s.

HUNTLY (MARQUIS OF)—

TRAVELS, SPORTS, AND POLITICS IN THE EAST OF EUROPE. With Illustrations by the Marchioness of Huntly. Large Crown 8vo, 12s.

INDUSTRIAL ARTS: Historical Sketches. With numerous Illustrations. Large crown 8vo, 3s.

IRELAND IN THE DAYS OF DEAN SWIFT. By J. B. DALY, LL.D. Crown 8vo, 5s.

IRISH ART OF LACEMAKING, A RENASCENCE OF THE. Illustrated by Photographic Reproductions of Irish Laces, made from new and specially designed Patterns. Introductory Notes and Descriptions. By A. S. C. Demy 8vo, 2s. 6d.

IRON (RALPH), (OLIVE SCHREINER)—

THE STORY OF AN AFRICAN FARM. New Edition. Crown 8vo, 1s.; in cloth, 1s. 6d.

ITALY, FROM THE ALPS TO MOUNT ETNA. Translated by FRANCES ELEANOR TROLLOPE, and Edited by THOMAS ADOLPHUS TROLLOPE. Illustrated with upwards of 100 full-page and 300 smaller Engravings. Imperial 4to.

JACKSON (FRANK G.), Master in the Birmingham Municipal School of Art—

DECORATIVE DESIGN. An Elementary Text Book of Principles and Practice. With numerous Illustrations. Crown 8vo, 7s. 6d.

JAMES (HENRY A.), M.A.—

HANDBOOK TO PERSPECTIVE. Crown 8vo, 2s. 6d.

PERSPECTIVE CHARTS, for use in Class Teaching. 2s.

JARRY (GENERAL)—

OUTPOST DUTY. Translated, with TREATISES ON MILITARY RECONNAISSANCE AND ON ROAD-MAKING. By Major-Gen. W. C. E. NAPIER. Third Edition. Crown 8vo, 5s.

JEANS (W. T.)—

CREATORS OF THE AGE OF STEEL. Memoirs of Sir W. Siemens, Sir H. Bessemer, Sir J. Whitworth, Sir J. Brown, and other Inventors. Second Edition. Crown 8vo, 7s. 6d.

JONES (CAPTAIN DOUGLAS), R.A.—

NOTES ON MILITARY LAW. Crown 8vo, 4s.

JONES. HANDBOOK OF THE JONES COLLECTION IN THE SOUTH KENSINGTON MUSEUM. With Portrait and Woodcuts. Large crown 8vo, 2s. 6d.

JUNKER (DR. WM.)—

TRAVELS IN AFRICA. With 38 Full-page Plates and 125 Illustrations in the Text and Map. Translated from the German by Professor KEANE. Demy 8vo, 21s.

KENNARD (EDWARD)—

NORWEGIAN SKETCHES: FISHING IN STRANGE WATERS. Illustrated with 30 beautiful Sketches. Second Edition. Oblong folio, 21s. A Set of Ten Hand-coloured Plates, £3; in Oak Frames, £4 10s.

Smaller Edition. 14s.

KENT (CHARLES)—

HUMOUR AND PATHOS OF CHARLES DICKENS. Crown 8vo, 6s.

KING (LIEUT.-COL. COOPER)—

GEORGE WASHINGTON. Large crown 8vo. [*In the Press.*

KLACZKO (M. JULIAN)—

TWO CHANCELLORS: PRINCE GORTCHAKOF AND PRINCE BISMARCK. Translated by MRS. TAIT. New and cheaper Edition, 6s.

KNOLLYS (MAJOR HENRY), R.A.—

SKETCHES OF LIFE IN JAPAN. With Illustrations. Large crown 8vo, 12s.

LACORDAIRE'S JESUS CHRIST; GOD; AND GOD AND MAN. Conferences delivered at Notre Dame in Paris. New Edition. Crown 8vo, 6s.

LAINÉ (J. M.), R.A.—

ENGLISH COMPOSITION EXERCISES. Crown 8vo, 2s. 6d.

LAING (S.)—

PROBLEMS OF THE FUTURE AND ESSAYS. Fifth Thousand. Demy 8vo, 3s. 6d.

MODERN SCIENCE AND MODERN THOUGHT. With a Supplementary Chapter on Gladstone's "Dawn of Creation" and Drummond's "Natural Law in the Spiritual World." Ninth Thousand. Demy 8vo, 3s. 6d.

A MODERN ZOROASTRIAN. Fourth Thousand. Demy 8vo, 3s. 6d.

LAVELEYE (ÉMILE DE)—

THE ELEMENTS OF POLITICAL ECONOMY. Translated by W. POLLARD, B.A., St. John's College, Oxford. Crown 8vo, 6s.

LANDOR (W. S.)—

LIFE AND WORKS. 8 vols.

VOL. 1. WALTER SAVAGE LANDOR. A Biography in Eight Books. By JOHN FORSTER. Demy 8vo, 12s.

VOL. 2. Out of print.

VOL. 3. CONVERSATIONS OF SOVEREIGNS AND STATESMEN, AND FIVE DIALOGUES OF BOCCACCIO AND PETRARCA. Demy 8vo, 14s.

VOL. 4. DIALOGUES OF LITERARY MEN. Demy 8vo, 14s.

VOL. 5. DIALOGUES OF LITERARY MEN (*continued*). FAMOUS WOMEN. LETTERS OF PERICLES AND ASPASIA. And Minor Prose Pieces. Demy 8vo, 14s.

VOL. 6. MISCELLANEOUS CONVERSATIONS. Demy 8vo, 14s.

VOL. 7. GEBIR, ACTS AND SCENES AND HELLENICS. Poems. Demy 8vo, 14s.

VOL. 8. MISCELLANEOUS POEMS AND CRITICISMS ON THEOCRITUS, CATULLUS, AND PETRARCH. Demy 8vo 14s.

LE CONTE (JOSEPH), Professor of Geology and Natural History in the University of California—

EVOLUTION AND ITS RELATIONS TO RELIGIOUS THOUGHT. Crown 8vo, 6s.

LEFÈVRE (ANDRÉ)—

PHILOSOPHY, Historical and Critical. Translated, with an Introduction, by A. W. KEANE, B.A. Large crown 8vo, 7s. 6d.

LE ROUX (H.)—

ACROBATS AND MOUNTEBANKS. With over 200 Illustrations by J. GARNIER. Royal 8vo, 16s.

LESLIE (R. C.)—

OLD SEA WINGS, WAYS, AND WORDS, IN THE DAYS OF OAK AND HEMP. With 135 Illustrations by the Author. Demy 8vo, 14s.

LIFE ABOARD A BRITISH PRIVATEER IN THE TIME OF QUEEN ANNE. Being the Journals of Captain Woodes Rogers, Master Mariner. With Notes and Illustrations by ROBERT C. LESLIE. Large crown 8vo, 9s.

A SEA PAINTER'S LOG. With 12 Full-page Illustrations by the Author. Large crown 8vo, 12s.

LETOURNEAU (DR. CHARLES)—

SOCIOLOGY. Based upon Ethnology. Large crown 8vo. 10s.

BIOLOGY. Translated by WILLIAM MACCALL. With 83 Illustrations. A New Edition. Demy 8vo, 3s. 6d.

LILLY (W. S.)—

ON RIGHT AND WRONG. Demy 8vo, 12s.

A CENTURY OF REVOLUTION. Second Edition. Demy 8vo, 12s.

CHAPTERS ON EUROPEAN HISTORY. With an Introductory Dialogue on the Philosophy of History. 2 vols. Demy 8vo, 21s.

ANCIENT RELIGION AND MODERN THOUGHT. Second Edition. Demy 8vo, 12s.

LITTLE (THE REV. CANON KNOX)—

THE CHILD OF STAFFERTON: A Chapter from a Family Chronicle. New Edition. Crown 8vo, boards, 1s.; cloth, 1s. 6d.

THE BROKEN VOW. A Story of Here and Hereafter. New Edition. Crown 8vo, boards, 1s.; cloth, 1s. 6d.

LITTLE (THE REV. H. W.)—

H. M. STANLEY: HIS LIFE, WORKS, AND EXPLORATIONS. Demy 8vo, 10s. 6d.

LLOYD (COLONEL E.M.), R.E., late Professor of Fortification at the Royal Military Academy, Woolwich—

VAUBAN, MONTALEMBERT, CARNOT: ENGINEER STUDIES. With Portraits. Crown 8vo, 5s

LLOYD (N. N.), late 24th Regiment—

ON ACTIVE SERVICE. Printed in Colours. Oblong 4to, 5s.

LIFE IN INDIA. Printed in Colours. 4to, 6s.

LONG (JAMES)—

DAIRY FARMING. To which is added a Description of the Chief Continental Systems. With numerous Illustrations. Crown 8vo, 9s.

LOW (WILLIAM)—

TABLE DECORATION. With 19 Full Illustrations. Demy 8vo, 6s.

LYTTON (ROBERT, EARL)—

POETICAL WORKS—

FABLES IN SONG. 2 vols. Fcap. 8vo, 12s.
THE WANDERER. Fcap. 8vo, 6s.
POEMS, HISTORICAL AND CHARACTERISTIC. Fcap. 6s.

McCOAN (J. C.)—

EGYPT UNDER ISMAIL: a Romance of History. With Portrait and Appendix of Official Documents. Crown 8vo, 7s. 6d.

MACDONALD (FREDERIKA)—

PUCK AND PEARL: THE WANDERINGS AND WONDERINGS OF TWO ENGLISH CHILDREN IN INDIA. By FREDERIKA MACDONALD. With Illustrations by MRS. IRVING GRAHAM. Second Edition. Crown 8vo, 5s.

MALLESON (COL. G. B.), C.S.I.—

PRINCE EUGENE OF SAVOY. With Portrait and Maps. Large crown 8vo, 6s.

LOUDON. A Sketch of the Military Life of Gideon Ernest, Freicherr von Loudon, sometime Generalissimo of the Austrian Forces. With Portrait and Maps. Large crown 8vo, 4s.

MALLET (ROBERT)—

PRACTICAL MANUAL OF CHEMICAL ASSAYING, as applied to the Manufacture of Iron. By L. L. DE KONINCK and E. DIETZ. Edited, with notes, by ROBERT MALLET. Post 8vo, cloth, 6s.

MARCEAU (SERGENT)—

REMINISCENCES OF A REGICIDE. Edited from the Original MSS. of SERGENT MARCEAU, Member of the Convention, ana Administrator of Police in the French Revolution of 1789. By M. C. M. SIMPSON, Author of the "Letters and Recollections of Julius and Mary Mohl." Demy 8vo, with Illustrations and Portraits, 14s.

MASKELL (ALFRED)—

RUSSIAN ART AND ART OBJECTS IN RUSSIA. A Handbook to the Reproduction of Goldsmiths' Work and other Art·Treasures. With Illustrations. Large crown 8vo, 4s. 6d.

MASKELL (WILLIAM)—

IVORIES: ANCIENT AND MEDIÆVAL. With numerous Woodcuts. Large crown 8vo, cloth, 2s. 6d.

HANDBOOK TO THE DYCE AND FORSTER COLLECTIONS. With Illustrations. Large crown 8vo, cloth, 2s. 6d.

MAUDSLAY (ATHOL)—

HIGHWAYS AND HORSES. With numerous Illustrations. Demy 8vo, 21s.

MECHELIN (SENATOR L.)—

FINLAND AND ITS PUBLIC LAW. Translated by CHARLES J. COOKE, British Vice-Consul at Helsingfors. Crown 8vo, 2s. 6d.

GEORGE MEREDITH'S WORKS.

A New and Uniform Edition. Crown 8vo, 3s. 6d. each.

DIANA OF THE CROSSWAYS.
EVAN HARRINGTON.
THE ORDEAL OF RICHARD FEVEREL.
THE ADVENTURES OF HARRY RICHMOND.
SANDRA BELLONI.
VITTORIA.
RHODA FLEMING.
BEAUCHAMP'S CAREER.
THE EGOIST.
THE SHAVING OF SHAGPAT; AND FARINA.

MERIVALE (HERMAN CHARLES)—

BINKO'S BLUES. A Tale for Children of all Growths. Illustrated by EDGAR GIBERNE. Small crown 8vo, 5s.

THE WHITE PILGRIM, and other Poems. Crown 8vo, 9s.

MILLS (JOHN), formerly Assistant to the Solar Physics Committee, and author of "Alternative Elementary Chemistry"—

ADVANCED PHYSIOGRAPHY (PHYSIOGRAPHIC ASTRONOMY). Designed to meet the Requirements of Students preparing for the Elementary and Advanced Stages of Physiography in the Science and Art Department Examinations, and as an Introduction to Physical Astronomy. Crown 8vo, 4s. 6d.

ELEMENTARY PHYSIOGRAPHIC ASTRONOMY. Crown 8vo, 1s. 6d.

ALTERNATIVE ELEMENTARY PHYSICS. Crown 8vo, 2s. 6d.

MILLS (JOHN) and NORTH (BARKER)—

QUANTITATIVE ANALYSIS (INTRODUCTORY LESSONS ON). With numerous Woodcuts. Crown 8vo, 1s. 6d.

HANDBOOK OF QUANTITATIVE ANALYSIS. Crown 8vo, 3s. 6d.

MOLESWORTH (W. NASSAU)—

HISTORY OF ENGLAND FROM THE YEAR 1830 TO THE RESIGNATION OF THE GLADSTONE MINISTRY, 1874. Twelfth Thousand. 3 vols. Crown 8vo, 18s.

ABRIDGED EDITION. Large crown, 7s. 6d.

MOLTKE (FIELD-MARSHAL COUNT VON)—

POLAND: AN HISTORICAL SKETCH. An Authorised Translation, with Biographical Notice by E. S. BUCHHEIM. Crown 8vo, 4s. 6d.

MORLEY (THE RIGHT HON. JOHN), M.P.—

RICHARD COBDEN'S LIFE AND CORRESPONDENCE. Crown 8vo, with Portrait, 7s. 6d.

Popular Edition. With Portrait. 4to, sewed, 1s. Cloth, 2s.

MUNTZ (EUGENE)—

RAPHAEL: his Life, Works, and Times. Illustrated with about 200 Engravings. A new Edition, revised from the Second French Edition by W. ARMSTRONG, B.A. Oxon. Imperial 8vo, 25s.

MURRAY (ANDREW), F.L.S.—

ECONOMIC ENTOMOLOGY. APTERA. With numerous Illustrations. Large crown 8vo, 7s. 6d.

NECKER (MADAME)—

THE SALON OF MADAME NECKER. By VICOMTE D'HAUSSONVILLE. 2 vols. Crown 8vo, 18s.

NESBITT (ALEXANDER)—

GLASS. With numerous Woodcuts. Large crown 8vo, cloth, 2s. 6d.

NEVINSON (HENRY)—

A SKETCH OF HERDER AND HIS TIMES. With a Portrait. Demy 8vo, 14s.

NICOL (DAVID)—

THE POLITICAL LIFE OF OUR TIME. Two vols. Demy 8vo, 24s.

NILSEN (CAPTAIN)—

LEAVES FROM THE LOG OF THE "HOMEWARD BOUND"; or, Eleven Months at Sea in an Open Boat. Crown 8vo, 1s.

NORMAN (C. B.)—

TONKIN; OR, FRANCE IN THE FAR EAST. With Maps. Demy 8vo, 14s.

O'BYRNE (ROBERT), F.R.G.S.—

THE VICTORIES OF THE BRITISH ARMY IN THE PENINSULA AND THE SOUTH OF FRANCE from 1808 to 1814. An Epitome of Napier's History of the Peninsular War, and Gurwood's Collection of the Duke of Wellington's Despatches. Crown 8vo, 5s.

O'GRADY (STANDISH)—

TORYISM AND THE TORY DEMOCRACY. Crown 8vo, 5s.

OLIVER (PROFESSOR D.), F.R.S., &c.—

ILLUSTRATIONS OF THE PRINCIPAL NATURAL ORDERS OF THE VEGETABLE KINGDOM, PREPARED FOR THE SCIENCE AND ART DEPARTMENT, SOUTH KENSINGTON. With 109 Plates. Oblong 8vo, plain, 16s.; coloured, £1 6s.

OLIVER (E. E.), Under-Secretary to the Public Works Department, Punjaub—

ACROSS THE BORDER; or, PATHAN AND BILOCH. With numerous Illustrations by J. L. KIPLING, C.I.E. Demy 8vo, 14s.

OXENHAM (REV. H. N.)—

MEMOIR OF LIEUTENANT RUDOLPH DE LISLE, R.N., OF THE NAVAL BRIGADE. Third Edition, with Illustrations. Crown 8vo, 7s. 6d.

PAYTON (E. W.)—

ROUND ABOUT NEW ZEALAND. Being Notes from a Journal of Three Years' Wandering in the Antipodes. With Twenty Original Illustrations by the Author. Large crown 8vo, 12s.

PERROT (GEORGES) and CHIPIEZ (CHARLES)—

A HISTORY OF ANCIENT ART IN SARDINIA, JUDÆA, SYRIA, AND ASIA MINOR. With 395 Illustrations. 2 vols. Imperial 8vo, 36s.

A HISTORY OF ANCIENT ART IN PHŒNICIA AND ITS DEPENDENCIES. Translated from the French by WALTER ARMSTRONG, B.A. Oxon. Containing 644 Illustrations in the text, and 10 Steel and Coloured Plates. 2 vols. Imperial 8vo, 42s.

PERROT (GEORGES) and CHIPIEZ (CHARLES)—Continued—

A HISTORY OF ART IN CHALDÆA AND ASSYRIA. Translated by WALTER ARMSTRONG, B.A. Oxon. With 452 Illustrations. 2 vols. Imperial 8vo, 42s.

A HISTORY OF ART IN ANCIENT EGYPT. Translated from the French by W. ARMSTRONG, B.A. Oxon. With over 600 Illustrations. 2 vols. Imperial 8vo, 42s.

PETERBOROUGH (THE EARL OF)—

THE EARL OF PETERBOROUGH AND MONMOUTH (Charles Mordaunt): A Memoir. By Colonel FRANK RUSSELL, Royal Dragoons. With Illustrations. 2 vols. demy 8vo. 32s.

PHŒNICIAN ART—

A HISTORY OF ANCIENT ART IN PHŒNICIA AND ITS DEPENDENCIES. By GEORGES PERROT and CHARLES CHIPIEZ. Translated from the French by WALTER ARMSTRONG, B.A. Oxon. Containing 644 Illustrations in the text, and 10 Steel and Coloured Plates. 2 vols. Imperial 8vo, 42s.

PILLING (WILLIAM)—

LAND TENURE BY REGISTRATION. Second Edition of "Order from Chaos," Revised and Enlarged. Crown 8vo, 5s.

PITT TAYLOR (FRANK)—

THE CANTERBURY TALES. Selections from the Tales of GEOFFREY CHAUCER rendered into Modern English, with close adherence to the language of the Poet. With Frontispiece. Crown 8vo, 6s.

POLLEN (J. H.)—

GOLD AND SILVER SMITH'S WORK. With numerous Woodcuts. Large crown 8vo, cloth, 2s. 6d.

ANCIENT AND MODERN FURNITURE AND WOODWORK. With numerous Woodcuts. Large crown 8vo, cloth, 2s. 6d.

POOLE (STANLEY LANE), B.A., M.R.A.S.—

THE ART OF THE SARACENS IN EGYPT. Published for the Committee of Council on Education. With 108 Woodcuts. Large crown 8vo, 4s.

POYNTER (E. J.), R.A.—

TEN LECTURES ON ART. Third Edition. Large crown 8vo, 9s.

PRINSEP (VAL), A.R.A.—

IMPERIAL INDIA. Containing numerous Illustrations and Maps. Second Edition. Demy 8vo, £1 1s.

PURCELL (the late THEOBALD A.), Surgeon-Major, A.M.D., and Principal Medical Officer to the Japanese Government)—

A SUBURB OF YEDO. With numerous Illustrations. Crown 8vo, 2s. 6d.

RADICAL PROGRAMME, THE. From the *Fortnightly Review*, with additions. With a Preface by the RIGHT HON. J. CHAMBERLAIN, M.P. Thirteenth Thousand. Crown 8vo, 2s. 6d.

RAE (W. FRASER)—

AUSTRIAN HEALTH RESORTS THROUGHOUT THE YEAR. A New and Enlarged Edition. Crown 8vo, 5s.

RAMSDEN (LADY GWENDOLEN)—

A BIRTHDAY BOOK. Illustrated. Containing 46 Illustrations from Original Drawings, and numerous other Illustrations. Royal 8vo, 21s.

RANKIN (THOMAS T.), C.E.—

SOLUTIONS TO THE QUESTIONS IN PURE MATHEMATICS (STAGES 1 AND 2) SET AT THE SCIENCE AND ART EXAMINATIONS FROM 1881 TO 1886. Crown 8vo, 2s.

RAPHAEL: his Life, Works, and Times. By EUGENE MUNTZ. Illustrated with about 200 Engravings. A New Edition, revised from the Second French Edition. By W. ARMSTRONG, B.A. Imperial 8vo, 25s.

REDGRAVE (GILBERT)—

OUTLINES OF HISTORIC ORNAMENT. Translated from the German. Edited by GILBERT REDGRAVE. With numerous Illustrations. Crown 8vo, 4s.

REDGRAVE (RICHARD)—

MANUAL OF DESIGN, compiled from the Writings and Addresses of RICHARD REDGRAVE, R.A. With Woodcuts. Large crown 8vo, cloth, 2s. 6d.

ELEMENTARY MANUAL OF COLOUR, with a Catechism on Colour. 24mo, cloth, 9d.

REDGRAVE (SAMUEL)—

A DESCRIPTIVE CATALOGUE OF THE HISTORICAL COLLECTION OF WATER-COLOUR PAINTINGS IN THE SOUTH KENSINGTON MUSEUM. With numerous Chromo-lithographs and other Illustrations. Royal 8vo, £1 1s.

REID (T. WEMYSS)—

THE LIFE OF THE RIGHT HON. W. E. FORSTER. With Portraits. Fourth Edition. 2 vols. Demy 8vo, 32s.
FIFTH EDITION, in one volume, with new Portrait. Demy 8vo, 10s. 6d.

RENAN (ERNEST)—

THE FUTURE OF SCIENCE. Demy 8vo.

HISTORY OF THE PEOPLE OF ISRAEL TILL THE TIME OF KING DAVID. Demy 8vo, 14s.

HISTORY OF THE PEOPLE OF ISRAEL. From the Reign of David up to the Capture of Samaria. Second Division. Demy 8vo, 14s.

RECOLLECTIONS OF MY YOUTH. Translated from the original French, and revised by MADAME RENAN. Crown 8vo, 8s.

REYNARDSON (C. T. S. BIRCH)—

SPORTS AND ANECDOTES OF BYGONE DAYS in England, Scotland, Ireland, Italy, and the Sunny South. With numerous Illustrations in Colour. Second Edition. Large crown 8vo, 12s.

DOWN THE ROAD: Reminiscences of a Gentleman Coachman. With Coloured Illustrations. Large crown 8vo, 12s.

RIANO (JUAN F.)—

THE INDUSTRIAL ARTS IN SPAIN. With numerous Woodcuts. Large crown 8vo, cloth, 4s.

RIBTON-TURNER (C. J.)—

A HISTORY OF VAGRANTS AND VAGRANCY AND BEGGARS AND BEGGING. With Illustrations. Demy 8vo, 21s.

ROBINSON (JAMES F.)—

BRITISH BEE FARMING. Its Profits and Pleasures. Large crown 8vo, 5s.

ROBINSON (J. C.)—

ITALIAN SCULPTURE OF THE MIDDLE AGES AND PERIOD OF THE REVIVAL OF ART. With 20 Engravings. Royal 8vo, cloth, 7s. 6d.

ROBSON (GEORGE)—

ELEMENTARY BUILDING CONSTRUCTION. Illustrated by a Design for an Entrance Lodge and Gate. 15 Plates. Oblong folio, sewed, 8s.

ROCK (THE VERY REV. CANON), D.D.—

TEXTILE FABRICS. With numerous Woodcuts. Large crown 8vo, cloth, 2s. 6d.

ROGERS (CAPTAIN WOODES), Master Mariner—

LIFE ABOARD A BRITISH PRIVATEER IN THE TIME OF QUEEN ANNE. Being the Journals of Captain Woodes Rogers, Master Mariner. With Notes and Illustrations by ROBERT C. LESLIE, Author of "A Sea Painter's Log." Large crown 8vo, 9s.

ROOSE (ROBSON), M.D., F.C.S.—

THE WEAR AND TEAR OF LONDON LIFE. Second Edition. Crown 8vo, sewed, 1s.

INFECTION AND DISINFECTION. Crown 8vo, sewed, 6d.

ROLAND (ARTHUR)—

FARMING FOR PLEASURE AND PROFIT. Edited by WILLIAM ABLETT. 8 vols. Crown 8vo, 5s. each.

DAIRY-FARMING, MANAGEMENT OF COWS, etc.

POULTRY-KEEPING.

TREE-PLANTING, FOR ORNAMENTATION OR PROFIT.

STOCK-KEEPING AND CATTLE-REARING.

DRAINAGE OF LAND, IRRIGATION, MANURES, etc.

ROOT-GROWING, HOPS, etc.

MANAGEMENT OF GRASS LANDS, LAYING DOWN GRASS, ARTIFICIAL GRASSES, etc.

MARKET GARDENING, HUSBANDRY FOR FARMERS AND GENERAL CULTIVATORS.

SARDINIAN ART: HISTORY OF ANCIENT ART IN SARDINIA, JUDÆA, SYRIA, AND ASIA MINOR. By GEORGES PERROT and CHARLES CHIPIEZ. With 395 Illustrations. 2 vols. Imperial 8vo, 36s.

SCIENCE AND ART: a Journal for Teachers and Scholars. Issued monthly. 3d. See page 38.

SCHREINER (OLIVE), (RALPH IRON)—

THE STORY OF AN AFRICAN FARM. Crown 8vo, 1s.; in cloth, 1s. 6d.

SCHAUERMANN (F. L.)—

HANDBOOK ON WOOD-CARVING. With numerous Illustrations. Imperial 8vo.

SCOTT (JOHN)—

THE REPUBLIC AS A FORM OF GOVERNMENT; or, The Evolution of Democracy in America. Crown 8vo, 7s. 6d.

SCOTT (LEADER)—

THE RENAISSANCE OF ART IN ITALY: an Illustrated Sketch. With upwards of 200 Illustrations. Medium quarto, 18s.

SCOTT-STEVENSON (MRS.)—

ON SUMMER SEAS. Including the Mediterranean, the Ægean, the Ionian, and the Euxine, and a voyage down the Danube. With a Map. Demy 8vo, 16s.

OUR HOME IN CYPRUS. With a Map and Illustrations. Third Edition. Demy 8vo, 14s.

OUR RIDE THROUGH ASIA MINOR. With Map. Demy 8vo, 18s.

SEEMAN (O.)—

THE MYTHOLOGY OF GREECE AND ROME, with Special Reference to its Use in Art. From the German. Edited by G. H. BIANCHI. 64 Illustrations. New Edition. Crown 8vo, 5s.

SETON-KARR (H. W.), F.R.G.S., etc.—

TEN YEARS' TRAVEL AND SPORT IN FOREIGN Lands; or, Travels in the Eighties. Second Edition, with additions and Portrait of Author. Large crown 8vo, 5s.

SHEPHERD (MAJOR), R.E.—

PRAIRIE EXPERIENCES IN HANDLING CATTLE AND SHEEP. With Illustrations and Map. Demy 8vo, 10s. 6d.

SHIRREFF (EMILY)—

A SHORT SKETCH OF THE LIFE OF FRIEDRICH FROBEL; a New Edition, including Fröbel's Letters from Dresden and Leipzig to his Wife, now first Translated into English. Crown 8vo, 2s.

HOME EDUCATION IN RELATION TO THE KINDERGARTEN. Two Lectures. Crown 8vo, 1s. 6d.

SHORE (ARABELLA)—

DANTE FOR BEGINNERS: a Sketch of the "Divina Commedia." With Translations, Biographical and Critical Notices, and Illustrations. With Portrait. Crown 8vo, 6s.

SIMKIN (R.)—

LIFE IN THE ARMY: Every-day Incidents in Camp, Field, and Quarters. Printed in Colours. Oblong 4to, 5s.

SIMMONDS (T. L.)—

ANIMAL PRODUCTS: their Preparation, Commercial Uses, and Value. With numerous Illustrations. Large crown 8vo, 7s. 6d.

SIMPSON (M. C. M.)—

REMINISCENCES OF A REGICIDE. Edited from the Original MSS. of Sergent Marceau, Member of the Convention, and Administrator of Police in the French Revolution of 1789. Demy 8vo, with Illustrations and Portraits, 14s.

SINGER'S STORY, A. Related by the Author of "Flitters, Tatters, and the Counsellor." Crown 8vo, sewed, 1s.

SINNETT (A. P.)—

ESOTERIC BUDDHISM. Annotated and enlarged by the Author. Sixth and cheaper Edition. Crown 8vo, 4s.

KARMA. A Novel. New Edition. Crown 8vo, 3s. 6d.

SINNETT (MRS.)—

THE PURPOSE OF THEOSOPHY. Crown 8vo, 3s.

SMITH (MAJOR R. MURDOCK), R.E.—

PERSIAN ART. With Map and Woodcuts. Second Edition. Large crown 8vo, 2s.

SMITH (S. THEOBALD)—

A RAMBLE IN RHYME IN THE COUNTRY OF CRANMER AND RIDLEY. Illustrated by HAROLD OAKLEY from Sketches by the Author. Crown 8vo, 2s. 6d.

STANLEY (H. M.): HIS LIFE, WORKS, AND EXPLORATIONS. By the Rev. H. W. LITTLE. Demy 8vo, 10s. 6d.

STOKES (MARGARET)—

EARLY CHRISTIAN ART IN IRELAND. With 106 Woodcuts. Demy 8vo, 7s. 6d.

STORY (W. W.)—

ROBA DI ROMA. Seventh Edition, with Additions and Portrait. Crown 8vo, cloth, 10s. 6d.

CASTLE ST. ANGELO. With Illustrations. Crown 8vo, 10s. 6d.

SUTCLIFFE (JOHN)—

THE SCULPTOR AND ART STUDENT'S GUIDE to the Proportions of the Human Form, with Measurements in feet and inches of Full-Grown Figures of Both Sexes and of Various Ages. By Dr. G. SCHADOW, Member of the Academies, Stockholm, Dresden, Rome, &c. &c. Translated by J. J. WRIGHT. Plates reproduced by J. SUTCLIFFE. Oblong folio, 31s. 6d.

SYMONDS (JOHN ADDINGTON)—

ESSAYS, SPECULATIVE AND SUGGESTIVE. 2 vols. Crown 8vo, 18s.

TAINE (H. A.)—

NOTES ON ENGLAND. Translated, with Introduction, by W. FRASER RAE. Eighth Edition. With Portrait. Crown 8vo, 5s.

TAIT (J. S.)—

WHO IS THE MAN? A Tale of the Scottish Border. Crown 8vo, 1s.; in cloth, 1s. 6d.

TANNER (PROFESSOR), F.C.S.—

HOLT CASTLE; or, Threefold Interest in Land. Crown 8vo, 4s. 6d.

JACK'S EDUCATION; OR, HOW HE LEARNT FARMING. Second Edition. Crown 8vo, 3s. 6d.

TAYLOR (EDWARD R.), Head Master of the Birmingham Municipal School of Art—

ELEMENTARY ART TEACHING: An Educational and Technical Guide for Teachers and Learners, including Infant School-work; The Work of the Standards; Freehand; Geometry; Model Drawing; Nature Drawing; Colours; Light and Shade; Modelling and Design. With over 600 Diagrams and Illustrations. Imperial 8vo, 10s. 6d.

TEMPLE (SIR RICHARD), BART., M.P., G.C.S.I.—

COSMOPOLITAN ESSAYS. With Maps. Demy 8vo, 16s.

THRUPP (GEORGE A.) and FARR (WILLIAM)—

COACH TRIMMING. With 60 Illustrations. Crown 8vo, 2s. 6d.

THRUPP (THE REV. H. W.), M.A.—

AN AID TO THE VISITATION OF THOSE DISTRESSED IN MIND, BODY, OR ESTATE. Crown 8vo, 3s. 6d.

TOPINARD (DR. PAUL)—

ANTHROPOLOGY. With a Preface by Professor PAUL BROCA. With 49 Illustrations. Demy 8vo, 3s. 6d.

TOVEY (LIEUT.-COL., R.E.)—

MARTIAL LAW AND CUSTOM OF WAR; or, Military Law and Jurisdiction in Troublous Times. Crown 8vo, 6s.

TRAHERNE (MAJOR)—

THE HABITS OF THE SALMON. Crown 8vo, 3s. 6d.

TRAILL (H. D.)—

THE NEW LUCIAN. Being a Series of Dialogues of the Dead. Demy 8vo, 12s.

TROLLOPE (ANTHONY)—

THE CHRONICLES OF BARSETSHIRE. A Uniform Edition, in 8 vols., large crown 8vo, handsomely printed, each vol. containing Frontispiece. 6s. each.

THE WARDEN and BARCHESTER TOWERS. 2 vols.
DR. THORNE.
FRAMLEY PARSONAGE.
THE SMALL HOUSE AT ALLINGTON. 2 vols.
LAST CHRONICLE OF BARSET. 2 vols.

LIFE OF CICERO. 2 vols. 8vo. £1 4s.

TROUP (J. ROSE)—

WITH STANLEY'S REAR COLUMN. With Portraits and Illustrations. Demy 8vo.

VANDAM (ALBERT D.)—

WE TWO AT MONTE CARLO. Second Edition. Crown 8vo, 1s.; in cloth, 1s. 6d.

VERON (EUGENE)—

ÆSTHETICS. Translated by W. H. ARMSTRONG. Large crown 8vo, 7s. 6d.

VERSCHOYLE (REV. J.), M.A.—

THE HISTORY OF ANCIENT CIVILISATION. A Handbook based upon M. Gustave Ducoudray's "Histoire Sommaire de la Civilisation." Edited by REV. J. VERSCHOYLE, M.A. With Illustrations. Large crown 8vo, 6s.

WALFORD (MAJOR), R.A.—

PARLIAMENTARY GENERALS OF THE GREAT CIVIL WAR. With Maps. Large crown 8vo, 4s.

WALKER (MRS.)—

UNTRODDEN PATHS IN ROUMANIA. With 77 Illustrations. Demy 8vo, 10s. 6d.

WALKER (MRS.)—Continued—

EASTERN LIFE AND SCENERY, with Excursions to Asia Minor, Mitylene, Crete, and Roumania. 2 vols., with Frontispiece to each vol. Crown 8vo, 21s.

WARD (JAMES)—

ELEMENTARY PRINCIPLES OF ORNAMENT. With 122 Illustrations in the text. 8vo, 5s.

WATSON (JOHN)—

BRITISH SPORTING FISHES. With Frontispiece. Crown 8vo, 3s. 6d.

WATSON (WILLIAM)—

LIFE IN THE CONFEDERATE ARMY: being the Observations and Experiences of an Alien in the South during the American Civil War. Crown 8vo, 6s.

WEGG-PROSSER (F. R.)—

GALILEO AND HIS JUDGES. Demy 8vo, 5s.

WELLS (HENRY P.)—

CITY BOYS IN THE WOODS; or, A Trapping Venture in Maine. With upwards of 100 Illustrations. Royal 8vo.

WHITE (WALTER)—

A MONTH IN YORKSHIRE. With a Map. Fifth Edition. Post 8vo, 4s.

A LONDONER'S WALK TO THE LAND'S END, AND A TRIP TO THE SCILLY ISLES. With 4 Maps. Third Edition. Post 8vo, 4s.

WILL-O'-THE-WISPS, THE. Translated from the German of Marie Petersen by CHARLOTTE J. HART. With Illustrations. Crown 8vo, 7s. 6d.

WORKING MAN'S PHILOSOPHY, A. By "ONE OF THE CROWD." Crown 8vo, 3s.

WORNUM (R. N.)—

ANALYSIS OF ORNAMENT: THE CHARACTERISTICS OF STYLES. An Introduction to the History of Ornamental Art. With many Illustrations. Ninth Edition. Royal 8vo, cloth, 8s.

WRIGHTSON (PROF. JOHN), M.R.A.C., F.C.S., &c.; President of the College of Agriculture, Downton, near Salisbury, &c., &c.

PRINCIPLES OF AGRICULTURAL PRACTICE AS AN INSTRUCTIONAL SUBJECT. With Geological Map. Second Edition. Crown 8vo, 5s.

FALLOW AND FODDER CROPS. Crown 8vo, 5s.

WORSAAE (J. J. A.)—

INDUSTRIAL ARTS OF DENMARK, FROM THE EARLIEST TIMES TO THE DANISH CONQUEST OF ENGLAND. With Maps and Woodcuts. Large crown 8vo, 3s 6d.

YOUNGE (C. D.)—

PARALLEL LIVES OF ANCIENT AND MODERN HEROES. New Edition. 12mo, cloth, 4s. 6d.

YOUNG OFFICER'S "DON'T"; or, Hints to Youngsters on Joining. 32mo, 1s.

SOUTH KENSINGTON MUSEUM SCIENCE & ART HANDBOOKS—*Continued.*

PLAIN WORDS ABOUT WATER. By A. H. CHURCH, M.A. Oxon. With Illustrations. Sewed, 6d.

ANIMAL PRODUCTS: their Preparation, Commercial Uses, and Value. By T. L. SIMMONDS. With Illustrations. 7s. 6d.

FOOD: Some Account of its Sources, Constituents, and Uses. By PROFESSOR A. H. CHURCH, M.A. Oxon. New Edition, enlarged. 3s.

ECONOMIC ENTOMOLOGY. By ANDREW MURRAY, F.L.S. APTERA. With Illustrations. 7s. 6d.

JAPANESE POTTERY. Being a Native Report. With an Introduction and Catalogue by A. W. FRANKS, M.A., F.R.S., F.S.A. With Illustrations and Marks. 2s. 6d.

HANDBOOK TO THE SPECIAL LOAN COLLECTION of Scientific Apparatus. 3s.

INDUSTRIAL ARTS: Historical Sketches. With Numerous Illustrations. 3s.

TEXTILE FABRICS. By the Very Rev. DANIEL ROCK, D.D. With numerous Woodcuts. 2s. 6d.

JONES COLLECTION IN THE SOUTH KENSINGTON MUSEUM. With Portrait and Woodcuts. 2s. 6d.

COLLEGE AND CORPORATION PLATE. A Handbook to the Reproductions of Silver Plate in the South Kensington Museum from Celebrated English Collections. By WILFRED JOSEPH CRIPPS, M.A., F.S.A. With Illustrations. 2s. 6d.

IVORIES: ANCIENT AND MEDIÆVAL. By WILLIAM MASKELL. With numerous Woodcuts. 2s. 6d.

ANCIENT AND MODERN FURNITURE AND WOOD-WORK. By JOHN HUNGERFORD POLLEN, M.A. With numerous Woodcuts. 2s. 6d.

MAIOLICA. By C. DRURY E. FORTNUM, F.S.A. With numerous Woodcuts. 2s. 6d.

THE CHEMISTRY OF FOODS. With Microscopic Illustrations. By JAMES BELL, Ph.D., &c., Principal of the Somerset House Laboratory.
Part I.—Tea, Coffee, Cocoa, Sugar, &c. 2s. 6d.
Part II.—Milk, Butter, Cheese, Cereals, Prepared Starches, &c. 3s.

MUSICAL INSTRUMENTS. By CARL ENGEL. With numerous Woodcuts. 2s. 6d.

MANUAL OF DESIGN, compiled from the Writings and Addresses of RICHARD REDGRAVE, R.A. By GILBERT R. REDGRAVE. With Woodcuts. 2s. 6d.

PERSIAN ART. By MAJOR R. MURDOCK SMITH, R.E. With Map and Woodcuts. Second Edition, enlarged. 2s.

CARLYLE'S (THOMAS) WORKS.

THE ASHBURTON EDITION.

An entirely New Edition, handsomely printed, containing all the Portraits and Illustrations, in Seventeen Volumes, demy 8vo, 8s. each.

THE FRENCH REVOLUTION AND PAST AND PRESENT. 2 vols.
SARTOR RESARTUS; HEROES AND HERO WORSHIP. 1 vol.
LIFE OF JOHN STERLING—LIFE OF SCHILLER. 1 vol.
LATTER-DAY PAMPHLETS—EARLY KINGS OF NORWAY—ESSAY ON THE PORTRAIT OF JOHN KNOX. 1 vol.
LETTERS AND SPEECHES OF OLIVER CROMWELL. 3 vols.
HISTORY OF FREDERICK THE GREAT. 6 vols.
CRITICAL AND MISCELLANEOUS ESSAYS. 3 vols.

LIBRARY EDITION COMPLETE.

Handsomely printed in 34 vols., demy 8vo, cloth, £15 3s

SARTOR RESARTUS. With a Portrait, 7s. 6d.

THE FRENCH REVOLUTION. A History. 3 vols., each 9s.

LIFE OF FREDERICK SCHILLER AND EXAMINATION OF HIS WORKS. With Supplement of 1872. Portrait and Plates, 9s.

CRITICAL AND MISCELLANEOUS ESSAYS. With Portrait. 6 vols., each 9s.

ON HEROES, HERO WORSHIP, AND THE HEROIC IN HISTORY. 7s. 6d.

PAST AND PRESENT. 9s.

OLIVER CROMWELL'S LETTERS AND SPEECHES. With Portraits. 5 vols., each 9s.

LATTER-DAY PAMPHLETS. 9s.

LIFE OF JOHN STERLING. With Portrait, 9s.

HISTORY OF FREDERICK THE SECOND. 10 vols., each 9s.

TRANSLATIONS FROM THE GERMAN. 3 vols., each 9s.

EARLY KINGS OF NORWAY; ESSAY ON THE PORTRAITS OF JOHN KNOX; AND GENERAL INDEX. With Portrait Illustrations. 8vo, cloth, 9s.

CHEAP AND UNIFORM EDITION.

23 vols., Crown 8vo, cloth, £7 5s.

THE FRENCH REVOLUTION: A History. 2 vols., 12s.

OLIVER CROMWELL'S LETTERS AND SPEECHES, with Elucidations, &c. 3 vols., 18s.

LIVES OF SCHILLER AND JOHN STERLING. 1 vol., 6s.

CRITICAL AND MISCELLANEOUS ESSAYS. 4 vols., £1 4s.

SARTOR RESARTUS AND LECTURES ON HEROES. 1 vol., 6s.

LATTER-DAY PAMPHLETS. 1 vol., 6s.

CHARTISM AND PAST AND PRESENT. 1 vol., 6s.

TRANSLATIONS FROM THE GERMAN OF MUSÆUS, TIECK, AND RICHTER. 1 vol., 6s.

WILHELM MEISTER, by Göethe. A Translation. 2 vols., 12s.

HISTORY OF FRIEDRICH THE SECOND, called Frederick the Great. 7 vols., £2 9s.

PEOPLE'S EDITION.

37 vols., small crown 8vo, 37s.; separate vols., 1s. each.

SARTOR RESARTUS. With Portrait of Thomas Carlyle.

FRENCH REVOLUTION. A History. 3 vols.

OLIVER CROMWELL'S LETTERS AND SPEECHES. 5 vols. With Portrait of Oliver Cromwell.

ON HEROES AND HERO WORSHIP, AND THE HEROIC IN HISTORY.

PAST AND PRESENT.

CRITICAL AND MISCELLANEOUS ESSAYS. 7 vols.

THE LIFE OF SCHILLER, AND EXAMINATION OF HIS WORKS. With Portrait.

LATTER-DAY PAMPHLETS.

WILHELM MEISTER. 3 vols.

LIFE OF JOHN STERLING. With Portrait.

HISTORY OF FREDERICK THE GREAT. 10 vols.

TRANSLATIONS FROM MUSÆUS, TIECK, AND RICHTER. 2 vols.

THE EARLY KINGS OF NORWAY; Essay on the Portraits of Knox.

Or in sets, 37 vols. in 18, 37s.

CHEAP ISSUE.

THE FRENCH REVOLUTION. Complete in 1 vol. With Portrait. Crown 8vo, 2s.

SARTOR RESARTUS, HEROES AND HERO WORSHIP, PAST AND PRESENT, AND CHARTISM. Complete in 1 vol. Crown 8vo, 2s.

OLIVER CROMWELL'S LETTERS AND SPEECHES. Crown 8vo, 2s. 6d.

CRITICAL AND MISCELLANEOUS ESSAYS. 2 vols. 4s.

WILHELM MEISTER. 1 vol. 2s.

SIXPENNY EDITION.

4to, sewed.

SARTOR RESARTUS. Eightieth Thousand.

HEROES AND HERO WORSHIP.

ESSAYS: BURNS, JOHNSON, SCOTT, THE DIAMOND NECKLACE.

The above in 1 vol., cloth, 2s. 6d.

DICKENS'S (CHARLES) WORKS.

ORIGINAL EDITIONS.

In demy 8vo.

THE MYSTERY OF EDWIN DROOD. With Illustrations by S. L. Fildes, and a Portrait engraved by Baker. Cloth, 7s. 6d.

OUR MUTUAL FRIEND. With Forty Illustrations by Marcus Stone. Cloth, £1 1s.

THE PICKWICK PAPERS. With Forty-three Illustrations by Seymour and Phiz. Cloth, £1 1s.

NICHOLAS NICKLEBY. With Forty Illustrations by Phiz. Cloth, £1 1s.

SKETCHES BY "BOZ." With Forty Illustrations by George Cruikshank. Cloth, £1 1s.

MARTIN CHUZZLEWIT. With Forty Illustrations by Phiz. Cloth, £1 1s.

DOMBEY AND SON. With Forty Illustrations by Phiz. Cloth, £1 1s.

DAVID COPPERFIELD. With Forty Illustrations by Phiz. Cloth, £1 1s.

BLEAK HOUSE. With Forty Illustrations by Phiz. Cloth, £1 1s.

LITTLE DORRIT. With Forty Illustrations by Phiz. Cloth, £1 1s.

THE OLD CURIOSITY SHOP. With Seventy-five Illustrations by George Cattermole and H. K Browne. A New Edition. Uniform with the other volumes, £1 1s.

BARNABY RUDGE: a Tale of the Riots of 'Eighty. With Seventy-eight Illustrations by George Cattermole and H. K. Browne. Uniform with the other volumes, £1 1s.

CHRISTMAS BOOKS: Containing—The Christmas Carol; The Cricket on the Hearth; The Chimes; The Battle of Life; The Haunted House. With all the original Illustrations. Cloth, 12s.

OLIVER TWIST and TALE OF TWO CITIES. In one volume. Cloth, £1 1s.

OLIVER TWIST. Separately. With Twenty-four Illustrations by George Cruikshank. Cloth, 11s.

A TALE OF TWO CITIES. Separately. With Sixteen Illustrations by Phiz. Cloth, 9s.

*** *The remainder of Dickens's Works were not originally printed in demy 8vo.*

DICKENS'S (CHARLES) WORKS.—*Continued.*

LIBRARY EDITION.

In post 8vo. With the Original Illustrations, 30 vols., cloth, £12.

			s.	d.
PICKWICK PAPERS	43 Illustrns.,	2 vols.	16	0
NICHOLAS NICKLEBY	39 ,,	2 vols.	16	0
MARTIN CHUZZLEWIT	40 ,,	2 vols.	16	0
OLD CURIOSITY SHOP & REPRINTED PIECES	36 ,,	2 vols.	16	0
BARNABY RUDGE and HARD TIMES	36 ,,	2 vols.	16	0
BLEAK HOUSE	40 ,,	2 vols.	16	0
LITTLE DORRIT	40 ,,	2 vols.	16	0
DOMBEY AND SON	38 ,,	2 vols.	16	0
DAVID COPPERFIELD	38 ,,	2 vols.	16	0
OUR MUTUAL FRIEND	40 ,,	2 vols.	16	0
SKETCHES BY "BOZ"	39 ,,	1 vol.	8	0
OLIVER TWIST	24 ,,	1 vol.	8	0
CHRISTMAS BOOKS	17 ,,	1 vol.	8	0
A TALE OF TWO CITIES	16 ,,	1 vol.	8	0
GREAT EXPECTATIONS	8 ,,	1 vol.	8	0
PICTURES FROM ITALY & AMERICAN NOTES	8 ,,	1 vol.	8	0
UNCOMMERCIAL TRAVELLER	8 ,,	1 vol.	8	0
CHILD'S HISTORY OF ENGLAND	8 ,,	1 vol.	8	0
EDWIN DROOD and MISCELLANIES	12 ,,	1 vol.	8	0
CHRISTMAS STORIES from "Household Words," &c.	14 ,,	1 vol.	8	0

Uniform with the above, 10s. 6d.

THE LIFE OF CHARLES DICKENS. By JOHN FORSTER. With Illustrations.

A NEW EDITION OF ABOVE, WITH THE ORIGINAL ILLUSTRATIONS, IN LARGE CROWN 8vo, 30 VOLS. IN SETS ONLY.

THE "CHARLES DICKENS" EDITION.

In Crown 8vo. In 21 vols., cloth, with Illustrations, £3 16s.

		s.	d.
PICKWICK PAPERS	8 Illustrations ...	4	0
MARTIN CHUZZLEWIT	8 ,, ...	4	0
DOMBEY AND SON	8 ,, ...	4	0
NICHOLAS NICKLEBY	8 ,, ...	4	0
DAVID COPPERFIELD	8 ,, ...	4	0
BLEAK HOUSE	8 ,, ...	4	0
LITTLE DORRIT	8 ,, ...	4	0
OUR MUTUAL FRIEND...	8 ,, ...	4	0
BARNABY RUDGE	8 ,, ...	3	6
OLD CURIOSITY SHOP	8 ,, ...	3	6
A CHILD'S HISTORY OF ENGLAND	4 ,, ...	3	6
EDWIN DROOD and OTHER STORIES	8 ,, ...	3	6
CHRISTMAS STORIES, from "Household Words" ...	8 ,, ...	3	6
SKETCHES BY "BOZ"	8 ,, ...	3	6
AMERICAN NOTES and REPRINTED PIECES ...	8 ,, ...	3	6
CHRISTMAS BOOKS	8 ,, ...	3	6
OLIVER TWIST	8 ,, ...	3	6
GREAT EXPECTATIONS...	8 ,, ...	3	6
TALE OF TWO CITIES	8 ,, ...	3	0
HARD TIMES and PICTURES FROM ITALY ...	8 ,, ...	3	0
UNCOMMERCIAL TRAVELLER	4 ,, ...	3	0

Uniform with the above.

		s.	d.
THE LIFE OF CHARLES DICKENS. Numerous Illustrations.	2 vols.	7	0
THE LETTERS OF CHARLES DICKENS	2 vols.	7	0

DICKENS'S (CHARLES) WORKS.—*Continued.*

THE ILLUSTRATED LIBRARY EDITION.

(WITH LIFE.)

Complete in 32 Volumes. Demy 8vo, 10s. each; or set, £16.

This Edition is printed on a finer paper and in a larger type than has been employed in any previous edition. The type has been cast especially for it, and the page is of a size to admit of the introduction of all the original illustrations.

No such attractive issue has been made of the writings of Mr. Dickens, which, various as have been the forms of publication adapted to the demands of an ever widely-increasing popularity, have never yet been worthily presented in a really handsome library form.

The collection comprises all the minor writings it was Mr. Dickens's wish to preserve.

SKETCHES BY "BOZ." With 40 Illustrations by George Cruikshank.

PICKWICK PAPERS. 2 vols. With 42 Illustrations by Phiz.

OLIVER TWIST. With 24 Illustrations by Cruikshank.

NICHOLAS NICKLEBY. 2 vols. With 40 Illustrations by Phiz.

OLD CURIOSITY SHOP and REPRINTED PIECES. 2 vols. With Illustrations by Cattermole, &c.

BARNABY RUDGE and HARD TIMES. 2 vols. With Illustrations by Cattermole, &c.

MARTIN CHUZZLEWIT. 2 vols. With 40 Illustrations by Phiz.

AMERICAN NOTES and PICTURES FROM ITALY. 1 vol. With 8 Illustrations.

DOMBEY AND SON. 2 vols. With 40 Illustrations by Phiz.

DAVID COPPERFIELD. 2 vols. With 40 Illustrations by Phiz.

BLEAK HOUSE. 2 vols. With 40 Illustrations by Phiz.

LITTLE DORRIT. 2 vols. With 40 Illustrations by Phiz.

A TALE OF TWO CITIES. With 16 Illustrations by Phiz.

THE UNCOMMERCIAL TRAVELLER. With 8 Illustrations by Marcus Stone.

GREAT EXPECTATIONS. With 8 Illustrations by Marcus Stone.

OUR MUTUAL FRIEND. 2 vols. With 40 Illustrations by Marcus Stone.

CHRISTMAS BOOKS. With 17 Illustrations by Sir Edwin Landseer, R.A., Maclise, R.A., &c. &c.

HISTORY OF ENGLAND. With 8 Illustrations by Marcus Stone.

CHRISTMAS STORIES. (From "Household Words" and "All the Year Round.") With 14 Illustrations.

EDWIN DROOD AND OTHER STORIES. With 12 Illustrations by S. L. Fildes.

LIFE OF CHARLES DICKENS. By John Forster. With Portraits. 2 vols. not separate.)

DICKENS'S (CHARLES) WORKS.—*Continued.*

THE POPULAR LIBRARY EDITION

OF THE WORKS OF

CHARLES DICKENS,

In 30 Vols., large crown 8vo, price £6; separate Vols. 4s. each.

An Edition printed on good paper, each volume containing 16 full-page Illustrations, selected from the Household Edition, on Plate Paper.

SKETCHES BY "BOZ."
PICKWICK. 2 vols.
OLIVER TWIST.
NICHOLAS NICKLEBY. 2 vols.
MARTIN CHUZZLEWIT. 2 vols.
DOMBEY AND SON. 2 vols.
DAVID COPPERFIELD. 2 vols.
CHRISTMAS BOOKS.
OUR MUTUAL FRIEND. 2 vols.
CHRISTMAS STORIES.
BLEAK HOUSE. 2 vols.
LITTLE DORRIT. 2 vols.
OLD CURIOSITY SHOP AND REPRINTED PIECES. 2 vols.
BARNABY RUDGE. 2 vols.
UNCOMMERCIAL TRAVELLER.
GREAT EXPECTATIONS.
TALE OF TWO CITIES.
CHILD'S HISTORY OF ENGLAND.
EDWIN DROOD AND MISCELLANIES.
PICTURES FROM ITALY AND AMERICAN NOTES.

HOUSEHOLD EDITION.

(WITH LIFE.)

In 22 Volumes. Crown 4to, cloth, £4 8s. 6d.

MARTIN CHUZZLEWIT, with 59 Illustrations, 5s.
DAVID COPPERFIELD, with 60 Illustrations and a Portrait, 5s.
BLEAK HOUSE, with 61 Illustrations, 5s.
LITTLE DORRIT, with 58 Illustrations, 5s.
PICKWICK PAPERS, with 56 Illustrations, 5s.
OUR MUTUAL FRIEND, with 58 Illustrations, 5s.
NICHOLAS NICKLEBY, with 59 Illustrations, 5s.
DOMBEY AND SON, with 61 Illustrations, 5s.
EDWIN DROOD; REPRINTED PIECES; and other Stories, with 30 Illustrations, 5s.
THE LIFE OF DICKENS. BY JOHN FORSTER. With 40 Illustrations, 5s.
BARNABY RUDGE, with 46 Illustrations, 4s.
OLD CURIOSITY SHOP, with 32 Illustrations, 4s.
CHRISTMAS STORIES, with 23 Illustrations, 4s.
OLIVER TWIST, with 28 Illustrations, 3s.
GREAT EXPECTATIONS, with 26 Illustrations, 3s.
SKETCHES BY "BOZ," with 36 Illustrations, 3s.
UNCOMMERCIAL TRAVELLER, with 26 Illustrations, 3s.
CHRISTMAS BOOKS, with 28 Illustrations, 3s.
THE HISTORY OF ENGLAND, with 15 Illustrations, 3s.
AMERICAN NOTES and PICTURES FROM ITALY, with 18 Illustrations, 3s.
A TALE OF TWO CITIES, with 25 Illustrations, 3s.
HARD TIMES, with 20 Illustrations, 2s. 6d.

DICKENS'S (CHARLES) WORKS.—*Continued.*

THE CABINET EDITION.

In 32 vols. small fcap. 8vo, Marble Paper Sides, Cloth Backs, with uncut edges, price Eighteenpence each.

Each Volume contains Eight Illustrations reproduced from the Originals.

In Sets only, bound in blue cloth, with cut edges, £2 8s.

CHRISTMAS BOOKS.
MARTIN CHUZZLEWIT, 2 vols.
DAVID COPPERFIELD, 2 vols.
OLIVER TWIST.
GREAT EXPECTATIONS.
NICHOLAS NICKLEBY, 2 vols.
SKETCHES BY "BOZ."
CHRISTMAS STORIES.
THE PICKWICK PAPERS, 2 vols.
BARNABY RUDGE, 2 vols.
BLEAK HOUSE, 2 vols.
AMERICAN NOTES AND PICTURES FROM ITALY.
EDWIN DROOD; AND OTHER STORIES.
THE OLD CURIOSITY SHOP, 2 vols.
A CHILD'S HISTORY OF ENGLAND.
DOMBEY AND SON, 2 vols.
A TALE OF TWO CITIES.
LITTLE DORRIT, 2 vols.
MUTUAL FRIEND, 2 vols.
HARD TIMES.
UNCOMMERCIAL TRAVELLER
REPRINTED PIECES.

CHARLES DICKENS'S CHRISTMAS BOOKS.

REPRINTED FROM THE ORIGINAL PLATES.

Illustrated by JOHN LEECH, D. MACLISE, R.A., R. DOYLE, C. STANFIELD, R.A., etc.

Fcap. cloth, 1s. each. Complete in a case, 5s.

A CHRISTMAS CAROL IN PROSE.
THE CHIMES: A Goblin Story.
THE CRICKET ON THE HEARTH: A Fairy Tale of Home.
THE BATTLE OF LIFE. A Love Story.
THE HAUNTED MAN AND THE GHOST'S STORY.

SIXPENNY REPRINTS.

READINGS FROM THE WORKS OF CHARLES DICKENS.

As selected and read by himself and now published for the first time. Illustrated.

A CHRISTMAS CAROL, AND THE HAUNTED MAN. By CHARLES DICKENS. Illustrated.

THE CHIMES: A GOBLIN STORY, AND THE CRICKET ON THE HEARTH. Illustrated.

THE BATTLE OF LIFE: A LOVE STORY, HUNTED DOWN, AND A HOLIDAY ROMANCE. Illustrated.

The last Three Volumes as Christmas Works, In One Volume, red cloth, 2s. 6d.

DICKENS'S (CHARLES) WORKS.—*Continued.*

A NEW EDITION, ENTITLED

THE PICTORIAL EDITION,

Now being issued in MONTHLY PARTS, royal 8vo, at

ONE SHILLING EACH.

Each Part will contain 192 pages of Letterpress, handsomely printed, and, besides full-page Plates on plate paper, about 24 Illustrations inserted in the Text.

The Edition will be completed in about THIRTY-SEVEN PARTS, of which Thirteen are now ready, and will contain in all—

UPWARDS OF NINE HUNDRED ENGRAVINGS.

The Volumes now ready are:

DOMBEY AND SON. 3s. 6d.
DAVID COPPERFIELD. 3s. 6d.
NICHOLAS NICKLEBY. 3s. 6d.
BARNABY RUDGE. 3s. 6d.
OLD CURIOSITY SHOP. 3s. 6d.
MARTIN CHUZZLEWIT. [*Now being issued in Parts.*

PROSPECTUSES AND SHOWCARDS ON APPLICATION.

THE TWO SHILLING EDITION.

Each Volume contains a Frontispiece. Crown 8vo, 2*s.*

The Volumes now ready are—

DOMBEY AND SON.
MARTIN CHUZZLEWIT.
THE PICKWICK PAPERS.
BLEAK HOUSE.
OLD CURIOSITY SHOP.
BARNABY RUDGE.

MR. DICKENS'S READINGS.

Fcap. 8vo, sewed.

CHRISTMAS CAROL IN PROSE. 1s.
CRICKET ON THE HEARTH. 1s.
CHIMES: A GOBLIN STORY. 1s.
STORY OF LITTLE DOMBEY. 1s.
POOR TRAVELLER, BOOTS AT THE HOLLY-TREE INN, and MRS. GAMP. 1s.

SOLUTIONS TO THE QUESTIONS IN PURE MATHEMATICS—Stages 1 and 2—SET AT THE SCIENCE AND ART EXAMINATIONS from 1881 to 1886. By THOMAS T. RANKIN, C.E., Rector of the Gartsherrie Science School, and West of Scotland Mining College. Crown 8vo, 2s.

SOLUTIONS TO THE QUESTIONS SET IN THE FOLLOWING SUBJECTS AT THE MAY EXAMINATIONS OF THE SCIENCE AND ART DEPARTMENT ANTERIOR TO 1887.

1. Animal Physiology ...	From	1881 to 1886.	
2. Hygiene	"	1884 to 1886	with Notes and Index.
3. Building Construction ...	"	1881 to 1886.	
4. Machine Construction ...	"	1881 to 1886.	
5. Agriculture	"	1881 to 1886.	
6. Magnetism and Electricity	"	1881 to 1886.	
7. Physiography	"	1881 to 1886.	
8. Sound, Light, and Heat	"	1881 to 1886.	

Each Subject will be dealt with in a Separate Volume, which will contain Complete Answers to the Elementary and Advanced Papers for the Years noted.

Price 1s. 6d. each.

ANSWERS TO QUESTIONS, 1887, 1888, & 1889.

1890 IN THE PRESS.

SOLUTIONS TO THE QUESTIONS *(Elementary and Advanced) set at the Examinations of the Science and Art Department of May,* 1887, 1888, *and* 1889, *are published as under, each subject and year being kept distinct, and issued in pamphlet form separately. Crown 8vo, 3d. each.*

1. ANIMAL PHYSIOLOGY. By J. H. E. Brock, M.D., B.S. (Lond.), F.R.C.S. (Eng.), D.P.H. (Univ. of Lond.)
2. BUILDING CONSTRUCTION. By H. Adams, C.E., M.I.M.E.
3. THEORETICAL MECHANICS. By J. C. Fell, M.I.M.E., E. Pillow, M.I.M.E., and H. Angel.
4. INORGANIC CHEMISTRY (Theoretical). By Rev. F. W. Harnett, M.A., and J. J. Pilley, F.C.S., F.R.M.S.
5. Ditto, ALTERNATIVE COURSE. By J. Howard, F.C.S.
6. MAGNETISM AND ELECTRICITY. By W. Hibbert, F.I.C., A.I.E.E.
7. PHYSIOGRAPHY. By W. Rheam, B.Sc., and W. S. Furneaux, F.R.G.S.
8. PRACTICAL PLANE AND SOLID GEOMETRY. By H. Angel.
9. ART—THIRD GRADE. PERSPECTIVE. By A. Fisher and S. Beale.
10. PURE MATHEMATICS. By R. R. Steel, F.C.S., and H. Carter, B.A.
11. MACHINE CONSTRUCTION AND DRAWING. By H. Adams, C.E., M.I.M.E.
12. PRINCIPLES OF AGRICULTURE. By H. J. Webb, Ph.D., B.Sc.
13. SOUND, LIGHT, AND HEAT. By C. A. Stevens and F. Carrodus, Assoc. N.S.S.
14. HYGIENE. By J. J. Pilley, F.C.S., F.R.M.S.
15. INORGANIC CHEMISTRY (Practical). By J. Howard, F.C.S.
16. APPLIED MECHANICS. For 1888 and 1889 only. By C. B. Outon, Wh.Sc.
17. ALTERNATIVE ELEMENTARY PHYSICS. By W. Hibbert, F.I.E., A.S.I.E. For 1889 only.

Price of each Pamphlet (dealing with both Elementary and Advanced Papers for a single year), 3d. net, postage included.

THE FORTNIGHTLY REVIEW.

Edited by FRANK HARRIS.

THE FORTNIGHTLY REVIEW is published on the 1st of every month, and a Volume is completed every Six Months.

The following are among the Contributors:—

ADMIRAL LORD ALCESTER.
GRANT ALLEN.
SIR RUTHERFORD ALCOCK.
AUTHOR OF "GREATER BRITAIN."
PROFESSOR BAIN.
SIR SAMUEL BAKER.
PROFESSOR BEESLY.
PAUL BOURGET.
BARON GEORGE VON BUNSEN.
DR. BRIDGES.
HON. GEORGE C. BRODRICK.
JAMES BRYCE, M.P.
THOMAS BURT, M.P.
SIR GEORGE CAMPBELL, M.P.
THE EARL OF CARNARVON.
EMILIO CASTELAR.
RT. HON. J. CHAMBERLAIN, M.P.
PROFESSOR SIDNEY COLVIN.
THE EARL COMPTON.
MONTAGUE COOKSON, Q.C.
L. H. COURTNEY, M.P.
G. H. DARWIN.
SIR GEORGE W. DASENT.
PROFESSOR A. V. DICEY.
PROFESSOR DOWDEN.
RT. HON. M. E. GRANT DUFF.
RIGHT HON. H. FAWCETT, M.P.
ARCHDEACON FARRAR.
EDWARD A. FREEMAN.
J. A. FROUDE.
MRS. GARRET-ANDERSON.
J. W. L. GLAISHER, F.R.S.
SIR J. E. GORST, Q.C., M.P.
EDMUND GOSSE.
THOMAS HARE.
FREDERIC HARRISON.
ADMIRAL SIR G. P. HORNBY.
LORD HOUGHTON.
PROFESSOR HUXLEY.
PROFESSOR R. C. JEBB.
ANDREW LANG.
E. B. LANIN.
EMILE DE LAVELEYE.
T. E. CLIFFE LESLIE.
W. S. LILLY.
MARQUIS OF LORNE.
PIERRE LOTI.
SIR JOHN LUBBOCK, BART., M.P.
THE EARL OF LYTTON.
SIR H. S. MAINE.
W. H. MALLOCK.
CARDINAL MANNING.
DR. MAUDSLEY.
PROFESSOR MAX MÜLLER.
GEORGE MEREDITH.
RT. HON. G. OSBORNE MORGAN, Q.C., M.P.
PROFESSOR HENRY MORLEY.
RT. HON. JOHN MORLEY, M.P.
WILLIAM MORRIS.
PROFESSOR H. N. MOSELEY.
F. W. H. MYERS.
F. W. NEWMAN.
PROFESSOR JOHN NICHOL.
W. G. PALGRAVE.
WALTER H. PATER.
RT. HON. LYON PLAYFAIR, M.P.
SIR HENRY POTTINGER, BART.
PROFESSOR J. R. SEELEY.
LORD SHERBROOKE.
PROFESSOR SIDGWICK.
HERBERT SPENCER.
M. JULES SIMON. (DOCTOR L'ACADEMIE FRANCAISE).
HON. E. L. STANLEY.
SIR J. FITZJAMES STEPHEN, Q.C.
LESLIE STEPHEN.
J. HUTCHISON STIRLING.
A. C. SWINBURNE.
DR. VON SYBEL.
J. A. SYMONDS.
SIR THOMAS SYMONDS. (ADMIRAL OF THE FLEET).
THE REV. EDWARD F. TALBOT (WARDEN OF KEBLE COLLEGE).
SIR RICHARD TEMPLE, BART.
HON. LIONEL A. TOLLEMACHE.
COUNT LEO TOLSTOI.
H. D. TRAILL.
PROFESSOR TYNDALL.
ALFRED RUSSELL WALLACE.
A. J. WILSON.
GEN. VISCOUNT WOLSELEY.
THE EDITOR.

ETC. ETC. ETC.

THE FORTNIGHTLY REVIEW *is published at 2s. 6d.*

CHAPMAN & HALL, LIMITED, 11, HENRIETTA STREET, COVENT GARDEN, W.C.

CHARLES DICKENS AND EVANS,] [CRYSTAL PALACE PRESS.